Clean Energy Production Technologies

Series Editors

Neha Srivastava, Department of Chemical Engineering and Technology, IIT (BHU) Varanasi, Varanasi, Uttar Pradesh, India

P. K. Mishra, Department of Chemical Engineering and Technology, IIT (BHU) Varanasi, Varanasi, Uttar Pradesh, India

The consumption of fossil fuels has been continuously increasing around the globe and simultaneously becoming the primary cause of global warming as well as environmental pollution. Due to limited life span of fossil fuels and limited alternate energy options, energy crises is important concern faced by the world. Amidst these complex environmental and economic scenarios, renewable energy alternates such as biodiesel, hydrogen, wind, solar and bioenergy sources, which can produce energy with zero carbon residue are emerging as excellent clean energy source. For maximizing the efficiency and productivity of clean fuels via green & renewable methods, it's crucial to understand the configuration, sustainability and techno-economic feasibility of these promising energy alternates. The book series presents a comprehensive coverage combining the domains of exploring clean sources of energy and ensuring its production in an economical as well as ecologically feasible fashion. Series involves renowned experts and academicians as volume-editors and authors, from all the regions of the world. Series brings forth latest research, approaches and perspectives on clean energy production from both developed and developing parts of world under one umbrella. It is curated and developed by authoritative institutions and experts to serves global readership on this theme.

Dan Bahadur Pal • Jay Mant Jha
Editors

Sustainable and Clean Energy Production Technologies

Springer

Editors
Dan Bahadur Pal
Department of Chemical Engineering
Birla Institute of Technology, Mesra
Ranchi, Jharkhand, India

Jay Mant Jha
Department of Chemical Engineering
Maulana Azad National Institute of Technology
Bhopal, Madhya Pradesh, India

ISSN 2662-6861 ISSN 2662-687X (electronic)
Clean Energy Production Technologies
ISBN 978-981-16-9134-8 ISBN 978-981-16-9135-5 (eBook)
https://doi.org/10.1007/978-981-16-9135-5

This Springer imprint is published by the registered company Springer Nature Singapore Pte Ltd.
The registered company address is: 152 Beach Road, #21-01/04 Gateway East, Singapore 189721, Singapore

Contents

About the Editors

Dan Bahadur Pal B. Tech, M. Tech, PhD has received his M.Tech. and Ph.D. in the field of Chemical Engineering from Indian Institute of Technology (BHU) Varanasi, Uttar Pradesh, India. Before that, he has completed his B. Tech. in Chemical Engineering from UPTU, Lucknow. Dr. Pal's research interest is nano-technology, catalysis, energy and environment and waste treatment, with a special focus on developing process and materials by using waste as raw materials. He also prefers to work on bio-waste processing and value addition. Dr. Pal published more than 45 publications in reputed journals and books along with 14 book chapters and edited 4 books.
Department of Chemical Engineering, Birla Institute of Technology, Mesra, Ranchi, Jharkhand, India

Jay Mant Jha B. Tech, M. Tech, PhD is a Chemical Engineering graduate with specialization in Industrial Pollution Abatement and Convective Heat Transfer. During his doctoral degree, he has investigated ultrafast cooling technology to improve the quality of steel. His research interest includes nano-technology, catalysis, energy and environment, waste management and heat transfer, with a special focus on developing process and materials by using waste as raw materials. He also prefers to work on bio-waste processing and value addition. To his credit, he has published more than 28 publications in reputed journals and book chapters.
Department of Chemical Engineering, Maulana Azad National Institute of Technology, Bhopal, Madhya Pradesh, India

Tiny Renewable Engines: Microalgae and Cyanobacteria as Biofuel Providers

M. M. Pathmalal and D. A. T. Madusanka

Abstract Petro-based fuel is getting depleted vastly with the current higher rate of consumption following the rapid development of industries. With the limited availability of petro fuels, global warming, and pollution due to petroleum-derived greenhouse gases (GHGs), the world's concern turns toward alternative, third-generation biofuel which are supposedly sustainable and eco-friendly. Third-generation biofuels include the fuel types produced with microalgae and cyanobacteria and are popular alternative energies. Most of the water bodies in tropical and subtropical regions in the world are being eutrophicated, due to the optimum light, temperature, and nutrients which accelerated the growth of microalgae and cyanobacteria. Blooms of microalgae and cyanobacteria adversely affect socially, economically, and environmentally landscapes profoundly. A limited number of studies in the world have revealed that the biomass of microalgae and cyanobacteria is an untapped treasure chest, as feedstocks for many biofuel types, especially their lipids, have a great potential for biodiesel. The trans-esterification process enhances the fuel properties and transforms lipids into their relevant lipid ester form, and oxygenated lipid ester molecules enhance the burning property of the fuel product. Most cyanobacteria and microalgae are *r* strategic microorganisms with greater reproduction and growth rate in even harsh environmental conditions forming cyanobacterial and microalgal blooms which makes a high potential to be used for biodiesel. Though biodiesel from cyanobacteria or microalgae lipids is a key alternative liquid fuel, it has many challenges in replacing the petrodiesel completely for compression-ignition (CI) engines without modifications, as a neat product. This

---

M. M. Pathmalal (✉)
Centre for Water Quality and Algae Research, Department of Zoology, Faculty of Applied Sciences, University of Sri Jayewardenepura, Gangodawila, Nugegoda, Sri Lanka

Faculty of Graduate Studies, University of Sri Jayewardenepura, Gangodawila, Nugegoda, Sri Lanka
e-mail: pathmalal@sjp.ac.lk

D. A. T. Madusanka
Centre for Water Quality and Algae Research, Department of Zoology, Faculty of Applied Sciences, University of Sri Jayewardenepura, Gangodawila, Nugegoda, Sri Lanka

D. B. Pal, J. M. Jha (eds.), *Sustainable and Clean Energy Production Technologies*, Clean Energy Production Technologies,
https://doi.org/10.1007/978-981-16-9135-5_1

chapter discusses the production of biodiesel from microalgae and cyanobacteria and the important challenges concerning the world context.

Keywords Microalgae · Cyanobacteria · Lipids · Biodiesel · Biodiesel blends

1 Introduction

Plants and animals which perished hundreds of millions of years ago are responsible for the formation of petroleum (Huang et al., 2012). Their organic matter decomposition forms petroleum after millions of years, and these resources have very high economic value because they are non-renewable energy sources that cannot be rapidly regenerated naturally on par with consumption.

Energy has become a significant factor in the present and essential to enhance the living standards directly and indirectly. Fossil fuels (coal, oil, and natural gas) play a dominant role in the present global energy system. At the end of the nineteenth century, the internal combustion engine (ICE) was invented, and that increases the demand dramatically for fossil fuel universally. Since that fossil energy has been a fundamental social, economic, industrial, and development-driven force for the industrial revolution. In the twentieth century, though the world moves toward modern applied technology, fossil fuel is still used as the main energy source that contributes more than 80% of the global energy needs (Huang et al., 2012; Madusanka & Manage, 2018a). However, due to the increments in the usage of fossil fuel with the populations rising, industrial development with modern technology, and unsustainable and unregulated use, the reserves of fossil fuel are being depleted rapidly. The estimated fossil fuel reserve depletion time for oil, natural gas, and coal are approximately 35, 37, and 107 years, respectively (Madusanka & Manage, 2018b; Gülüm & Bilgin, 2015).

Air pollution and global warming have been increasing gradually along with the combustion of fossil fuel that mainly produces carbon-based gasses such as carbon dioxide, carbon monoxide, etc. (Zhou & Thomson, 2009; Mahmudul et al., 2017). Due to the rapid development of industries with the usage of modern technology, the limited availability of fossil fuels (Rangabhashiyam et al., 2017), and the rising footprint of global warming and pollution due to greenhouse gases (GHGs) and other pollutants, therefore, the world is increasingly shifting toward alternative, sustainable, and eco-friendly energy productions (Halim et al., 2011).

Presently, many studies are being conducted on renewable energies such as wind (Shezan & Ishraque, 2019), ocean thermal energy conversion (OTEC) (Song, 2019), tidal (Park, 2017), solar power (Tidwell et al., 2018), geothermal water (Ghazvini et al., 2019), and biofuel (Chisti, 2007; Huang et al., 2012; Madusanka & Manage, 2018a, 2018b). In recent times, biofuels are considered as one of the best available alternative energy sources for future requirements, which could be replaced with diesel-like liquid fossil fuels (Atadashi et al., 2012; Liu et al., 2012; Raj & Bhandari, 2017). Biofuel is simply defined as any fuel that is derived from biomass including plants, algae, or animal materials and is carbon neutral which means there is no net release of carbon dioxide to the environment, as the biomasses consume carbon dioxides from the onset through carbon fixation or the Calvin cycle (Yew et al., 2019). Apart from carbon neutrality, biofuel's lesser toxicity footprint and renewability ensure economic and environmental sustainability (Madusanka & Manage, 2018b).

1.1 Biofuel Types

Regarding the biomass type, biofuel is categorized into first-, second-, third-, and fourth-generation biofuel. Considering liquid biofuels, the first-generation biofuels are synthesized from edible feedstocks such as corn, rapeseed, sugar cane, and soybean, whereas the second-generation biofuels are produced from non-edible feedstocks such as *Jatropha*, switchgrass, corn cobs, animal waste, paper waste, food waste, etc. (Mubarak et al., 2015; Mubarak et al., 2016; Ranjith Kumar et al., 2015). Based on the context of "food vs. energy," the first-generation fuels are considered as less sustainable with the fuel demand and food crisis. Furthermore, both first- and second-generation biofuels are considered less sustainable due to the limited availability of cultivable land for edible and non-edible feedstocks, except the waste products (Mubarak et al., 2015). The waste products such as cooking waste oil and paper waste have a higher potential to be utilized for biodiesel and bioethanol production in any part of the world.

Biofuel types that belong to the third generation are derived from microalgae and cyanobacteria (blue-green algae), and the metabolic engineering of photosynthetic organisms based biofuel production belongs to fourth-generation (Daroch et al., 2013; Jung et al., 2013). Considering the land availability for plantation or culture of oil-rich biomasses, the biomass harvest per hectare is greater for microalgae and cyanobacteria than the first- and second-generation plants as microalgae and cyanobacteria can be cultured three-dimensionally. Therefore, based on several key factors, fast growth rate, rapid reproduction, limited land availability, and the rate of carbon conversion, microalgae and cyanobacteria are considered as a higher potential feedstock for bioenergy production, and studies on microalgae liquid biofuel production had begun in the mid-1980s (Huang et al., 2010; Shuba & Kifle, 2018).

1.2 Biodiesel

The world's biodiesel production has been increasing day by day, and despite 2020, COVID-19 pandemic situation, in year 2019, the total biodiesel consumption was over 40 billion liters, an 11.7% increment compared to year 2018 (Mizik & Gyarmati, 2021). It seems biofuels are the most successful alternative instead of fossil fuel and can increase the contribution up to 25% of the world energy requirement (Madusanka & Manage, 2018b; Shuba & Kifle, 2018). Biofuel consists of a variety of types that includes biodiesel, bioethanol, biogas, biohydrogen, biochar, etc. (Madusanka & Manage, 2018b; Weerasinghe et al., 2021). Out of the biofuel types mentioned, the liquid form of biodiesel and bioethanol has gained extra attention since their formation is similar to petrodiesel and replacing them day by day. Most developed countries have already changed their policies according to the green concept that most of the future development plans rely on it. According to the American Society for Testing and Materials (ASTM), biodiesel is defined as a fuel composed of mono-alkyl esters of long-chain fatty acids derived from renewable biomass oils or fats meeting the requirements of the D7651 standard. Chemically, methyl-esterified fatty acids (fatty acid methyl esters) are considered "biodiesel" (A.S.T.M., 2008).

As an oxygenated fuel, biodiesel generally contains 10 wt.% of oxygen that improves burning efficiency and reduction of particulate matter and other gaseous pollutants (such as carbon monoxide) (Huang et al., 2012; Madusanka & Manage, 2018a; Pandey et al., 2019). Apart from that, biodiesel appears to be an attractive and successful alternative energy source for several reasons including sustainable supply, low sulfur content (Dash & Lingfa, 2018; Xie et al., 2019; Simsek, 2020), high flashpoint, and biodegradability (Chen et al., 2019).

In the early and mid-nineteenth century, vegetable oil had been used as an alternative to diesel in emergencies (Huang et al., 2010). Averagely, the carbon chain of diesel molecules contains about 15 carbon atoms, whereas plant oil contains 9–22 carbons. However, vegetable oil is not a successful alternative fuel because of its higher price than petroleum fuels, and due to higher viscosities of oil, as it does not comply with the diesel properties (Chisti, 2007).

The use of neat biodiesel for normal compression-ignition (CI) engines without modifications is questionable, but "biodiesel blends" which means mixing biofuel with petrodiesel according to a certain proportion is one of the solutions. Especially lower blends (blends below 20%) make a move forward along with the biodiesel without engine modifications. Theoretically, biodiesel-diesel blends decrease the environmental impacts and the usage of petrodiesel by its relevant percentage (i.e., 20% biodiesel-diesel blend reduces the usage of petrodiesel by 20% and the environmental impacts by 20%). Madusanka and Manage (2018b) recorded production of biodiesel from the bloom of cyanobacterium *Microcystis aeruginosa* and achieved 6% yield from biodiesel-petrodiesel blend (B6) with satisfied Sri Lankan Ceypetco standard fuel properties. The results of the study showed that even with a small percentage of biodiesel, it enhances the fuel properties of petrodiesel especially higher gross calorific value and lower sulfur content.

2 Microalgae and Cyanobacteria

Cyanobacteria and microalgae are simple cellular photosynthetic organisms that are observed under the microscope and found in both marine and freshwater environments (Da Rós et al., 2012; Madusanka & Manage, 2018a; Olufemi et al., 2020; Idroos & Manage, 2014). Unlike microalgae, cyanobacteria are a prominent phytoplankton group in eutrophic freshwater bodies. All cyanobacteria are photosynthetic and contain only chlorophyll-α (Wijesekara & Manage, 2017). The morphological diversity of cyanobacteria ranges from unicellular to small colonies of cells and simple or branched filamentous forms (Da Rós et al., 2012; Idroos & Manage, 2014; Wijesekara & Manage, 2017).

Microalgae and cyanobacteria can produce 1 ton of algae biomass by consuming 1.83 tons of CO_2 in the atmosphere (Chisti, 2019), and due to the simple structure and high cell surface/volume ratio, it leads to a higher rate of nutrient assimilation, green algae growth is approximately 100 times greater than terrestrial plants, and most of the microalgal species commonly double their biomass within 24 h; that period reduces to about 3.5 h during their exponential growth phase (Manage et al., 2000; Chisti, 2007; Idroos & Manage, 2014; Pacheco et al., 2020).

Microalgae and cyanobacteria, mainly sunlight-driven organisms, are producing a wide range of bioactive compounds that include vitamins, pigments, carotenoids, ketones, and lipids and are useful as feedstocks for biofuels, foods, feeds, nitrogen-fixing biofertilizers, and in bioremediation applications (Chisti, 2007; Da Rós et al., 2012; Halim et al., 2012; Wijesekara & Manage, 2017; Andrade & Andrade, 2018; Sudhakar et al., 2019; Molino et al., 2020; Pacheco et al., 2020).

2.1 *Microalgae and Cyanobacteria for Biodiesel Production*

As a source for different types of natural compounds, microalgae and cyanobacteria have diversified applications, such as human and animal food (Kaga et al., 2021), aquaculture (Lin et al., 2019), chemicals and pharmaceuticals (Rizwan et al., 2018), fertilizers (Rizwan et al., 2018), cosmetics (Rizwan et al., 2018), and fuels (Urtubia et al., 2016; Madusanka & Manage, 2018a). About a decade ago, microalgae-based biofuel production wasn't commercialized to large-scale production as debates existed regarding the microalgae life cycle impacts of large-scale microalgae-based biofuel production, especially the impact on water usage.

Microalgal and cyanobacterial renewable biofuels include methane (biogas) that is produced through anaerobic digestion of algal biomass (Perendeci et al., 2019), photobiologically produced biohydrogen (Mubarak et al., 2015; Hossain, 2017), biochar (Mubarak et al., 2015), and lipids provisioning a great potential as a feedstock for biodiesel (Gouveia et al., 2017; Madusanka & Manage, 2018a, 2018b; Mofijur et al., 2019).

Due to high photosynthetic efficiency, high biomass production, rapid lipid accumulation (Halim et al., 2012), and fast growth rate than other conventional crop products, fatty acid-rich microalgae and cyanobacteria (Huang et al., 2010) are considered as a good source of oil that is possible for transformation into biofuel, especially carbon-neutral biodiesel (Huang et al., 2010; Madusanka & Manage, 2018a, 2018b).

Nowadays, commercial scaled-up biodiesel production has been introduced with modern technology to overcome the emerged issues. Raceways and photobioreactors are the most popular methods for large-scale microalgae biomass production (Kumar & Banerjee, 2019). The factors such as the continuous growth with a broad range of environmental conditions (such as salinity, pH, and chemical composition), the higher production yield with the higher photosynthetic ability, and ease to grow in any land type (either arable or non-arable) make microalgae and cyanobacteria as a potential biomass for biofuel production (Chisti, 2007; Smith et al., 2010; Rashid et al., 2013).

Depending on species, conditions, and other environmental factors, microalgae and cyanobacteria produce many different compounds that include kinds of lipids, hydrocarbons, and other complex oils such as phospholipids, glycolipids, etc. (Huang et al., 2010; Kolmakova & Kolmakov, 2019; Sánchez-Bayo et al., 2020). Considering the presence of oil types, all the algal oils aren't satisfactory for the production of biodiesel. However, it was recorded that almost all oil-rich microalgae or cyanobacteria species commonly contain suitable fatty acid types, such as palmitic acids (Da Rós et al., 2012; Madusanka & Manage, 2018a).

The growth of microalgae or cyanobacteria depends on carbon source, light, and vital nutrients such as nitrogen and phosphorus (Yen et al., 2019). Microalgae can utilize carbon dioxide or bicarbonates (as inorganic carbon) and glucose and acetate (as organic carbon) for lipid production (Huang et al., 2010) and have more production per unit area than conventional oil crops (Chisti, 2008) which means that microalgae do not need much space compared to other conventional oil crops to produce the same quantity of lipids (Chisti, 2007). Therefore, biofuels which are produced from microalgal and cyanobacterial biomass are considered as the most promising alternative fuels for the production of oil per acre which is higher (100 times more) in microalgae and cyanobacteria than the other types of conventional oil plants like Castro (Mubarak et al., 2015). The carbon number of microalgal and cyanobacterial fatty acids mostly ranges from 12 to 22, and that can be either neutral or polar lipid molecules and either saturated or unsaturated (Huang et al., 2010; Mubarak et al., 2015).

2.2 *Fatty Acids in Microalgae for Biodiesel Production*

Chemically, saturated fatty acids have no double bonds, while unsaturated fatty acids have at least one double bond in the hydrocarbon chain (Nelson et al., 2008). Medina et al. (1998) reported that all or most all the microalgae- and

Fig. 1 (**a**) Triacylglycerol (neutral lipid). (**b**) Phospholipid (polar lipid). R′, R″, and R‴ represent fatty acid chains

cyanobacteria-unsaturated fatty acids are cis-isomers and never exceed six double bonds along with the carbon chain.

Overall, lipids are categorized into two groups, neutral and polar lipids, based on the head group polarity (Arif et al., 2020). Neutral lipids (like acylglycerols and free fatty acids) contain a carboxylate group at the end of the fatty acid that binds with a neutral group. Polar lipid molecules found in cell membranes (such as phospholipids and glycolipids) contain a charged head group (Halim et al., 2012) (Fig. 1). Triacylglycerol consists of three long-chain fatty acids which are bound to the glycerol backbone chemically (Moser, 2009).

The two main reasons for neutral acylglycerols attracting more attention for commercial-scale biodiesel production are as follows:

(1) The industrial-scale alkaline-catalyzed trans-esterification is designed to process acylglycerols and has limited efficiencies on other lipid fractions, such as polar lipids and free fatty acids.
(2) Acylglycerols generally contain a lower degree of unsaturation than other lipid fractions (such as polar lipids), and they produce fatty acid methyl esters (FAMEs) with higher oxidation stability (Halim et al., 2012).

Unlike other vegetable oils or animal fat, microalgal fatty acids consist of polyunsaturated carbon chain structure that contains more than one double bond (e.g., eicosapentaenoic acid (EPA), C20:5n-3, and docosahexaenoic acid (DHA), C22:6n-3), which are highly susceptible for oxidation that reduces the acceptability of using their fatty acid methyl esters as biodiesel (Rangabhashiyam et al., 2017). Iodine value indicates the total availability of unsaturated fatty acid and the free fatty acid amount that is a major limitation for biodiesel production where the global standards, EN 14214 and EN 14213, require the value to not exceeding 120 and 130 g iodine/100 g of biodiesel, respectively (Chisti, 2007). Neutral lipid molecules such as acylglycerols (tri-, di-, monoglycerols) are involved in high energy storage and are the main materials (Halim et al., 2012; Huang et al., 2010) which are desirable for commercial-scale biodiesel production (Chen et al., 2018). The acid value of the lipids indicates the availability of free fatty acids (FFAs), and higher

FFA causes saponification during the trans-esterification and reduces the available reactive catalysts in media decreasing the esterification efficiency.

Compared to microalgae, the fatty acid composition of cyanobacteria is simpler. Both Da Rós et al. (2012) and Madusanka and Manage (2018a) reported that most of the fatty acids of cyanobacterium *Microcystis* were saturated (about 50%) and the major saturated fatty acids were palmitic acid (C16:0), lauric acid (C12:0), and myristic acid (C14:0), and dominant unsaturated fatty acids were oleic acid (C18:1) and linoleic acid (C18:2), while erucic acid (C22:1), linolenic acid (C18:3), and palmitoleic acid (C16:1) were in lower amounts.

2.3 Blooms of Microalgae and Cyanobacteria in the Aquatic Environment

The oil recovery from biomass and the conversion of oil to biodiesel aren't affected by culturing systems (Chisti, 2007). Therefore, the naturally available biomass of microalgae or cyanobacteria is profitable for biodiesel production without the cultivation cost, and the cost of raw materials for biodiesel production is also economical. However, in natural environments, microalgae and cyanobacteria are not available as monocultures which can be problematic for the sustainable production of biodiesel; thus the lipid composition of microalgae or cyanobacteria vary from species to species (Manage et al., 2000; De Souza Silva et al., 2014; Idroos & Manage, 2014; Wijesekara & Manage, 2017). However, some of the blooms have extreme densities of particular cyanobacteria or microalgae, due to the eutrophication process that has a great potential to take as the feedstock for biodiesel production (Madusanka & Manage, 2018a).

Extreme densities of cyanobacteria or microalgae bloom naturally occur due to the eutrophication process, which is due to the high amounts of nitrate and phosphate in the environment (Idroos & Manage, 2014; Idroos et al., 2016; Sethunga & Manage, 2010). Eutrophication is a naturally occurring phenomenon but accelerated in hundreds or thousands by anthropogenic activities (Manage, 2019). Freshwater ecosystems which are eutrophicated may contain a higher abundance of phytoplankton biomass that include cyanobacteria, microalgae, chlorococcales, or dinoflagellates (Chu & Rienzo, 2013). Most of the water bodies in tropical and subtropical countries are being polluted, and this leads to eutrophication with excess amounts of available nitrogen and phosphorus (Sethunga & Manage, 2010). Cyanobacteria are the most common in freshwater blooms, while dinoflagellates form marine blooms frequently, and cyanobacteria form blooms occasionally.

In urbanized areas get polluted human-induced nutrients introduction that accelerates the natural eutrophication process and enhances the algae blooms' intensity, density. and frequency (Chu & Rienzo, 2013; Patel et al., 2019). Both freshwater and marine water are known to be changed by eutrophication, and freshwater eutrophication affects society environmentally and economically as well. Both point and non-point sources (including fertilizers containing agricultural runoff, industrial wastewater, sewage effluents, domestic wastewater, and pet wastes discarded) adversely affect the aquatic environments that collapse the existing equilibrium between phytoplankton and the other aquatic life (including flora and fauna) that leads to the eutrophication (Chu & Rienzo, 2013; Idroos & Manage, 2014; Idroos et al., 2016; Sethunga & Manage, 2010; Wijewickrama & Manage, 2019). Especially in tropical countries, most of the stagnant water bodies easily get eutrophicated with less fluctuation of sunlight. The uplifted algal mat covers the water's surface and prevents sunlight penetration to the lower portions of the water bodies. Not all the microalgae and cyanobacteria form blooms, but some species produce bloom with the optimum conditions for eutrophication. Some of the bloom-forming cyanobacteria are toxic that produce hepatotoxins, cytotoxins, neurotoxins, and endotoxins (Chu & Rienzo, 2013; Wijewickrama & Manage, 2019). Utilizing cyanobacteria or microalgae blooms for commercial biodiesel production is yet to be evaluated to maintain a continuous supply for the current global demand. As lake management strategies along with lake restorations, blooms can be harvested and utilized for biodiesel production. The production may vary with the capacity of the particular water body and the lipid content of the bloomed cyanobacteria or microalgae species. Madusanka and Manage (2018a, 2018b) reported cyanobacterium *Microcystis aeruginosa* as the dominant cyanobacteria in a hyper-eutrophic lake in Sri Lanka and used the bloom to produce biodiesel. The study has highlighted that the use of microalgae or cyanobacteria blooms isn't sustainable with the current fuel requirement and the bloom regeneration rate.

2.4 Culturing of Microalgae and Cyanobacteria for Biodiesel Production

Naturally available microalgae or cyanobacteria blooms seem unsustainable, and the biomass varies with the external environmental factors. To make sure biodiesel production from microalgae or cyanobacteria is sustainable, cultivation of naturally available microalgae and cyanobacterial blooms is more important to obtain sustainable and sufficient feedstock. Photobioreactors (PBRs) (Chen et al., 2011; Khalekuzzaman et al., 2020) and algae raceways (Delrue et al., 2012; Koley et al., 2019) are the major techniques for culturing microalgae or cyanobacteria.

In photoautotrophic cultivation, microalgae and cyanobacteria produce energy through the photosynthesis process that converts inorganic carbon (such as carbon dioxide, bicarbonates) into carbohydrates (e.g., glucose) with the presence of water and sunlight (Yen et al., 2019). Some of the microalgae species utilize organic carbon (such as glucose) directly with or without sunlight that is called heterotrophic cultivation (Huang et al., 2017; Yen et al., 2019). The heterotrophic microalgae require a carbon source such as sugar for their metabolic activities, while autotrophic microalgae require light, nutrients, and CO_2 (Huang et al., 2017). The accumulation of lipids in heterotrophic microalgae is higher (around 55%), while autotrophic microalgae accumulate 20–50% (Huang et al., 2017). Some microalgae genera such as *Chlorella*, *Scenedesmus*, *Tetraselmis*, and *Nitzschia* can switch between phototrophic metabolism and heterotrophic metabolism according to the environmental condition (Pacheco et al., 2020). The photoautotrophic cultivation of microalgae is the most suitable feedstock for biofuel production since heterotrophic cultivation would increase the additional feedstock production cost. Furthermore, a biofuel derived from photoautotrophic microalgae or cyanobacteria ensures the "no net release of CO_2," whereas though fuel combustion produces CO_2, photosynthetic microalgae utilize CO_2 from the environment (Chisti, 2007; Huang et al., 2017; Yen et al., 2019).

Based on the requirement, there are various types of PBRs, and they can be constructed either indoor or outdoor. Comparatively, outdoor PBR systems are having a relatively low operational cost as they use sunlight as the light source, while indoor systems should be facilitated with optimum light conditions. Though the operational cost is higher with indoor PBRs, the microalgae harvest would be the highest considering the algae concentration and the operation time (outdoor PBRs take light for approx. 12 h day^{-1}, while indoor PBRs take light for almost 24 h day^{-1}).

PBRs are the most suitable culturing technique, as the biomass production rate is higher compared to the required land area. In the comparison of PBRs with raceways, though the construction cost for PBRs is higher than algae raceways, the cultivation of microalgae and cyanobacteria is more efficient with PBRs. Therefore, PBRs are popular in the commercial-scale biomass cultivations (Yen et al., 2019).

Chisti (2007) compared several culturing variables such as volumetric production, areal production, and required area that need to produce a particular production, extractable lipid yield, etc. for PBRs and raceway ponds. According to the given data, PBRs gain more economical sustainability rather than raceway ponds. Especially the volumetric production of PBRs was 1.535 kg m^{-3} day^{-1}, while the value for raceway ponds was 0.117 kg m^{-3} day^{-1}, and Chisti (2007) shows that there is a significant difference in the extractable lipid yield (assuming that 70% by wt. in oil in biomass), 136.9 m^3 ha^{-1} and 99.4 m^3 ha^{-1} for PBRs and raceway ponds, respectively.

2.5 *Culturing of Microalgae and Cyanobacteria Using Wastewater*

Microalgae and cyanobacteria can be utilized to remediate wastewater (Rawat et al., 2011; Pires et al., 2013). Media of wastewater facilitates favorable conditions to cyanobacteria and microalgae that allow for a fast growth and reproduction rate. Wastewater is used water that contains dissolved and suspended solids such as nutrients, microbes, and toxic organic and inorganic compounds (Mahagamage &

Manage, 2014; Pacheco et al., 2020). Organic matter and NH_4^+ oxidize with the O_2 produced by photosynthetic microalgae, while both autotrophic and heterotrophic metabolism of algal-bacterial consortia can support an enhanced nutrient recovery (Posadas et al., 2017).

The use of microalgae and cyanobacteria for wastewater treatment is a novel and multidisciplinary approach that allows reduction of COD, nitrate, phosphate, heavy metals, and other xenobiotic chemicals in wastewater by providing growth media for microalgae and cyanobacteria. The raceways method is the most ideal as wastewater contains a lack of oxygen and a higher capacity rate of raceways. Microalgae harvested after the wastewater treatment can be used as value-added products because the microalgae are rich in carbohydrates, proteins, and lipids (Udaiyappan et al., 2017).

3 Biodiesel Production from Microalgae and Cyanobacteria Biomass

The basic steps for the microalgae and cyanobacteria biodiesel production include obtaining biomass (either cultured or natural), pre-treatment process (including dewatering, thickening, and drying processes), lipid extraction, lipid trans-esterification, and finally biodiesel separation which is depicted in Fig. 2.

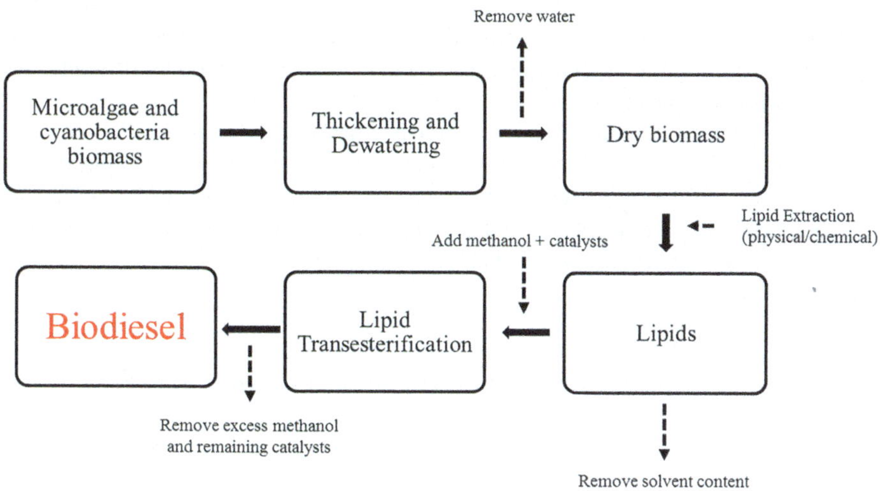

Fig. 2 Schematic diagram for biodiesel production from microalgae and cyanobacteria

3.1 *Pre-Treatment for Microalgae and Cyanobacteria Biomass*

As lipid molecules are located inside microalgae or cyanobacteria cells, lipid globules can be found in the cytoplasm or bound with the cell membranes as phospholipids or glycolipids (Kadir et al., 2018; Madusanka & Manage, 2018a; Onumaegbu et al., 2018).

The biomass harvesting process is one of the major steps in biodiesel production, and based on the method, the cost and efficiency are varied. Averagely, the industrial-scale microalgae harvesting processes account for about 20% of the total cost (Barros et al., 2015; Yew et al., 2019). The concentration of cyanobacteria biomass is one of the lipid extraction pre-treatment procedures that reduces water consistency and enhances the extraction of lipids and improves the quality of the product (Al-Hattab et al., 2015). Several techniques are being employed for harvesting biomass. Both thickening and dewatering methods include physical harvesting methods, chemical harvesting methods, auto and bioaccumulation, auto-flocculation, bio-flocculation, electrophoresis, and gravity sedimentation (Al-Hattab et al., 2015; Barros et al., 2015; Yin et al., 2020). The thickening process increases the cell concentration of the microalgae suspension. The microalgae surface is negatively charged, and the cells carry algogenic organic matters (AOMs) that keep cells in a dispersed state (Danquah et al., 2009). Commercially dewatering can be performed by centrifugation or filtration. After the dewatering process, the dewatered biomass (also called "dewatered cake") is subjected to drying to improve the downstream process (lipid extraction).

Typically, the two-step pre-treatment method is the simplest way to perform the pre-treatments on the laboratory scale. In that procedure, the concentrate completely dries under solar light or oven conditions and then is ground to a fine powder (Halim et al., 2012; Madusanka & Manage, 2018a) using a grinder or mortar and pestle. Other than these, freeze-drying, air-flow drying, etc. are potential techniques that can be used for pre-treatment, but the cost relevant to each method is different and subjected to requirement.

Extraction pre-treatments are used to enhance the lipid extraction rate (Lee et al., 2010; Roux et al., 2017; Lakshmikandan et al., 2020) that includes breaking the cell barriers either physically or chemically to expose the lipids to the outside, reduction of particulate size (Halim et al., 2012), and reduction of water content in oil to enhance the trans-esterification efficiency.

3.2 *Lipid Extraction from Microalgae and Cyanobacteria Biomass*

As given in Mubarak et al. (2015), lipid extraction from microalgae and cyanobacteria can be either chemical or mechanical. Chemical methods may include solvent extraction method, supercritical CO_2 extraction, ionic liquids as solvents for extraction, etc., and the mechanical methods may include expeller or press method, ultrasound-assisted extraction method, microwave-assisted extraction method, etc.

On the laboratory scale, chemical-based lipid extraction methods are important, and fatty acids are soluble in non-polar organic solvents. The lipid extraction process for biodiesel production from microalgae needs to be not only lipid specific but also need to reduce the co-extraction of non-lipid contaminants such as proteins and carbohydrates and should be selected for desirable neutral lipid fractions (e.g., acylglycerols) (Halim et al., 2012; Sheng et al., 2011). Madusanka and Manage (2018a) used the Soxhlet extraction method for lipid extraction from cyanobacteria.

Industrially, lipid extraction is a costly but important process, and the economical production of microalgae biodiesel depends on the energy utilized for the processing of biomass and the type of lipid extraction process (Kim et al., 2013). The lipid extraction method should be effective and energetically efficient to be profitable.

3.2.1 Mechanical Methods for Microalgae Lipid Extraction

Microalgae or cyanobacteria lipids locate either bound to the cell membrane or as lipid globules in the cytoplasm. Lipid globules contain lipid produced by cell metabolic activities, while the cell membrane contains lipid molecules as the structural compound. Available percentages vary with the species and the environmental condition. Mechanical methods are easy to operate but are slow and require larger quantities of biomass and sometimes require an excessive power supply. Mubarak et al. (2015) reported that the expeller or press method shows up to 75% lipid extraction efficiency. Even though the press method gives a comparatively larger result, the thick cell walls reduce the release of intra-lipid available inside the cell (Lee et al., 2010).

Ultrasound-assisted lipid extraction and microwave-assisted lipid extraction are two techniques that enhance lipid extraction as microalgae cell walls get damaged and exposed with both ultrasound and microwaves (Selvakumar & Sivashanmugam, 2019).

During the ultrasound method, intense sonication of liquid generates higher-pressure and lower-pressure cycles. It produces vacuum bubbles during the low-pressure cycle and collapses further during the high-pressure cycles, which is called "cavitation" (Mubarak et al., 2015; Kumar & Banerjee, 2019). With the cavitation, microalgae cell membrane gets broken mechanically and improves the lipid extractability. Suali and Sarbatly (2012) recorded that the ultrasound technique enhances the lipid extraction by 50–500% with tenfold reduction of extraction time.

The microwave-mediated method uses a microwave to heat (non-contact heat) which penetrates the biomass and heat uniformly which allows super-quality lipid yield with the reduced extraction time (Chen et al., 2011; Selvakumar & Sivashanmugam, 2019).

3.2.2 Chemical Solvent Methods for Microalgae Lipid Extraction

The principle behind the solvent extraction method of lipids is a basic chemistry concept of "like dissolving like" (Halim et al., 2012; Mubarak et al., 2015). Static and dynamic methods are two different solvent extraction techniques.

Static Solvent Extraction Method

The static solvent extraction method is the easiest way to perform the lipid extraction, but the efficiency is quite doubtable, thus in this technique, lipid extraction is based on diffusion-driven principle. When the gradient of lipid concentrations in between the microalgal cell and the organic solvent is higher, the extraction becomes more rapid, and when the concentration gradient is lower, the lipid extraction rate slows down (Mubarak et al., 2015). No further lipid transfer from the cell to the solvent will take place when lipid concentration between the organic solvent and the cellular matrices reach partitions equilibrium (Halim et al., 2012).

Dynamic Solvent Extraction Method (Soxhlet Extraction Method)

The Soxhlet extraction method is cheap and easy to execute, but the use of organic hazardous chemicals (such as n-hexane) and longer operation time are the major drawbacks of this method (Mubarak et al., 2015). In laboratory-scale analysis, Soxhlet extraction is the most suitable lipid extraction method.

The Soxhlet apparatus consists of three main parts: the round bottom flask which is continuously heated, the middle part that holds thimble, and the condenser for continuous cooling and condenses gaseous solvent into the liquid phase.

In dynamic solvent extraction, during the solvent evaporation and condensation cycles, the Soxhlet apparatus refreshes the organic solvent continuously with biomass and enhances the lipid extraction from biomass while minimizing the solvent consumption (Wang & Weller, 2006) where one cycle is equal for the static state for a certain time and more cycles allow a higher extraction percentage. Higher temperatures increase the solubility of the solvents (Mubarak et al., 2016). But in Soxhlet extraction, higher temperature increases the solvent cycles (evaporation and condensation), but the solvent solubility is not considerably affected and does not contribute to the lipid yield significantly.

The thimble locates in the middle part of the Soxhlet extractor that avoids the dried biomass from being moved away by the organic solvent, and therefore it serves as a filter to remove cell debris (Halim et al., 2012).

Non-polar Solvents for Dynamic Lipid Extraction

According to the Halim et al. (2012), organic solvent extraction incorporates five steps including penetration of organic solvent through the cell membrane into the cytoplasm, interactions with neutral lipids by van der Waals forces, the formation of solvent-lipid complexes, diffusion back across the cell membrane according to the concentration gradient, and static solvent film formation surrounding the cell due to interactions between the cell wall and bulk organic solvents (Loyao Jr et al., 2018; Shin et al., 2018). Non-polar solvents such as n-hexane and dimethyl ether mainly interact with neutral lipid molecules in the cytoplasm (Madusanka & Manage, 2018a) (Fig. 3).

Mixed Solvents for Dynamic Lipid Extraction

Preparation of mixed solvents includes a mixture of polar solvents and non-polar solvents according to a proportion. Some of the neutral lipids locate in the microalgae, while polar lipids interact with the cell membrane. Strong hydrogen and electrostatic bonds keep polar lipids bound to the anchor proteins of the cell membrane (Loyao Jr et al., 2018; Madusanka & Manage, 2018b). The van der Waals forces that formed between non-polar organic solvents and neutral lipids of the complex are inadequate to break these membrane-based lipid-protein associations. But polar organic solvents (such as methanol, isopropanol, ethanol, etc.) can break the lipid-protein complex by forming hydrogen bonds with the polar lipids in the complexes (Halim et al., 2012; Loyao Jr et al., 2018) (Fig. 4).

Unlike non-polar organic solvents, the mixture of polar/non-polar organic solvent mixture can extract neutral lipids in the cytoplasm as well as polar lipids in

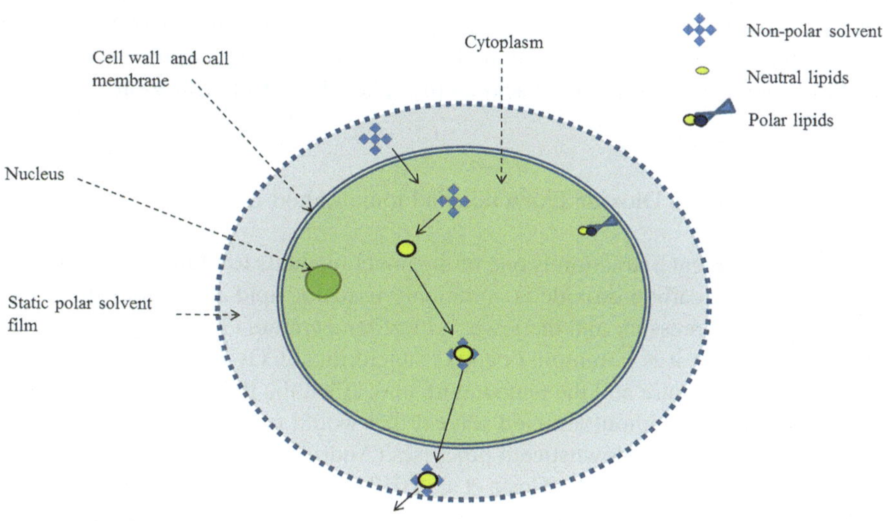

Fig. 3 Non-polar lipid extraction

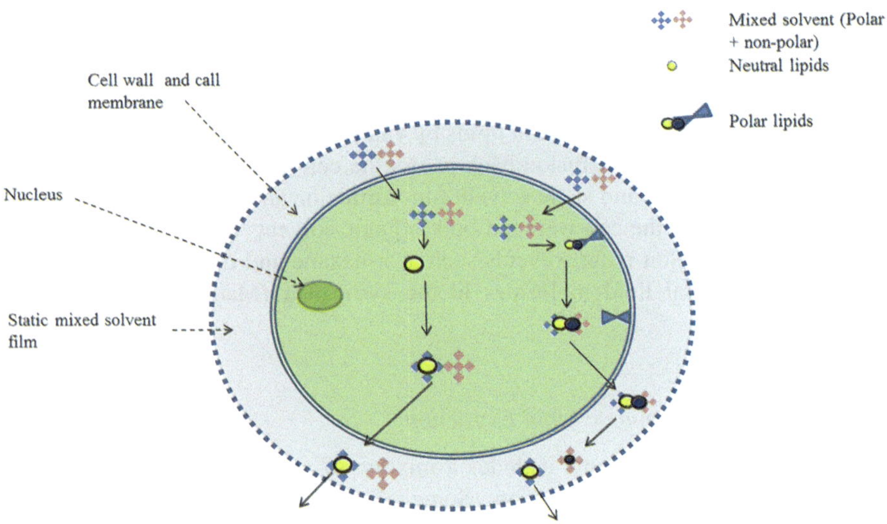

Fig. 4 Mixed solvent-lipid extraction

association with the cell membrane (Halim et al., 2012). In the case of mixed solvents for lipid extraction, the non-polar organic solvents surround the lipid complex and form a weaker bond (van der Waals associations) with neutral lipids in complexes, while polar organic solvents surround the lipid complex and form hydrogen bonds with the polar lipid complex (Loyao Jr et al., 2018). The hydrogen bonds are strong enough to disrupt the bond between lipid-protein complexes in cell membranes.

However, the mixed solvents may lead to co-extraction, such as pigments, ketones, and aldehyde groups (Halim et al., 2012; Loyao Jr et al., 2018; Madusanka & Manage, 2018b), and could result in difficulties in downstream processes such as lipid trans-esterification and fatty acid methyl ester (FAME) purification.

Supercritical Carbon Dioxide Extraction and Ionic Liquid Extraction

Supercritical solvent extraction is one of the novel methods used in lipid extraction, and supercritical carbon dioxide is commonly used for lipid extractions due to its moderate critical pressure and the lower critical temperature (7.4 MPa and 31.1 °C, respectively); thus it is commonly called "supercritical CO_2 extraction" ($SCCO_2$). The extraction pressure and the temperature may affect the fluid, and it can produce crude lipid extracts without a mixed solvent that would reduce the production cost and is advantageous for downstream processes (Andrich et al., 2005; Santana et al., 2012; Al-hattab et al., 2015; Da Costa et al., 2019). Halim et al. (2012) compared the performance of $SCCO_2$ extraction and the n-hexane extraction (both static and dynamic) and concluded that hexane solvent extraction was less efficient than

$SCCO_2$ that achieves higher lipid yield and less operating time with no significant difference in recorded FAME composition (Andrich et al., 2005; Couto et al., 2010).

Ionic liquid extraction is considered to be suitable for lipid extraction due to its non-volatility, thermal stability, and synthetic flexibility (Mubarak et al., 2015; Kim et al., 2013) but has not been used widely due to less cost-effectiveness.

3.3 *Trans-Esterification*

Three moles of methanol react with one mole of triglyceride (fatty acid) and result in one mole of glycerol and three moles of fatty acid methyl esters (FAMEs). This reaction is accelerated with the presence of catalyzing (chemicals (acids or bases), enzymes (lipase enzyme), etc.) (Chisti, 2007; Cheirsilp & Louhasakul, 2013). For commercial-scale production, chemical-based trans-esterification is used, which is more convenient. Considering the enzymes, it is needed to provide optimum storing temperature and other conditions that increase the production cost indirectly.

Chisti (2007) cited that alkali-based trans-esterification is faster (about 4000 times) than acid-based trans-esterification. But acid-catalyzed trans-esterification is important in the trans-esterification of free fatty acids (FFAs) (Madusanka & Manage, 2018a). Alkali catalyses cause saponification during the trans-esterification with the presence of FFA and cause a change in the fuel properties. Table 1 compares the advantages and disadvantages associated with different trans-esterification methods.

Table 1 Common trans-esterification methods and their relevant advantages and disadvantages (source: Chisti, 2007)

Type of trans-esterification	Advantages	Disadvantages
Chemical-based catalysts (acid- and alkali-based trans-esterification)	(a) The reaction can be controlled (b) Suitable for large-scale production (c) Production cost is low (d) Remained methanol can be reused (e) Higher production efficiency	(a) Energy consumption as it needs relatively higher temperature (50–60 °C) (b) Methanol recovery adds additional cost (c) Generates wastewater
Enzyme-based trans-esterification	(a) No pollution occurs (b) Needs relatively small amount of methanol	(a) Conversion of short-length fatty acids is limited (b) Reaction efficiency is affected by the chemicals available in the reaction medium

4 Properties of Microalgae and Cyanobacteria Biodiesel

Madusanka and Manage (2018b) reported fuel properties of biodiesel blend (B6) produced from *Microcystis* sp. and compared the obtained results with ASTM standards to determine the fuel quality and Ceypetco (Ceylon Petroleum Corporation) specification (local standard) for auto diesel to assess the usability instead of petrodiesel. Table 2 shows the fuel properties of *Microcystis* biodiesel B6 blend according to the obtained results of Madusanka and Manage (2018b).

Due to extreme properties of neat biodiesel (B100) such as high flashpoint, viscosity, etc., they cannot be used directly to operate normal compression-ignition (CI) diesel engines without any engine or biodiesel modifications (Madusanka & Manage, 2018b). Engine modifications are quite expensive. Therefore, biodiesel blends are important to overcome this problem (Knothe, 2006). Lower blends (<B20) of biodiesel are allowed for normal diesel engines without modifications

Table 2 Fuel properties of *Microcystis* biodiesel B6 blend (source: Madusanka & Manage, 2018b)

Parameter	Test method	Units	Limits of ASTM D7467 standard	B6 biodiesel blend from *Microcystis* lipids	Auto diesel[a]
Biodiesel content	ASTM D7371	% Vol.	6–20	6	–
Lubricity	ASTM D6079	μm	520 max	405	ns
Viscosity	ASTM D445	cSt	1.9–4.1	2.83	1.5–5
Flashpoint	ASTM D93	°C	52 min	<30	60 min
Density	ASTM D1298/ 4052	kg/m^3	ns	831.0	820–860
Sulfur content[b] S15 S500 S5000	ASTM D4294	wt.%	0.0015 (15 ppm) 0.050 (500 ppm) 0.5 (5000 ppm)	0.231 (2310 ppm)	0.3 (3000 ppm)
Gross calorific value	ASTM D240	kCal/kg	Ns	11,180	10,500
Cloud point	ASTM D2500	°C	Ns	6.0	Ns
Cold flow properties (CFPP)	ASTM D6371	°C	Ns	4.0	10 max

ns Not specified

[a]Ceypetco auto diesel specifications

[b]The fuel grades S15, S500, and S5000 refer to the maximum sulfur content allowed in the fuel expressed in ppm by weight and represent maximum sulfur content of 0.0015 wt.%, 0.05 wt. %, and 0.5 wt. %, respectively

instead of petrodiesel. Also, biodiesel blends may achieve some advanced qualities like equipment compatibility, cold weather operability, performance, and cost rather than B100 biodiesel. Contaminants such as remaining unreacted acylglycerols (mono-, di-, and triacylglycerol), partial un-separated glycerols, free fatty acids, residual alcohols, and catalysts may result in engine deposition, filter clogging, and fuel property deterioration (Knothe, 2006). Therefore proper refining and removing particles in the product are essential before utilization.

The composition of FAMEs and the availability of saturated and unsaturated FAMEs directly affect the biodiesel quality (Madusanka & Manage, 2018b). Commonly, higher content of polyunsaturated fatty acids and their methyl esters results in poor volatility, high viscosity, low oxidation stability, and the tendency for gum formation (Halim et al., 2011; Ryckebosch et al., 2014) that lead to burning difficulties and engine depositions which reduce the engine efficiency and fuel properties of biodiesel (Chisti, 2007) while enhancing the cold flow properties (low cloud point and low pour point) (Halim et al., 2012). With the temperature decrease, saturated fatty acid chains are packed and form tight semi-crystalline structures that freeze at a much lower temperature, while unsaturated fatty acids are prevented due to bends formed by the cis-double bonds (Lang et al., 2001). Therefore, biodiesel or biodiesel blends must comply with either ASTM standards or EN standards.

Many structural features such as the availability of heteroatoms, length of the carbon backbone, and degree of unsaturation (availability of double bonds) influence the biodiesel lubricity (Knothe & Steidley, 2005b). Two oxygen atoms in the head group enhance biodiesel lubricity. Increased chain length and degree of unsaturation show superior lubricity that enhances the fuel properties. Knothe and Steidley (2005a) reported that kinematic viscosity is useful in monitoring the fuel flow property of biodiesel that is mainly considered before determining the use of biodiesel as an alternative fuel instead of neat oils or fats. A high viscous fuel could lower the performance of the fuel flow system. The use of neat vegetable oil as diesel fuel has been considered unsatisfactory and impractical due to lower fuel flow properties (Moser, 2009).

The kinematic viscosity of biodiesel is influenced by structural features, such as double-bond availability and orientation, chain length, and type of ester head group (Moser, 2009). Longer-chain length and larger ester head groups increase the kinematic viscosity. The studies of Knothe and Steidley (2005a) provided evidence for higher kinematic viscosities cause operational defects such as engine depositions. The kinematic viscosities of FAMEs are much lower than their fatty acids and slightly higher than petrodiesel. Flashpoint is considered a safety limitation and serves to restrict the amount of alcohol in the biodiesel fuel (Knothe, 2006). As mentioned in Moser (2009), biodiesel contaminated with methanol, even with a small fraction ($<1\%$), fails to meet the biodiesel standards for the flashpoint that is considered as the safety limit for the fuel product. Therefore, contaminants such as methanol, water, catalyst, glycerol, FFA, soap, and metals may cause to lower the flashpoint. Therefore, proper purification of the product is very important in downstream processes.

The gross calorific value represents the higher value of the heat of combustion. As Moser (2009) defined, the heat of combustion is the thermal energy released by the burning of fuel. Oxygen composition and the hydrocarbon ratio are the major factors that influence the energy content of biodiesel or biodiesel blends (Moser, 2009). As Knothe and Steidley (2005b) assessed, the heat of combustion is directly proportional to chain length. Longer chains contain more carbons with a similar number of oxygen atoms that increase the gross calorific value, and lower carbon-to-hydrogen ratios (more hydrogen) exhibit greater energy content. The saturated FAMEs have lower C:H ratios, while unsaturated FAMEs of similar chain length have higher C:H ratios.

As reported in Madusanka and Manage (2018a) and De Oliveira et al. (2005), the gross calorific value for *Microcystis* FAMEs B6 blend is 11,180 kcal/kg (46.81 MJ/kg), and gross calorific value for soya bean biodiesel B20 blend was 10468.45 kcal/kg (43.83 MJ/kg). The higher percentage of saturated FAMEs may increase the gross calorific value (Madusanka & Manage, 2018b).

Moser (2009) documented that both cloud point and cold flow properties (CFPP, cold filter plug point) are parameters of low-temperature operability. Cloud point is defined as the specific temperature level that the larger crystals fuse together and form agglomerations that prevent the pouring of fluid.

According to several works of literature, many types of microalgae and cyanobacteria have a great potential to produce biodiesel. According to Minowa et al.'s (1995) experimental data, the good-quality biodiesel was synthesized from *Dunaliella tertiolecta*, and the viscosity and calorific value were 1.5–3.3 cSt and 36 MJ/kg, respectively. Also, Xu et al. (2006) produced high-quality biodiesel from *Chlorella protothecoides* and was characterized by a viscosity (at 40 °C) of 5.2 cSt., a high heating value (calorific value) of 41 MJ/kg, and a density of 864 kg/m^3. A viscosity (at 40 °C) of 2.83 cSt, a density of 831 kg/m^3, and a calorific value of 46.81 MJ/kg resulted in *Microcystis*-based biodiesel in Madusanka and Manage (2018b).

5 Potentials and Challenges of the Production of Biodiesel

Though biodiesel production has the potential to be used instead of petrodiesel, it mostly adheres with the technological developments of a few countries and thus is only popular among certain countries only. Especially, developed countries like the USA, Australia, the UK, and China and developing countries such as India produce biodiesel from many biomass types that contribute more than 50% of the global biodiesel production. Introducing novel technologies in biodiesel production to potential countries would be economically and socially beneficial both regionally and internationally.

Biodiesel production consists of the cost of the raw materials (fat or oil) and the downstream processing cost. The raw materials' cost accounts for a considerable amount (> 40%) of the total biodiesel production cost (Bhati, 2020). Though there is a wide variety of low-cost oils and fats available (such as restaurant waste and animal fats), the content of a large amount of free fatty acids (FFAs) is the major problem in low-cost oils and fats. The conversion of FFAs into relative FAMEs is difficult through alkali trans-esterification and increases soap formation during the trans-esterification process. Soap formation reduces the available reactive catalysts in the medium that also slow down the whole trans-esterification process (Demirbas, 2003; Madusanka & Manage, 2018b). As these are organic chemical transformations, it is very hard to predict an exact outcome or exact result.

Wastewater generation is increased day by day that most industries dispose of wastewater to the environment either partially treated or untreated that cause adverse environmental issues as well as social and health issues. Naturally, microalgae and cyanobacteria have great potential to absorb pollutants, nutrients, chemicals that available in wastewater and use them for their catabolism and completely degrade or partially degrade chemical components while increasing their biomasses (Rawat et al., 2011; Pires et al., 2013). This integration would provide a sustainable green solution for wastewater management and the fuel crisis of the future.

The wastewater generated in the production of biodiesel is another challenge as it contains many chemical and biological residues that could be polluted natural water bodies. Unreacted catalysts, oil compounds such as free fatty acids, non-oil compounds such as carbohydrates, and proteins may enter the natural waters and will cause eutrophication indirectly. Raw microalgae or cyanobacteria can be introduced accidentally to the natural environment and can be adversely affected ecologically and environmentally.

6 Conclusion

Though studies on sustainable biodiesel production are one of the emerging themes all around the world and had been popular in the past two decades, some of the challenges remain to be resolved. As cyanobacteria and microalgae have favorable lipid compositions, many studies have shown that microalgae and cyanobacteria are potential feedstock for biodiesel production. Cost for feedstock and production costs are significant factors that are crucial in biodiesel production on a commercial scale. Naturally available microalgae and cyanobacteria have been given much less attention for commercial biodiesel production rather than culturing microalgae and cyanobacteria using PBRs and raceways due to less sustainability. Wastewater-mediated microalgae and cyanobacteria cultures have a sustainable point of view due to providing feedstock for biodiesel production while remediating wastewater simultaneously. The lipid extraction method and lipid trans-esterification method are the main contributors to the total cost of biodiesel production. Lower blends (<20) have a higher potential to be used as an alternative fuel without engine modifications. However, biodiesel from microalgae and cyanobacteria is carbon neutral, renewable, and environment-friendly; so, it's a challenge to completely replace petrodiesel for normal CI engines, without modifying it with a neat biofuel (B100).

References

A.S.T.M. (2008). D6751, Standard Specification for Bio-Diesel Fuel (B100) Blend Stock for Distillate Fuels. West Conshohocken, PA: ASTM. Automotive Fuels-Fatty-Acid Methyl Esters (FAME) for Diesel Engines—Requirements and Test Methods.

Al-Hattab, M., Ghaly, A., & Hammoud, A. (2015). Microalgae harvesting methods for industrial production of biodiesel: Critical review and comparative analysis. *Journal of Fundamentals of Renewable Energy and Applications, 5*, 2.

Andrade, C. J., & Andrade, L. M. (2018). Microalgae for bioremediation of textile wastewater. An overview. *MOJ Food Processing and Technology, 6*(5), 432–433.

Andrich, G., Nesti, U., Venturi, F., Zinnai, A., & Fiorentini, R. (2005). Supercritical fluid extraction of bioactive lipids from the microalga Nannochloropsis sp. *European Journal of Lipid Science and Technology, 107*(6), 381–386.

Arif, M., Bai, Y., Usman, M., Jalalah, M., Harraz, F. A., Al-Assiri, M. S., Li, X., Salama, E. S., & Zhang, C. (2020). Highest accumulated microalgal lipids (polar and non-polar) for biodiesel production with advanced wastewater treatment: Role of lipidomics. *Bioresource Technology, 298*, 122299.

Atadashi, I. M., Aroua, M. K., Aziz, A. A., & Sulaiman, N. M. N. (2012). The effects of water on biodiesel production and refining technologies: A review. *Renewable and Sustainable Energy Reviews, 16*(5), 3456–3470.

Barros, A. I., Gonçalves, A. L., Simões, M., & Pires, J. C. (2015). Harvesting techniques applied to microalgae: A review. *Renewable and Sustainable Energy Reviews, 41*, 1489–1500.

Bhati, R. (2020). Algae biodiesel: Fundamentals and future prospects. In *Biotechnology business-concept to delivery* (pp. 95–108). Springer.

Cheirsilp, B., & Louhasakul, Y. (2013). Industrial wastes as a promising renewable source for production of microbial lipid and direct transesterification of the lipid into biodiesel. *Bioresource Technology, 142*, 329–337.

Chen, C. Y., Yeh, K. L., Aisyah, R., Lee, D. J., & Chang, J. S. (2011). Cultivation, photobioreactor design and harvesting of microalgae for biodiesel production: A critical review. *Bioresource Technology, 102*(1), 71–81.

Chen, J., Li, J., Dong, W., Zhang, X., Tyagi, R. D., Drogui, P., & Surampalli, R. Y. (2018). The potential of microalgae in biodiesel production. *Renewable and Sustainable Energy Reviews, 90*, 336–346.

Chen, Y. A., Liu, P. W. G., Whang, L. M., Wu, Y. J., & Cheng, S. S. (2019). Biodegradability and microbial community investigation for soil contaminated with diesel blending with biodiesel. *Process Safety and Environmental Protection, 130*, 115–125.

Chisti, Y. (2007). Biodiesel from microalgae. *Biotechnology Advances, 25*(3), 294–306.

Chisti, Y. (2008). Biodiesel from microalgae beats bioethanol. *Trends in Biotechnology, 26*(3), 126–131.

Chisti, Y. (2019). Introduction to algal fuels. In *Biofuels from algae* (pp. 1–31). Elsevier.

Chu, T. C., & Rienzo, M. J. (2013). Bloom forming cyanobacteria and other phytoplankton in northern New Jersey freshwater bodies. *International Perspectives on Water Quality Management and Pollutant Control*, 1–23.

Couto, R. M., Simoes, P. C., Reis, A., Da Silva, T. L., Martins, V. H., & Sánchez-Vicente, Y. (2010). Supercritical fluid extraction of lipids from the heterotrophic microalga Crypthecodinium cohnii. *Engineering in Life Sciences, 10*(2), 158–164.

Da Costa, W. A., Bezerra, F. W. F., De Oliveira, M. S., de Aguiar Andrade, E. H., dos Santos, A. P. M., Cunha, V. M. B., dos Santos, D. C. S., da Silva Banna, D. A. D., Teixeira, E., & de Carvalho Junior, R. N. (2019). Supercritical CO2 extraction and transesterification of the residual oil from industrial palm kernel cake with supercritical methanol. *The Journal of Supercritical Fluids, 147*, 179–187.

Da Rós, P. C., Silva, C. S., Silva-Stenico, M. E., Fiore, M. F., & de Castro, H. F. (2012). Microcystis aeruginosa lipids as feedstock for biodiesel synthesis by enzymatic route. *Journal of Molecular Catalysis B: Enzymatic, 84*, 177–182.

Danquah, M. K., Ang, L., Uduman, N., Moheimani, N., & Forde, G. M. (2009). Dewatering of microalgal culture for biodiesel production: Exploring polymer flocculation and tangential flow filtration. *Journal of Chemical Technology Biotechnology International Research in Process Environmental & Clean Technology, 84*(7), 1078–1083.

Daroch, M., Geng, S., & Wang, G. (2013). Recent advances in liquid biofuel production from algal feedstocks. *Applied Energy, 102*, 1371–1381.

Dash, S. K. and Lingfa, P. (2018). An overview of biodiesel production and its utilization in diesel engines. In *IOP Conference Series: Materials Science and Engineering* (Vol. 377, No. 1, p. 012006). IOP Publishing.

De Oliveira, D., Di Luccio, M., Faccio, C., Dalla Rosa, C., Bender, J. P., Lipke, N., Amroginski, C., Dariva, C. & De Oliveira, J. V. (2005). Optimization of alkaline transesterification of soybean oil and castor oil for biodiesel production. *In Twenty-Sixth Symposium on Biotechnology for Fuels and Chemicals* (pp. 553–560). Humana Press.

De Souza Silva, A. P. F., Costa, M. C., Lopes, A. C., Neto, E. F. A., Leitao, R. C., Mota, C. R., & dos Santos, A. B. (2014). Comparison of pretreatment methods for total lipids extraction from mixed microalgae. *Renewable Energy, 63*, 762–766.

Delrue, F., Setier, P. A., Sahut, C., Cournac, L., Roubaud, A., Peltier, G., & Froment, A. K. (2012). An economic, sustainability, and energetic model of biodiesel production from microalgae. *Bioresource Technology, 111*, 191–200.

Demirbas, A. (2003). Biodiesel fuels from vegetable oils via catalytic and non-catalytic supercritical alcohol transesterifications and other methods: A survey. *Energy Conversion and Management, 44*(13), 2093–2109.

Ghazvini, M., Sadeghzadeh, M., Ahmadi, M. H., Moosavi, S., & Pourfayaz, F. (2019). Geothermal energy use in hydrogen production: A review. *International Journal of Energy Research, 43*(14), 7823–7851.

Gouveia, L., Oliveira, A. C., Congestri, R., Bruno, L., Soares, A. T., Menezes, R. S., & Tzovenis, I. (2017). Biodiesel from microalgae. In *Microalgae-based biofuels and bioproducts* (pp. 235–258). Woodhead Publishing.

Gülüm, M., & Bilgin, A. (2015). Density, flash point and heating value variations of corn oil biodiesel–diesel fuel blends. *Fuel Process Technol, 134*, 456–464.

Halim, R., Danquah, M. K., & Webley, P. A. (2012). Extraction of oil from microalgae for biodiesel production: A review. *Biotechnology Advances, 30*(3), 709–732.

Halim, R., Gladman, B., Danquah, M. K., & Webley, P. A. (2011). Oil extraction from microalgae for biodiesel production. *Bioresource Technology, 102*(1), 178–185.

Hossain, M. F. (2017). Production of clean energy from cyanobacterial biochemical products. *Strategic Planning for Energy and the Environment, 36*(3), 6–23.

Huang, D., Zhou, H., & Lin, L. (2012). Biodiesel: an alternative to conventional fuel. *Energy Procedia, 16*, 1874–1885.

Huang, G., Chen, F., Wei, D., Zhang, X., & Chen, G. (2010). Biodiesel production by microalgal biotechnology. *Applied Energy, 87*(1), 38–46.

Huang, Q., Jiang, F., Wang, L., & Yang, C. (2017). Design of photobioreactors for mass cultivation of photosynthetic organisms. *Engineering, 3*(3), 318–329.

Idroos, F. S., Hettiarachchi, I. U., & Manage, P. M. (2016). Optimization of a standard method for enumeration of total cell counts of colonial Microcystis aeruginosa in environmental samples collected from Boralasgamuwa Lake, Sri Lanka. *Vidyodaya Journal of Science, 18*.

Idroos, F. S., & Manage, P. M. (2014). Seasonal occurrence of microcystin-LR with respect to physico-chemical aspects of Beira lake water. *International Journal of Multidisciplinary Studies, 1*(2), 27–37.

Jung, K. A., Lim, S. R., Kim, Y., & Park, J. M. (2013). Potentials of macroalgae as feedstocks for biorefinery. *Bioresource Technology, 135*, 182–190.

Kadir, W. N. A., Lam, M. K., Uemura, Y., Lim, J. W., & Lee, K. T. (2018). Harvesting and pre-treatment of microalgae cultivated in wastewater for biodiesel production: A review. *Energy Conversion and Management, 171*, 1416–1429.

Kaga, Y., Kuda, T., Taniguchi, M., Yamaguchi, Y., Takenaka, H., Takahashi, H., & Kimura, B. (2021). The effects of fermentation with lactic acid bacteria on the antioxidant and anti-glycation properties of edible cyanobacteria and microalgae. *LWT, 135*, 110029.

Khalekuzzaman, M., Kabir, S. B., Islam, M. B., Datta, P., Alam, M. A., & Xu, J. (2020). Enhancing microalgal productivity and quality by different colored photobioreactors for biodiesel production using anaerobic reactor effluent. *Biomass Conversion and Biorefinery*, 1–13.

Kim, J., Yoo, G., Lee, H., Lim, J., Kim, K., Kim, C. W., Park, M. S., & Yang, J. W. (2013). Methods of downstream processing for the production of biodiesel from microalgae. *Biotechnology Advances, 31*(6), 862–876.

Knothe, G. (2006). Analyzing biodiesel: Standards and other methods. *Journal of the American Oil Chemists' Society, 83*(10), 823–833.

Knothe, G., & Steidley, K. R. (2005a). Kinematic viscosity of biodiesel fuel components and related compounds. Influence of compound structure and comparison to petrodiesel fuel components. *Fuel, 84*(9), 1059–1065.

Knothe, G., & Steidley, K. R. (2005b). Lubricity of components of biodiesel and petrodiesel. The origin of biodiesel lubricity. *Energy & Fuels, 19*(3), 1192–1200.

Koley, S., Mathimani, T., Bagchi, S. K., Sonkar, S., & Mallick, N. (2019). Microalgal biodiesel production at outdoor open and polyhouse raceway pond cultivations: A case study with Scenedesmus accuminatus using low-cost farm fertilizer medium. *Biomass and Bioenergy, 120*, 156–165.

Kolmakova, A. A., & Kolmakov, V. I. (2019). Amino acid composition of green microalgae and diatoms, cyanobacteria, and zooplankton. *Inland Water Biology, 12*(4), 452–461.

Kumar, S. J., & Banerjee, R. (2019). Enhanced lipid extraction from oleaginous yeast biomass using ultrasound assisted extraction: A greener and scalable process. *Ultrasonics Sonochemistry, 52*, 25–32.

Lakshmikandan, M., Murugesan, A. G., Wang, S., Abomohra, A. E. F., Jovita, P. A., & Kiruthiga, S. (2020). Sustainable biomass production under CO_2 conditions and effective wet microalgae lipid extraction for biodiesel production. *Journal of Cleaner Production, 247*, 119398.

Lang, X., Dalai, A. K., Bakhshi, N. N., Reaney, M. J., & Hertz, P. B. (2001). Preparation and characterization of bio-diesels from various bio-oils. *Bioresource Technology, 80*(1), 53–62.

Lee, J. Y., Yoo, C., Jun, S. Y., Ahn, C. Y., & Oh, H. M. (2010). Comparison of several methods for effective lipid extraction from microalgae. *Bioresource Technology, 101*(1), S75–S77.

Lin, W. R., Tan, S. I., Hsiang, C. C., Sung, P. K., & Ng, I. S. (2019). Challenges and opportunity of recent genome editing and multi-omics in cyanobacteria and microalgae for biorefinery. *Bioresource Technology, 291*, 121932.

Liu, X., Ye, M., Pu, B., & Tang, Z. (2012). Risk management for Jatropha curcas based biodiesel industry of Panzhihua prefecture in Southwest China. *Renewable and Sustainable Energy Reviews, 16*(3), 1721–1734.

Loyao, A. S., Jr., Villasica, S. L. G., Peña, P. L. L. D., & Go, A. W. (2018). Extraction of lipids from spent coffee grounds with non-polar renewable solvents as alternative. *Industrial Crops and Products, 119*, 152–161.

Madusanka, D. A. T., & Manage, P. M. (2018a). Optimizing a solvent system for lipid extraction from cyanobacterium Microcystis spp.: Future perspective for biodiesel production. *Journal of National Sciences Foundation, 46*(2), 219–227.

Madusanka, D. A. T., & Manage, P. M. (2018b). Potential utilization of Microcystis sp. for biodiesel production: Green solution for future energy crisis. *Asian Journal of Microbiology, Biotechnology and Environmental Science., 19*(2), 143–149.

Mahagamage, M. G. Y. L., & Manage, P. M. (2014). Water quality index (CCME-WQI) based assessment study of water quality in Kelani River Basin, Sri Lanka. *International Journal of Environment and Natural Resources, 1*, 199–204.

Mahmudul, H. M., Hagos, F. Y., Mamat, R., Adam, A. A., Ishak, W. F. W., & Alenezi, R. (2017). Production, characterization and performance of biodiesel as an alternative fuel in diesel engines–a review. *Renewable and Sustainable Energy Reviews, 72*, 497–509.

Manage, P. (2019). Cyanotoxins: A hidden cause of chronic kidney disease of unknown etiology (CKDu) in Sri Lanka-a review. *Sri Lanka Journal of Aquatic Science, 24*(1), 1–10.

Manage, P. M., Kawabata, Z., & Nakano, S. I. (2000). Algicidal effect of the bacterium Alcaligenes denitrificans on Microcystis spp. *Aquatic Microbial Ecology, 22*(2), 111–117.

Medina, A. R., Grima, E. M., Giménez, A. G., & González, M. I. (1998). Downstream processing of algal polyunsaturated fatty acids. *Biotechnology Advances, 16*(3), 517–580.

Minowa, T., Yokoyama, S. Y., Kishimoto, M., & Okakura, T. (1995). Oil production from algal cells of Dunaliella tertiolecta by direct thermochemical liquefaction. *Fuel, 74*(12), 1735–1738.

Mizik, T., & Gyarmati, G. (2021). Economic and sustainability of biodiesel production—A systematic literature review. *Clean Technologies, 3*(1), 19–36.

Mofijur, M., Rasul, M. G., Hassan, N. M. S., & Nabi, M. N. (2019). Recent development in the production of third generation biodiesel from microalgae. *Energy Procedia, 156*, 53–58.

Molino, A., Mehariya, S., Di Sanzo, G., Larocca, V., Martino, M., Leone, G. P., Marino, T., Chianese, S., Balducchi, R., & Musmarra, D. (2020). Recent developments in supercritical fluid extraction of bioactive compounds from microalgae: Role of key parameters, technological achievements and challenges. *Journal of CO2 Utilization, 36*, 196–209.

Moser, B. R. (2009). Biodiesel properties, and feedstocks. *Society for In Vitro Biology, 45*(3), 229–266.

Mubarak, M., Shaija, A., & Suchithra, T. V. (2015). A review on the extraction of lipid from microalgae for biodiesel production. *Algal Research, 7*, 117–123.

Mubarak, M., Shaija, A., & Suchithra, T. V. (2016). Optimization of lipid extraction from Salvinia molesta for biodiesel production using RSM and its FAME analysis. *Environmental Science and Pollution Research, 23*(14), 14047–14055.

Nelson, D. L., Lehninger, A. L., & Cox, M. M. (2008). *Lehninger principles of biochemistry*. Macmillan.

Olufemi, B., Sulaimon, S., & Arikawe, A. (2020). Optimum production and characterization of biodiesel from Microcystis Aeruginosa algae. *Journal of Engineering Studies and Research, 26*(4), 96–105.

Onumaegbu, C., Mooney, J., Alaswad, A., & Olabi, A. G. (2018). Pre-treatment methods for production of biofuel from microalgae biomass. *Renewable and Sustainable Energy Reviews, 93*, 16–26.

Pacheco, D., Rocha, A. C., Pereira, L., & Verdelhos, T. (2020). Microalgae water bioremediation: Trends and hot topics. *Applied Sciences, 10*(5), 1886.

Pandey, A., Srivastava, S., & Kumar, S. (2019). Isolation, screening and comprehensive characterization of candidate microalgae for biofuel feedstock production and dairy effluent treatment: A sustainable approach. *Bioresource Technology, 293*, 121998.

Park, Y. H. (2017). Analysis of characteristics of dynamic tidal power on the west coast of Korea. *Renewable and Sustainable Energy Reviews, 68*, 461–474.

Patel, R., de Oliveira, A., Newby, R., & Chu, T. (2019). Flow cytometric analysis of freshwater cyanobacteria: A case study. *Water, 11*(7), 1422.

Perendeci, N. A., Yılmaz, V., Taştan, B. E., Gökgöl, S., Fardinpoor, M., Namlı, A., & Steyer, J. P. (2019). Correlations between biochemical composition and biogas production during anaerobic digestion of microalgae and cyanobacteria isolated from different sources of Turkey. *Bioresource Technology, 281*, 209–216.

Pires, J. C. M., Alvim-Ferraz, M. C. M., Martins, F. G., & Simões, M. (2013). Wastewater treatment to enhance the economic viability of microalgae culture. *Environmental Science and Pollution Research, 20*(8), 5096–5105.

Posadas, E., Alcántara, C., García-Encina, P.A., Gouveia, L., Guieysse, B., Norvill, Z., Acién, F.G., Markou, G., Congestri, R., Koreiviene, J. and Muñoz, R. (2017). Microalgae cultivation in wastewater. *In Microalgae-based biofuels and bioproducts* (pp. 67–91).

Raj, S., & Bhandari, M. (2017). Comparison of methods of production of biodiesel from Jatropha Curcas. *Journal of Biofuels, 8*(2), 58–80.
Rangabhashiyam, S., Behera, B., Aly, N., & Balasubramanian, P. (2017). Biodiesel from microalgae as a promising strategy for renewable bioenergy production-a review. *Journal of Environment & Biotechnology Research, 6*(4), 260–269.
Ranjith Kumar, R., Hanumantha Rao, P., & Arumugam, M. (2015). Lipid extraction methods from microalgae: A comprehensive review. *Frontiers in Energy Research, 2*, 61.
Rashid, N., Rehman, S. U., & Han, J. I. (2013). Rapid harvesting of freshwater microalgae using chitosan. *Process Biochemistry, 48*(7), 1107–1110.
Rawat, I., Ranjith Kumar, R., Mutanda, T., & Bux, F. (2011). Dual role of microalgae: Phycoremediation of domestic wastewater and biomass production for sustainable biofuels production. *Applied Energy, 88*, 3411–3424.
Rizwan, M., Mujtaba, G., Memon, S. A., Lee, K., & Rashid, N. (2018). Exploring the potential of microalgae for new biotechnology applications and beyond: A review. *Renewable and Sustainable Energy Reviews, 92*, 394–404.
Roux, J. M., Lamotte, H., & Achard, J. L. (2017). An overview of microalgae lipid extraction in a biorefinery framework. *Energy Procedia, 112*, 680–688.
Ryckebosch, E., Bruneel, C., Termote-Verhalle, R., Muylaert, K., & Foubert, I. (2014). Influence of extraction solvent system on extractability of lipid components from different microalgae species. *Algal Research, 3*, 36–43.
Sánchez-Bayo, A., Morales, V., Rodrı́guez, R., Crespo, G. V., & Bautista, L. F. (2020). Effect of cultivation variables on biomass composition and growth of microalgae and cyanobacteria. *Molecules, 25*, 2834.
Santana, A., Jesus, S., & Larrayoz, M. A. (2012). Supercritical carbon dioxide extraction of algal lipids for the biodiesel production. *Procedia Engineering, 42*, 1755–1761.
Selvakumar, P., & Sivashanmugam, P. (2019). Ultrasound assisted oleaginous yeast lipid extraction and garbage lipase catalyzed transesterification for enhanced biodiesel production. *Energy Conversion and Management, 179*, 141–151.
Sethunga, H., & Manage, P. M. (2010). Contamination status of algae toxins microcystins in some selected water bodies in Sri Lanka. *International Environmental Symposium*, 25.
Sheng, J., Vannela, R., & Rittmann, B. E. (2011). Evaluation of methods to extract and quantify lipids from Synechocystis PCC 6803. *Bioresource Technology, 102*(2), 1697–1703.
Shezan, S. A. and Ishraque, M. F. (2019). Assessment of a micro-grid hybrid wind-diesel-battery alternative energy system applicable for offshore islands. In 2019 5th International Conference on Advances in Electrical Engineering (ICAEE) (pp. 457–462). IEEE.
Shin, H. Y., Shim, S. H., Ryu, Y. J., Yang, J. H., Lim, S. M., & Lee, C. G. (2018). Lipid extraction from Tetraselmis sp. microalgae for biodiesel production using hexane-based solvent mixtures. *Biotechnology and Bioprocess Engineering, 23*(1), 16–22.
Shuba, E. S., & Kifle, D. (2018). Microalgae to biofuels:'Promising'alternative and renewable energy, review. *Renewable and Sustainable Energy Reviews, 81*, 743–755.
Simsek, S. (2020). Effects of biodiesel obtained from canola, sefflower oils and waste oils on the engine performance and exhaust emissions. *Fuel, 265*, 117026.
Smith, V. H., Sturm, B. S., Denoyelles, F. J., & Billings, S. A. (2010). The ecology of algal biodiesel production. *Trends in Ecology & Evolution, 25*(5), 301–309.
Song, Y. (2019). A study of OTEC application on deep-sea FPSOs. *Journal of Marine Science and Technology, 24*(2), 466–478.
Suali, E., & Sarbatly, R. (2012). Conversion of microalgae to biofuel. *Renewable and Sustainable Energy Reviews, 16*(6), 4316–4342.
Sudhakar, M. P., Kumar, B. R., Mathimani, T., & Arunkumar, K. (2019). A review on bioenergy and bioactive compounds from microalgae and macroalgae-sustainable energy perspective. *Journal of Cleaner Production, 228*, 1320–1333.

Tidwell, J. H., Tidwell, A., & Nelson, S. (2018). Surveying the solar power gap: Assessing the spatial distribution of emerging photovoltaic solar adoption in the state of Georgia, USA. *Sustainability, 10*(11), 4117.

Udaiyappan, A. F. M., Hasan, H. A., Takriff, M. S., & Abdullah, S. R. S. (2017). A review of the potentials, challenges and current status of microalgae biomass applications in industrial wastewater treatment. *Journal of Water Process Engineering, 20*, 8–21.

Urtubia, H. O., Betanzo, L. B., & Vásquez, M. (2016). Microalgae and cyanobacteria as green molecular factories: Tools and perspectives. *Algae-Organisms for Imminent Biotechnology*, 1–27.

Wang, L., & Weller, C. L. (2006). Recent advances in extraction of nutraceuticals from plants. *Trends in Food Science & Technology, 17*(6), 300–312.

Weerasinghe, W. M. L. I., Madusanka, D. A. T., & Manage, P. M. (2021). Isolation and identification of cellulase producing and sugar fermenting bacteria for second-generation bioethanol production. *International Journal of Renewable Energy Development, 10*(4), 699–711.

Wijesekara, W. A. M. A., & Manage, P. M. (2017). In vitro screening of antibacterial antifungal and cytotoxicity activities in crude extract of freshwater cyanobacterium Oscillatoria sp. *Journal of Trophical Forestry and Environment, 7*(2), 78–90.

Wijewickrama, M. M., & Manage, P. M. (2019). Accumulation of microcystin-LR in grains of two Rice varieties (Oryza sativa L.) and a leafy vegetable, Ipomoea aquatica. *Toxins, 11*(8), 432.

Xie, Q., Cai, L., Xia, F., Liang, X., Wu, Z., Liu, Y., Li, X., Lu, M., Nie, Y., & Ji, J. (2019). High vacuum distillation for low-sulfur biodiesel production: From laboratory to large scale. *Journal of Cleaner Production, 223*, 379–385.

Xu, H., Miao, X., & Wu, Q. (2006). High quality biodiesel production from a microalga Chlorella protothecoides by heterotrophic growth in fermenters. *Journal of Biotechnology, 126*(4), 499–507.

Yen, H. W., Hu, I. C., Chen, C. Y., Nagarajan, D., & Chang, J. S. (2019). Design of photobioreactors for algal cultivation. In *Biofuels from algae* (pp. 225–256). Elsevier.

Yew, G. Y., Lee, S. Y., Show, P. L., Tao, Y., Law, C. L., Nguyen, T. T. C., & Chang, J. S. (2019). Recent advances in algae biodiesel production: From upstream cultivation to downstream processing. *Bioresource Technology Reports, 7*, 100227.

Yin, Z., Zhu, L., Li, S., Hu, T., Chu, R., Mo, F., Hu, D., Liu, C., & Li, B. (2020). A comprehensive review on cultivation and harvesting of microalgae for biodiesel production: Environmental pollution control and future directions. *Bioresource Technology, 301*, 122804.

Zhou, A., & Thomson, E. (2009). The development of biofuels in Asia. *Applied Energy, 86*, S11–S20.

Different Energy Management Strategies for Clean Energy

Anjini Kumar Tiwary, Prashant Kumar Singh, Amit Kumar Tiwari, and Rajeev Kumar

Abstract Energy production is viewed today as one of the main culprits in polluting the atmosphere with respect to greenhouse emissions. Several works are being done to make energy production environment-friendly. Several approaches to protect the environment from this devil is to check the leakage at different levels. Since energy is one of the major cost factors in any manufacturing unit, its leakage is sealed as a major target area for effective cost control. Different energy management programme has currently carried out in many industries as a tool for this purpose. This paper reviews how such energy management programme is being used to make the energy consumption most efficient and thus support the environment.

Keywords Auditing energy · Clean energy · Energy optimization · Energy management

1 Introduction

The important factor of the whole world is energy. The definition of energy varies in accordance to the domain of subject. In economic aspect, energy acts as a fuel, which is a source of energy (Martina's, 2005). Fuel is the important factor other than food

A. K. Tiwary (✉)
Department of ECE, Birla Institute of Technology, Mesra, Ranchi, Jharkhand, India
e-mail: aktiwary@bitmesra.ac.in

P. K. Singh
University College of Engineering and Technology, Vinoba Bhave University, Hazaribagh, Jharkhand, India

A. K. Tiwari
Department of Chemical Engineering, Birla Institute of Technology, Mesra, Ranchi, Jharkhand, India

R. Kumar
Department of Mechanical Engineering, Birla Institute of Technology, Mesra, Ranchi, Jharkhand, India

D. B. Pal, J. M. Jha (eds.), *Sustainable and Clean Energy Production Technologies*, Clean Energy Production Technologies,
https://doi.org/10.1007/978-981-16-9135-5_2

which directly affects human. In earlier days, fire is the only source of energy as discovered earlier. Later fossil fuels turn the human civilization in every sector like industry, agriculture and environment. The first industrial revolution is started in year 1800 with the invention of stream engines powered by coal. In year 1870, new sources (electricity, gas and oil) of energy were invented. Another technical revolution in year 1922 was made with the electrification of manufacturing system (Ford & Crowther, 1922). With the invention of nuclear power, third industrial era started in year 1969. The first, second and third industrial era started with the production of stream power, electric power and industrial automation, respectively (Schwab, 2016). The fourth industrial era is started with the integration of process of production with energy which is an alternative, and the examples are geothermal energy, sun and wind (Sentryo, 2017).

In nature, numerous sources of energy are present. Generally, two major sources, viz. renewable and nonrenewable energies, are considered traditional sources of energy. The example of nonrenewable energy resource is hydrocarbon, oil, coal, natural gas and nuclear. They are limited and require a lot of time to refill, whereas renewable resources fill in a very short time span, e.g. solar, wind, geothermal, hydro, biomass and marine (Andrea, 2014). In recent year the unconventional sources are growing sources of energy such as carbon capture, waste to energy, etc.

Coal, a carbon-intensive fossil fuel, is an organic rock made up of carbon, hydrogen and oxygen (6, WCA, 2009). In low-graded coals, the highest level is moisture, whereas the lowest level or percentage is carbon content, while high-graded coals are those coals where the carbon contents are high and moisture level is low. The high- and low-graded coals are harder and softer, respectively. Nearly 40% of global power plant uses coal as a source of energy (6, WCA, 2009).

Fuels are a mixture of hydrocarbons which are extracted and refined from the rock strata.

The inflammable mixture of gaseous hydrocarbons is the natural gas. The natural gas mainly contains methane (more than 80%) with ethane, propane, butane, etc. Shale gas is commonly found in stratified sedimentary rocks or tight sandstones. Till 2015, it was observed that many countries have the ability to produce shale gas, and due to the availability of resources, it can be produced by the countries like Canada, the United States, Argentina and China (7, EIA, 2015). The range of methane and oxygen content in coal mine methane, which is another form of natural gas, is 25–60% and 5–12% (CE, 2017).

By using nuclear fusion and nuclear fission reactions, nuclear energy can be generated. This nuclear energy is appearing as heat. The nuclear fission can be natural or artificially using fissile materials. The nuclear plants are generating nearly 11% of the world electricity.

Hydroelectric power plant was made in the year of 1879 and is known as Niagara Falls (IHA, 2016). Hydropower is renewable and emission-free. Nearly 71% of renewable electricity is from hydropower.

It is evident that bioenergy technologies can transform biomass into electricity and heat energies. Bioenergy uses the organic substances or biomass for the production of energy. The forests and agriculture residues are important sources for the production of bioenergy. In renewable sources of energy, bioelectricity keeps its third position; and presently liquid biofuels are frequently utilized in the transportation system. There are many other examples of bioenergy products, e.g. biogas, charcoal, etc. A good amount of mixture of methane and carbon dioxide can also be produced from biomass. It is a universal truth that the sun is our primary source of energy which helps in energy production through various routes. The collector panels receive the solar power, i.e. the energy of the sun, and convert it into a kind of energy. In the desert area, large solar panels are used to collect enough power. These powers can be utilized in homes for hot water, cooling and electricity. In the whole world, limited geographical areas are there which get enough power from the sun. The availability of sun power is dependent on the seasons and weather. High investment is required to produce solar electricity, so it has not reached its optimum potential yet.

Another common power is wind power. Mostly large turbines are used to utilize the maximum available wind which is a source of power. Large turbines turn a generator to produce electricity. The installation of large turbines requires huge investment. The wind speed is non-uniform. Its speed varies from time to time. So, the generation of power is affected, and hence wind power is not an ideal choice.

Geothermal energy is produced beneath the earth, within the crust of earths. The radioactive particles have slow delay which produces high temperature continuously. This available high temperature heated the rocks, and this heated rock heats up the water within the earth. Hot water produces steam, and this steam is utilized to rotate the turbines. The rotating turbines generate the power and hence produce the geothermal energy. This energy is sustainable, clean and environment-friendly. During ancient times this geothermal energy is utilized for bathing purpose. This geothermal energy can be utilized in an industrial purpose or in residential work in small unit. The pumping of water and steam through the reservoir leads the plants to less emissions. The geothermal energy is produced in limited places throughout the world. Creating underground new reservoirs may increase the risk of an earthquake. So, the new reservoir is generally avoided. The Geysers, California, USA, is the location of the largest power plants in the world.

Another energy source is hydrogen, and it can be utilized as the source of fuel in the world. These fuels are utilized in various applications, e.g. vehicles, homes and industries. It is a renewable energy. Water is the most common element available on earth. In water, hydrogen is available in two-thirds of the proportions. This hydrogen once separated with other elements utilized is as the source of energy. It is not harmful for the environment and for the humans since its emission is not giving any wrong effect.

Tidal energy is another important renewable source of energy. The tides are produced in the ocean once the water waves in the ocean get an optimum height. To generate enough energy from the high tidal force, the altitude of the tide needs to be 5 metre minimum greater than the lowest tide. The kinetic energy, which is generated due to the high and low movements of the tides, finally can be converted into electrical energy. Even with the low speed of tides, tidal energy produces large energy. Mostly in coastal areas, tidal power is generated. Since coastal areas are limited, tidal energy generation is also limited based on the sites. Huge investment is required because of high-civil-construction power plants.

In the oceans, the wave energy is generally produced from the water waves, since the ocean is governed by the gravity of the moon, which makes chains of wave power. The wave energy is a renewable and environment-friendly source of energy. The transformation of wave energy into electric power can be done with several different approaches, e.g. dam structures, floor-anchored devices or below the water surface. It can be exploited in the coastal regions, and we can make the country to be self-dependent for fuels. Thus, the extra burdens on the countries may be minimized. But the generation of wave energy in oceans can disturb the watercraft which may be either private or commercial. It also disturbs the ecosystem of the marine. Since the wave is dependent on the wavelength, it can create the noise and visual disturbances.

In hydropower, the waterpower is utilized to move generators and the production of electricity. So, the major dam provides hydropower to an electric station. The important aspect of hydropower plant is the dam; thus the life span of dam is utmost. Major absolution work should always be functional, and it needs a huge amount of money. The hydropower is pollution-free. It involves no waste or toxic gases; hence it is not harmful for the environment.

Biomass energy is generated from the organic material. The whole world is using biomass energy. In plants, photosynthesis process takes place. Plants capture the sun's energy during the presence of the sun. Chlorophyll, which is available in plants, traps the energy from the sun by well-known process 'photosynthesis'. After burning the plants, the available H_2O and CO_2 are liberated into the atmosphere. Fuels in the industry and in homes are generally biomass energy used for the purpose of cooking and heating. It includes the waste of crops, wood chips, animal wastes, trees, yard clippings, etc. Here, the collection of fuel involved grind. In the atmosphere, a huge amount of carbon dioxide is released with the biomass energy. The biomass fuels cause air pollution when indoor cooking takes place. So, proper ventilation is required when biomass energy is used for indoor purposes to avoid any health problem. Unfeasible utilization of biomass causes the deformation of greenery, and thus it degrades the present environment.

The major renewable source of energy is nuclear power which is available at present in almost each country. Whether nuclear power is safe or whether it is energy-efficient while considering its waste, it is a topic to debate. By using nuclear reactions, this energy is created, and it is then collected and finally used by the power generators. At present there are moratoriums on its use since the research is going in the aspect of their safety and their waste materials it produces after their use. The uranium atoms are splitting, and the process of splitting is called nuclear fission. The splitting of the uranium creates heat and electricity. The formation of uranium is long ago. In the crust of earths, the atom uranium is found. The processing of conversion of uranium into the fuel for nuclear power plant is very difficult and too expensive. Thorium can be used as a fuel, which is an abundant element. The source of fuel from the fast-neutron reactors is ample enough that its scarcity is nearly impossible under practical scenarios.

The examples of fossil fuels are natural gases, coal and oil. It provides power using coal and oil in most of the world. Many products are made from oil. The most used products are gasoline. Another more common product is natural gas which is mostly used for heating purposes. The vehicle, which is powered by natural gas, is also more common nowadays easily seen on streets. Once fossil fuels are used, there is a heavy destructive refutation that pollutes the environment. Fossil fuels are in limited quantity, and their consumption should be done in an efficient manner.

2 Energy Management

Energy management is the utilization and technology applications in an efficient manner. The proper planning and its operation are made for the production and its consumption so that the organization should be energy-efficient (VDI, 2007). The principal goals of energy management are conservation and maintenance of resources, minimum budget, the extension of equipment life, lowering of maintenance cost and prevention of climate from undesired change as well as ensuring the easy and ingrained access of energy spectrum for all. Energy audit is the initial step towards a successful energy management programme. It evaluates the energy which is utilized by all building and system process. The auditor initiates at the utility metres. The energy auditor also locates each and every energy source coming into a facility. The auditor identifies energy streams for each fuel, locates those energy streams into different discrete functions, determines the efficiency of each of those functions and then finally makes a note on the energy and options for cost savings.

2.1 Different Approaches for Management of Energy

There are several different methods intended to promote energy management and their efficient way to utilize the energy use.

2.1.1 Standard of Energy Management and Auditing Energy

ISO 50001 and ISO 50002 are the international standards for energy management and energy, respectively. Standard EN 16247 is the European energy audit standard. In year 2011, the International Organization for Standardization (ISO) designed a standard 'ISO 50001'. This standard was designed to facilitate the different organizations and industries with the need of an energy management system (EnMS). It directs various regional and national energy management standards, regulations and its specifications which include Denmark, China, Ireland, Europe, etc. All organizations of any size are easily accepting the standard ISO 50001. Those organizations are either public or private sector. For the services and manufacturing, it is basically a classical approach of management system standard. The design is done in a way so that it could be merged with other systems that are mainly dealing with the quality and environmental management systems. It can be implemented either individually or in a group. With the concept of plan-do-check-act (PDCA), several standards designed such as ISO 14001 and ISO 9001 are environment and quality management standard, respectively. The standard ISO 50001 is not target oriented for energy performance improvement. This is designed in such a way that it can be tuned to any organizations or the regulatory authorities. This shows that any private sector or public sector can implement the standard with ISO 50001 with its self-energy policy and set a performance improvement of energy continuously in terms of its budget resources and its capacity (ISO, 2011). Those industries that have adopted a standardized EnMS for a wider energy management programme have generally come out with savings at par with what they have accomplished through their self-designed systems. The best model of this is Ireland; in Ireland, almost all the companies are following the standard EnMS. Companies are also involved in LIEN programme of the government through which their energy performance improved is markable. Those companies are not new in the programme of energy management. Those firms already set a standard for energy saving over the last 10 years without using any EnMS system (IEA/IIP, 2012).

2.1.2 Software for Energy Audit

There are several software available for energy audit and related works like Microsoft Visio, eQuest 3–5, eVALUator, DIAL.ux, MATLAB 6.5 and MotoMaster 4. Nordenaudit is a software used for the industrial energy audit purpose. This software is launched in 2012. In this software, the unit process concept is used. In unit process concept, a structure is established for the balance of energy in the map of audit.

2.1.3 Systems for Energy Monitoring

Different energy monitoring devices and systems are available in the markets, which may be related to a specific installation for specific systems. These monitoring devices and systems monitor the entire energy operation of a company where directly or indirectly energy is involved. It also analyses the energy uses and compares their uses at regular intervals. By monitoring this operation, energy uses are minimized.

2.1.4 Process Integration

In industry the system is more complex with different units involved in it, e.g. reactor, separator, boiler and heat exchanger. These different units are connected to each other, and during operation they interact with each other. Since here several units are cascaded to each other, analysing such system is more complex. Hence systemic methods are used to minimize the energy use. This systemic method is called process integration. This process consists of quite a few methods used for the system solutions of processes and also associates with different sub-processes. As per the definition given by IEA, the process integration is a 'systematic and general methods for designing integrated production systems, ranging from individual processes to total sites, with special emphasis on the efficient use of energy and reducing environmental effects' (Gundersen, 2000). Three examples of methods like pinch analysis, mathematical programming and energy analysis are the methods that are involved in process integration. According to ASHRAE and the organization of EN 50002, this type of arrangement comes under level III audit. Level III audits are in-depth studies of the data of level II energy audit, which is utilized for the optimization of the energy systems.

There are various important reasons of utilization of process integration methods. Some are listed below:

1. Sub-process improvement does not mean that improved entire system.
2. If a system is more complex, then with the help of systematic method, there is always a greater probability to finding the solution.
3. To achieve different goals, there should be always comparisons of solutions and trade-off between the clash targets.
4. Try to find out the result of the strategic issues, changes, new processes and investments.

2.1.5 Pinch Inspection

Pinch analysis is a very specific and systematic tool; this tool is more commonly used among the three process integration tools. According to Linnhoff et al. (1982), it is mostly used to determine the aspects of heat integration which is called process integration between the processes. In the analysis of pinch, the streams which are to be heated should be coupled with the streams which are to be cooled. Heat exchangers are used in between these hot and cold streams. Using a pinch analysis method, the theoretical minimum possible value of external heat requirement for a network of the heat exchanger can be made. The heat exchanger network is the external boilers or cooling towers through which either heat is added to the system or the amount of heat that is to be removed to cool off the system. The determination of quantity of heat that can be interchanged within the process is possible. Different curves are evolved through which the system can be investigated in several ways. The minimum possible external heating and cooling requirement could be determined theoretically by using a composite curve diagram, whereas the same information can be retrieved by using the grand composite curves (GCC). After the analysis of foreground and background in grand composite curve, the integration of present system with the process can be made. Apart from this GCC, there are other valuable curves that are there. Among them pinch curve is the important one (Nordman & Berntsson, 2005).

2.1.6 Energy Optimization

According to Rardin (1998), optimization or mathematical programming is another process integration method that is commonly used for energy optimization. During optimization, the question is imparted with an objective function and constraints. The objective function should either be maximized or minimized, e.g. minimizing cost or maximizing the profit of any firm or CO_2 emissions from energy use. As per the requirement, a multi-objective function can be used; in a multi-objective function, different objectives are weighted against each other at the same time. In the objective function, different variables are time, temperatures and pressures. During the optimization, the search for the values of the variables is made, and they incite the values of the objective functions. These variables are also added in the constraints. It allows the variables to assume certain value and exclude other values. During the analysis of energy system, the number of constraints is varying. In industrial energy system analysis, this number can go up to thousands, and the problem should be either linear or nonlinear. In nonlinear constraints, they are linearized with the help of integers. The problem is formulated with objective function and constraints, and it can be solved by using an algorithm.

2.1.7 Decision Support Tool

The decision support tool is reMIND. This software is based on the analysis method of industrial energy systems, i.e. MIND. Mixed-integer linear programming based this tool in programming nonlinear functions to the linear one with the help of integers. Apart from industrial energy system model, another system model can be designed provided its mathematical relations are developed. This tool can be utilized in the issues of the best or most favourable operating strategy and improved structures of the entire process, etc. The results of the application of this tool may include boiler operation, storage and turbines, cost of the system, materials and energy flows and other investments. The dynamics can be determined once a specific time is broken into smaller time intervals (Karlsson, 2011).

2.1.8 Energy Analysis

Energy analysis deals with reducing the loss of energy and consequently maximizes the amount of work output with the efficient utilization of the energy. Hence, this analysis identifies and evaluates the cause of deficiencies in the energy processes. It also tells how to improve the system deficiencies. In this analysis, due to system irreversibility, energy within the system is all time more than the energy out of the system. The efficiency of energy and its losses can be examined by properly evaluating the energy at different points in a series of process of energy conversion (Gong, 2004).

2.1.9 Manufacturing Simulation

A simulation is an emulation of actual processes or systems over time. The depth of understanding any system is clearer with the application of manufacturing simulation. In this process, a proper understanding of different parameters like identifying bottlenecks, reduce downtimes, reduce operating costs and preparation of quick operation schedules are important to broader aspects. But, in the process integration toolbox, the simulation tool is not available.

2.1.10 Investment Calculation Tools

In the audit system, investment can also be checked by analysing the net present value, straight payback, life cycle cost and initial rate of return parameters and then tells about whether it is profitable or non-profitable.

2.1.11 Energy Services

The energy services are a market-driven approach through which improved energy efficiency in the industry can be attained. This service can drive the individual or a country to the most developed country. This energy service company is represented as ESCO (energy service company). The collective name for the different form of energy services is called energy services. The target of each of these services is to make use of energy more efficiently. Direct and indirect energy services are the two energy services. Operational services, as well as services for financing investments in more energy-efficient technologies and processes, and advisory services are the direct and indirect energy services, respectively. Energy audits, analysis of energy invoices, hourly data for a company's electricity consumption and identification of energy efficiency measures are some of the examples of indirect energy services. The operation, maintenance and hands-on support for implementation of energy efficiency measures are the examples of direct energy services (Bergmasth & Strid, 2004). Energy services have been used to enhance energy efficiency and are used in the public sector or in personal uses. The revenue collection of energy service companies is based on how energy efficiency has changed along with fixed remuneration.

2.1.12 Energy Efficiency Networks

The way of organizing energy management in a group form is called energy efficiency networks. The structured way of a company for improving energy efficiency in stimulated time frame, initiating with the auditing of energy. The energy networks model was created in Switzerland in the late 1980s. After that, Germany and Sweden also started the energy efficiency network in their own country. In the country Sweden, Hackefors model are used. This model is on the concept of networking which is a network approach. This network mesh is a certification of a standard management group of environmental and quality systems. Hackefors model was started during the mid-1990s (Altea, 2018). One of the most important drivers in an energy efficiency network is that the cost of energy management is consequently minimized, as several companies share the cost of the energy efficiency network coordinator. If the energy efficiency network functions well, the coordinator of the energy is 'in-sourced' energy manager.

2.1.13 Database and Real Energy Efficiency Measures

The energy managers and energy auditors can analyse the current energy efficiency measures with the database of earlier data and find out the important points of their improvement. Comparative analysis reduces the cost of transaction in lieu of the energy audit. Since the energy auditors are directly not involved in the production process in the company, this database helps improve the quality of energy audit and ensures that the proposed measures taken are reliable. The proposed measures are reliable. Using the proposed measures, the quality of energy audit report improved. The database is useful for customers, in order to evaluate, review and validate from several energy service firms. Nearly greater than 50% of the industrial energy of Swedish is coming in the database of NEAD. NEAD includes the industrial policy programme and voluntary agreement programme with energy audit programme. The industrial policy programme covers from years 2005 to 2015. The improvement of energy efficiency programme in the electricity industry is associated with voluntary agreement programme (Energy Efficiency Program, PEE). Nearly 2300 measures of industrial energy efficiency are the actions of PEE in his first two programmes from years 2005 to 2014. Nearly 6200 energy efficiency measures are reported for energy audit policy programme between 2010 and 2015 along with PFE. In the aspect of promoting the national energy audit policy programmes, there are several tools for energy audits. The major energy audit components are obtaining comparable results that can then be used to evaluate, analyse, and improve the programme. An energy auditor, who has enough experience, never replaces the tools of practical energy efficiency database with the new one. He never proposes the new scheme without these database tools which is most useful and energy-efficient.

2.1.14 Training and Education

A company's proficiency development depends on the training of every individual of the company. The training can be organized in several steps for its proper understanding. So, it is important that those persons directly involved in energy issues will get training in the first phase and the persons who are not directly involved will get the training in the stages (Caffal, 1996).

3 The Standard of Energy Management

The energy management system is based on some model known as plan-do-check-act (PDCA). In year 2011, the international standard for energy management is ISO 50001. The European and Swedish standard is SS EN 16001 and SS 62 77 50, respectively. In Sweden, the first energy management standard was set in year 2000, and the standard is SS 62 77 50. Since it is part of the standard of Danish, it will be converted to align with the Swedish national industrial policy programme. There are several standards which relate to the energy efficiency of the industry. Some of them are listed below:

1. EN 16212:2012 (the efficiency of energy and its saving measurement)—This method is a top-down and bottom-up method.
2. EN 13187 (the buildings' thermal performance—detection based on qualitative analysis of the building and its thermal irregularities)—This method uses infrared.
3. EN 15603:2008 (the performance of energy in a building)—It defines the rating of energy and the complete use of energy.
4. EN 15193:2007 (the performance of energy in a building)—For lighting purposes, this energy is required.
5. EN 12464–1:2011 (workplaces lighting and light)—Mainly used for indoor workplaces.
6. EN 16231:2012—The method used was benchmarking of energy efficiency.
7. ISO 50003:2014 (need for professionals which provides certification of EnMS, auditing of EnMS)—The method is EnMS.
8. ISO 50006:2014 (EnMS)—Measuring the performance of energy with the help of the indicators of energy performance and the baselines of energy.

4 Comparison between Method and Result Management

Internally the complete energy management work is based on two pillars: (a) method management and (b) result management. Altogether these two make this EnMS completely successful. In the result management system, a clear target is set for getting improved energy efficiency. Since it is the important aspect in the process of advancement towards EnMS, it centres about a particular person, and it is mostly successful. However, if the concerned person changes his workplace, then the result deviates. In the process of method management, more freedom is provided in comparison to result management. With more freedom employee makes his or her own routines and responsibility of improvement of energy efficiency. Hence the employee is more responsible about his or her own created routine and more involved towards EnMS. Once the person changes his position, the routines are still with the organization, which is more useful. In result management, previous experience is not useful. If a person is moving on the sand, his path is not permanent. As the wave of air comes, the paths of the previous runner vanish completely. In method management programme, the paths made by the previous persons are useful for the others. With the route of the previous runner or person's footprint, the others will follow the same path. There is more consumption of time. Normally both methods are useful for a company when it is adopted in-house for energy management programme. The proper balance of these two methods is helpful in accordance to the company's work culture, previous experience of work and the available knowledge of the seniors.

5 Monitoring Progress of Energy Efficiency Improvement

Energy Key Performance Indicator (e-KPI) is the fundamental method of benchmarking process of energy efficiency improvement. According to thermodynamics energy cannot be consumed. The process of e-KPI is to monitor the progress. The specific use of energy divided by products which are used for a particular time period is called specific energy consumption (SEC). The review paper of a researcher Lawrence et al. in year 2019 on the topic of SEC found that this terminology is not defined properly and it is the huge deviations in the actual magnitude of SEC.

6 Different Obstacles for the Implementation of Energy Efficiency

Several advantages of energy management systems are elaborated in the present chapter. What is the need of government incentives to promote the energy management programme? Why any organization is not identifying without the absence of government incentives? In year 1995, Oates et al. explained that companies should keep attention of their cost reduction in the perfect competitive economy by looking the different options which is useful of the companies. Different reported literatures explained this question empirically like in Ambec et al. in year 2013 (Ambec et al., 2013). Lanoie et al. in year 2011 (Lanoie et al., 2011) hired a database which contains observations of many countries which are in number 7. He found that regulations of environment slightly offset the costs of using the environmental policies. Since the regulations of environment do excite the innovations of environment only, it is equally essential to discriminate between measures where huge public goods benefits and strong private benefit goods. Since energy efficiency comes under the second category, it is essential to find out why such measures are not carefully analysed. Individual countries have narrowly defined sectors even then there is a large variation in management quality among different companies. This is the indication for the capacity of different companies to find out opportunities through which productivity and profitability should be increased. This is nearly same in all the different firms like as iron sector or steel sector, etc. The process of data collection may not be available; this is one of the major problems. In some firms, this data collection for the relevant information should be there, but there is no such communication link between the data collection staffs and the responsible management staffs who are taking necessary decisions. In some other firms, maybe the communication link between the two is available, but again this energy efficiency improvement programme is not implemented in a meaningful manner due to other problems like capital allocation within the firm which is not promoting such investments or split responsibility. If the advantages associated with a given measure are in one cost centre, while the costs are behaved in another one, then the decision will be sub-optimal. Behavioural, technical and organizational obstacles are the other barriers of energy management system. These barriers restrict the companies to take part in energy efficiency investments and actions. An EnMS is a process which helps reduce many of these obstacles to energy efficiency. The same obstacles also affect the uptake of EnMS. Lack of general awareness of energy efficiency and its advantages is also a barrier of EnMS. An EnMS advises companies to their standing of their energy use and assess the financial possibility of relevant energy efficiency measures. These measures enable the documentation and clarification of energy use and aligned potential energy savings. Several literatures report various barriers to adopt energy management measures (IEA/IIP, 2012; Worrel, 2011) which are described below:

- Barriers from organization—without support system from their senior management of their firms—slow down the adoption of energy efficiency measures. Companies always focus on their production activity and their improvement, and they expense their core money on these things. Companies are least concern about their energy efficiency.
- Restricted knowledge of energy efficiency since the companies are not updating the advancement of energy-saving strategies.
- Observed technicality and its operational risks which are aligned with its execution of energy-efficient practices. This is because of lack of knowledge of energy reduction process which is mostly in practice to the major projects of business.
- Subsidized rate of energy in some regions restricts the companies to give less incentive to reduce their energy consumption. Here companies are getting the energy with very less cost, so they are not concerned with this.

- Distinguished professional and functional limitations within the organization create poor understanding. It should be helpful for the project of energy efficiency. For example, the bills made by the company's staff are another people and the staff who is purchasing and maintaining the equipment's are different.
- Any companies will primarily focus on the growth of their company by investing capital investment on their production capacity and revenue generation. Thus, limited finances restrict the companies to invest in energy efficiency measures.
- The observed directness of EnMS implementation is the requirement of the hardware installation. These instruments can be utilized to determine the energy and their emission streams. This is a particular case of SMEs.

In large industrial companies, several drive forces typically direct decision-makers. So senior management considers the common drivers while making the fresh investments or going with the other opportunities in business. The fresh opportunities in business arise with the reasons of energy and productivity efficiency (Reinaud & Goldberg, 2011):

- Companies' financial imperious.
- To attain compliance of environment, the policy should be the burden.
- Within the company the energy-saving opportunity knowledge.
- The concerned of the company towards energy efficiency and the environment.
- The demands of the public and market to improve the company's environmental or energy performance.

The effective way of stimulating the engagement and agreement of senior authorities for implementing an EnMS includes the development of a package of policies or an overarching energy management programme that addresses the above range of drivers, applicable to the sector and country.

7 Energy Management Program

The prime objectives of energy management programmes are to decrease industrial energy use and reduce greenhouse gas emissions through the implementation of an energy management system (EnMS). These programmes can also help achieve other objectives if properly designed and hence saving to its more productive use. The important activity of an energy management programme is EnMS. The success of EnMS is dependent on several factors like appropriate drivers, support mechanisms and linkages with other policies. Along with its effectivity, it is equally important that the development of an energy management programme is coordinated with national and/or regional low-carbon strategies. Thus, this energy management programme is equivalent to a tool that gives to the practical realization of the strategy. An industry should encourage energy management programmes through different drives. The drives include regulation, mandatory requirement and incentives which stimulate the industry. Different support systems are required for the

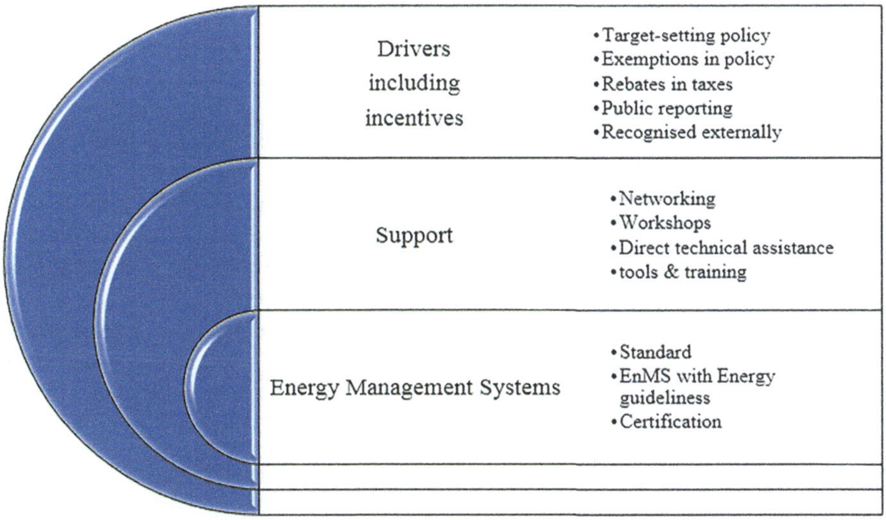

Fig. 1 Elements of energy management programmes (source: IEA/IIP, 2012)

initial implementation of an EnMS and its integration into any organization (IEA/IIP, 2012). The different support systems include technical assistance, capacity building and training, benchmarking tools and guidance. Figure 1 shows how the element makes well-designed energy management programmes.

A government should always provide sufficient support and incentives to the industry, and appropriate approaches are listed below based on government and business-related factors as reported in literatures (Global Superior Energy Performance Partnership, 2014):

- Energy intensity and level of energy consumptions in the industrial sector and small, and medium-sized enterprises (SMEs) to large companies' ratio.
- As long as companies exist on the global market and it is driven.
- Country's culture for management systems (earlier uplifting of other ISO management systems, such as ISO 9001 and ISO 14001).
- The government measures regarding possibility and culture in those areas.
- The energy service companies and equipment supplier's status.
- The financial and human resources available in the companies.
- The relation between the customer and supplier.
- Minimum track of carbon and improved efficiency of energy are the major needs for the environment performance.

The important elements listed above, and the factor of the company enables the drive of energy management programmes. These factors are flexible instruments which can be utilized to change policy and evolve industry and hence continue its relevance and its effectiveness. Many international organizations such as the Institute for Industrial Productivity (IIP), Global Superior Energy Performance (GSEP), the International Energy Agency (IEA) and UNIDO give detailed guidance on EnMS programme development. The above organizations strongly suggested the development of the EnMS programme of the government. It requires large energy concern industry and also encourages other energy users of the company for the confirmation of ISO 50001 and its equivalent management protocol (IEA, 2011). The Global Superior Energy Performance, Institute for Industrial Productivity, United Nations Industrial Development Programme and Energy Management Action Network (EMAK) are some of the international organizations promoting EnMS. IIP supports EnMS adoption on the ground in India, China and the United States, and it compiles and develops the best global practice information. Clean energy is the key concept of environment. For the improvement of EnMS in the industry and different commercial places, the government helps through the GSEP made. In industry the promotion of energy management is done with a network of policymakers that was built by the EMAK.

8 Incentives to Drive EnMS Uplift

Any financial support system is designed and adapted in such a manner to avoid potential unintended consequences that can be associated with the provision of subsidies or any incentives. These financial benefits by the government associated with energy efficiency improvement give broad energy efficiency policies that have a strong focus on lifting additional private sector finance in order to ensure sufficient finance for relevant energy efficiency in their countries. An EnMS requires minimum equipment investment and personnel time as reported (Therkelsen et al., 2013). Such types of financial support will help and uplift the drive of EnMS. If energy management programme should be designed in accordance to industrial needs, then it is a very cost-effective instrument for the governments to use in incentivizing industrial energy efficiency. The government can subsidize partially energy audit cost for the industry or even at no cost. To induce participation, incentives for EnMS are designed after which companies understand the benefits of EnMS and actively participate. The success of this programme is correlated with the provision of appropriate resources during its implementation stage. As the companies benefits of EnMS and cost-saving, they can achieve, further financial incentives are no longer required as reported in several literatures (McKane et al., 2009; Reinaud & Goldberg, 2012). Once a company executes the EnMS, then the resulting knowledge that demonstrates and qualifies the advantages of energy efficiency has the ability to afterwards encourage the private companies or small companies to invest in the identified opportunities, especially those that are more capital intensive. By using the standard method provided by the EnMS, energy saving and cost saving could help banks better assess the risks and returns of these projects if documenting and quantifying energy use are there. However, incentives for EnMS are only effective when they are adapted to the real needs of the industry. They are designed to motivate also private sector companies. The market variation is large, and regulation force of different countries increases the range of incentives to uplift of EnMS. EnMS incentives can improve the involvement of the industry in implementing energy efficiency, but in few cases, it unintentionally leads to restricting private sector investment support that would have occurred naturally in the absence of the incentive. Once the incentives are not utilized properly or not well adapted to the real demand of the industry, then it results in one of the possible hazards. Financial support is useful for the EnMS uplift if and only if the earlier direct technical support is provided; otherwise lack of knowledge does not uplift the EnMS programme.

8.1 Different Types of Typical Incentives for Industry Implementing EnMS

- Exemptions from policies in exchange for energy management programme participation, such as from an energy or carbon tax.
- Equipment cost which is used for energy saving should be on rebate price or some tax exemption which is not directly associated with energy management system.
- Audit cost of energy should be at a subsidized rate; recognition programme based on implementation and mandatory implementation of EnMS.
- The performance of a company should be in public to promote world-class behaviour.
- All such programme should be direct and with full technical help.

1. Canada (subsidy based on certification)—The industrial companies which adopt EnMS as per the standard of ISO 50001 and the other energy-linked things should be given cost-shared assistance. Maximum of 50% incentive should be provided for the proper implementation of ISO 50001.
2. China (pricing on electricity is differential)—The government has implemented the electricity bill as follows: the company who is more energy efficient should have differential price of electricity, and the company who is less energy efficient have to pay electricity price at higher cost.
3. Denmark (tax rebate is based on certification)—several investigations are needed to get the certification of energy management, i.e. standard EN16001, and energy-efficient projects should be in profit with that all projects should be payback horizon within 4 years to demand for the tax rebate of electricity.
4. Finland (subsidy on energy audit)—In year 1992, those companies who is interested in energy audit should be given nearly 40 to 50% subsidy, and their expectation was at the end of 2011; all should have done at least once.
5. Germany (based on certification tax rebate is there)—Those companies who got an EnMS certificate after verification can apply and get the energy tax rebate.
6. Ireland (direct technical help)—Companies who did the agreement with the Irish energy agency are getting a support manager. This support manager

generally gives advice one on one. In every year the support system is boosted in terms of technologies, initiatives and the specific domain of one's interest to promote EnMS programme through these members.

7. United States (subsidy on energy audit programme)—The departments of energy are providing free system energy who are small- and medium-sized enterprises (SMEs).

9 Support System for EnMS Implementation

The practical resources, which support the implementation of energy management system, are training, technical assistance, case study, benchmarking tools and practice sharing. These are the important tools to support system for EnMS. The companies which include these resources in their energy management programme will certainly enhance the energy efficiency with its proper utilization. Benchmarking process in companies deals with these parameters compared to processes, its operations and their systems. It leads to the best operation in comparison with the other similar companies or sites. In the sector of the iron and steel, companies who can set a standard by adopting EnMS in their firms put them as an example. They can be considered as the merits of energy efficiency in their equivalent ones. The two important tools are energy reviews and cost-benefit analysis. These two tools are universally accepted tools. These tools have many advantages with respect to energy efficiency measures. They are useful to the companies who are new in the area of EnMS and even useful for those who want to improve their present energy management processes. Since these programmes are adopted by the industry and it is always supported and funded by the government agencies, the valuable information is always available. There are a number of successful stories from the companies which tells the importance of EnMS. There are a different number of best practice materials available in the market for use. Those valuable experiences should be integrated into the present or new management programmes of energy.

9.1 Energy Management Practitioner's Toolbox

There should be a robust platform for internationally accepted EnMS tools. But there is no place where they host a robust platform which is accepted universally. The information resources and training materials, the Global Superior Energy Performance (GSEP) Partnership and the Institute for Industrial Productivity (IIP) have agreed to design and develop the toolbox for the initial phase of an energy management practitioner. The government and the private sector both will use and keep the available toolbox. The toolbox projects have collected nearly more than 400 tools, and some of them are for resources information, comparison of case studies, tools for software, energy-saving calculators and their methodologies, technology guidelines, training materials and other resources. In year 2015, the available toolbox is made public through an interactive website. These tools and resources will be used by the industry and mechanical persons which lead their steps towards successful EnMS and supply chain management.

10 Conclusion

The different sources of energy generation, renewable and nonrenewable, are elaborated. The prime objectives of energy management are conservation of resources, saving budget, extended equipment life, low maintenance cost and preventing climate change as well as ensuring easy and ingrained access for all the energy spectra. Energy audit is the initial step towards a successful energy management programme. The different approach and standard for energy management is explained. Comparative analysis is done between the method and result management. Different obstacles in the path of EnMS, its monitoring process, support system and different incentive schemes are explained for energy efficiency improvement.

11 Acknowledgement

First I would like to thank the editors for providing a chapter space for me in the book entitles *Sustainable Energy Production Technologies/Recent Technologies for the Production of Clean Energy* (edited by Dr. Dan Bahadur Pal and Dr. Jay Mant Jha). I'd also like to thank the many people who have helped me learn and practise both the art and science of energy management throughout the years. Additionally, I want to extend my special thanks to my co-authors Dr. Amit Kumar Tiwari, Dr. Prashant Kumar Singh and Dr. Rajeev Kumar, for their valuable contribution in this book chapter 'Review on Different Energy Management Strategies'.

I would also want to extend my special thanks to our Prof. S. S. Solanki, Prof. V. R. Gupta and Prof. Nisha Gupta (Department of ECE, Birla Institute of Technology, Mesra, Ranchi) for their full enthusiastic support and motivation throughout my career. Finally, I would like to thank my better-half, Shweta, for tolerating my incessant disappearances into my home. A lifelong partner makes both the journey and destination worthwhile.

References

Altea, 2018. The Hackefors-model. https://altea.se/en/the-hackefors-model/ (accessed 02.01.19).

Ambec, S., Cohen, M. A., Elgie, S., & Lanoie, P. (2013). The porter hypothesis at 20: Can environmental regulation enhance innovation and competitiveness? *Review of Environmental Economics and Policy, 7*(1), 2–22.

Andrea, A (2014). Nonrenewable and renewable energy resources. Retrieved from https://ww2.kqed.org/quest/2014/02/13/nonrenewable-and-renewable-energy-resources-2/

Bergmasth, M., Strid, M. (2004). Energitjänster på en avreglerad marknad - för en effektivare energianvändning? (Energy Services in a deregulated market—for a more efficient use ¨ of energy?), Doctoral thesis. School of Economics and Commercial Law, 2004. Gothenburg University, Gothenburg, Sweden [in Swedish].

Caffal, C. (1996). Energy Management in Industry. Centre for the Analysis and Dissemination of Demonstrated Energy Technologies (CADDET), Sittard, The Netherlands, Analysis Series 17.

CE. (2017). Working mine methane. Retrieved from https://www.clarke-energy.com/coal-gas/working-mine-methane/.

EIA. (February 13, 2015). Shale gas and tight oil are commercially produced in just four countries. Retrieved from https://www.eia.gov/todayinenergy/detail.php?id¼19991.

Ford, H., & Crowther, S. (1922). *My life and work*. Doubleday Page & Company.

Global Superior Energy Performance Partnership (2014). Models for driving energy efficiency nationally using energy management.

Gong, M. (2004). Using exergy an optimization models to improve industrial energy systems towards sustainability, Linko¨ping Studies in Science and Technology, Dissertation No. 868, ISBN 91-7373-933-2.

Gundersen, T., 2000. A process integration primer e implementing agreement on process integration. International Energy Agency Report, SINTEF Energy Research, Trondheim, Norway.

IHA. (2016). A brief history of hydropower. Retrieved from https://www.hydropower.org/a-briefhistory-of-hydropower.

International Energy Agency (IEA) (2011). 25 Energy efficiency policy recommendations.

International Energy Agency/Institute for Industrial Productivity (IIP) (2012). Energy management programmes for industry, gaining through saving, IEA policy pathway series, IEA/OECD, Paris, France.

International Organization for Standardization (ISO (2011)), Win the energy challenge with ISO 50001. DSTI/SU/SC(2014)14/FINAL 29.

Karlsson, M. (2011). The MIND method: A decision support for optimization of industrial energy systems: Principles and case studies. *Applied Energy, 88*, 577589.

Lanoie, P., Laurent-Lucchetti, J., Johnstone, N., & Ambec, S. (2011). Environmental policy, innovation and performance: New insights on the porter hypothesis. *Journal of Economics & Management Strategy, 20*, 803–842.

Linnhoff, B., Townsend, D. W., Boland, D., Hewitt, G. F., Thomas, B. E. A., Guy, A. R., et al. (1982). *A user guide on process integration for the efficient use of energy*. IChemE. Revised edition published in 1994.

Martina's, K. (2005). Energy in physics and in economy. *Interdisciplinary Description of ComplexSystems, 3*(2), 44e58.

McKane, A., Desai, D., Matteini, M., Meffert, W., Williams, R., & Risser, R. (2009). *Thinking globally: How ISO 15001—Energy management can make industrial energy efficiency standard practice*. Lawrence Berkeley National Laboratory.

Nordman, R., Berntsson, T. (2005). Advanced pinch technology based composite curves for evaluating the usable excess heat potential. The 18th International Conference on Efficiency, Cost, Optimization, Simulation and Environmental Impact of Energy Systems. Trondheim, Norway, June 2222.

Rardin, R. L. (1998). *Optimization in operations research*. Prentice-Hall Inc.

Reinaud, J. and Goldberg, A. (2011). The boardroom perspective: how does energy efficiency policy influence decision making in industry?, International Energy Agency and Institute for Industrial Productivity for the IEA Energy efficiency series.

Reinaud, J. and Goldberg, A. (2012). Promoting Energy Management Systems through Energy Efficiency Programmes, Incentives and Support—Lessons Learnt from Evaluations in Denmark, Ireland and Sweden.

Schwab, K. (2016). The fourth industrial revolution: What it means and how to respond. Retrievedfrom https://www.weforum.org/agenda/2016/01/the-fourth-industrial-revolution-what-itmeans-and-how-to-respond/

Sentryo. (2017). The 4 industrial revolutions. Retrieved from https://www.sentryo.net/the-4-industrial-revolutions/.

Therkelsen, P., Sabouni, R., McKane, A., and Scheihing, P. (2013). Assessing the costs and benefits of the superior energy performance program, 2013 ACEEE Summer.

VDI. (2007). Energy management—terms and definitions. Berlin, Germany.

WCA. (2009). *The coal resources: A comprehensive overview of coal.* World Coal Association.

Worrel, E. (2011), Barriers to energy efficiency: International case studies on successful barrier removal, working paper 14/2011, Development Policy, Statistics and Research Branch, United Nations Industrial Development Organization. Vienna.

Recycle of Plastic Waste to Liquid Fuel: A Sustainable Energy Production

Satarupa Pattanayak, Nirupama Prasad, and Sumit Kumar Jana

Abstract The rapid rate of urbanization of global plastic production has increased over the years. A significant volume of plastic waste has resulted in this rise. In this context, the book chapter has been conducted to contribute to the creation of successful worldwide plastic waste policy approaches and to the conversion of this waste into fuel. For the high energy demand compensation, the researchers turned their attention to the method of energy recovery. The conversion of plastic waste into energy has been developed through comprehensive research and technological growth. Conversion of plastic waste into oil using pyrolysis process had a great potential. Plastics are mainly produced by petroleum. Thus, oil produced from plastic wastes had a good calorific value and is comparable to the commercial fuel. For each type of plastic, this chapter examined the pyrolysis process and its key process parameters that affected the product quality, such as diesel oil, gases, and char. Furthermore, several points of view were also explored in this book to maximize the development of liquid oil for each plastic. This graded waste plastic was mixed with various catalysts on different occasions and fed into the experimental set-up, which included a pyrolysis reactor, condenser, and collector. The feed was kept at a temperature of around 450–550 °C and atmospheric pressure. The produced vapors were passed through a condenser, and the final liquid fuel was collected in a jar. Various quality tests were performed on the obtained fuel, and the obtained results were compared with the existing commercial counterparts.

Keywords Plastic waste · Chemical recycle · Reactor · Pyrolysis · Catalyst

---

S. Pattanayak
Department of Chemistry, Birla Institute of Technology, Mesra, Ranchi, India

N. Prasad · S. K. Jana (✉)
Department of Chemical Engineering, Birla Institute of Technology, Mesra, Ranchi, India
e-mail: sumitkrjana@bitmesra.ac.in

D. B. Pal, J. M. Jha (eds.), *Sustainable and Clean Energy Production Technologies*, Clean Energy Production Technologies,
https://doi.org/10.1007/978-981-16-9135-5_3

1 Introduction

Plastic waste pyrolysis is one of the efficient techniques of chemical recycling. In the present days, one has to be very cautious regarding the use of petroleum and other energy sources. The fossil fuel reserves are depleting day by day. Municipal wastes contain a high amount of plastic wastes, which really creates problem like choking of sewage systems. If pyrolyzed the waste material generates a fuel oil which may be used in diesel engines. In this way a good amount of foreign currency may also be saved as the import of crude petroleum may be reduced.

There are various methods to overcome this problem. Blending of plastic with virgin material to some extent solves the problem. However, blending reduces the quality of the material; articles made from blended plastics will never retain the same elasticity and strength as those are prepared from virgin plastic. Considerable fraction of the plastics in municipal waste ends up in land filling. The main hindrance to the implementation of plastic waste recycling is the inhomogeneity of plastic wastes.

Plastic consists of malleable synthetic or semi-synthetic organic compounds that can be easily molded into any shape. These are usually high-molecular-mass organic polymers. They are synthetic substances that are mostly derived from petrochemicals. However, many are derived from renewable sources. These plastics have good plasticity, i.e., able to bend irreversibly without cracking. These materials have a wide range of applications from simple household items to sophisticated industrial equipments. These materials have many advantages such as light weight, low cost, easy processing, flexibility, corrosion-free, and water resistance. In most of their former uses, they have already replaced many conventional materials, such as wood, stone, horn and bone, leather, paper, metal, glass, and ceramics.

An intense search for alternative fuels for internal combustion engines has resulted from the rapid depletion of petroleum fuel and environmental problems. The conversion of waste substances into energy is one of the methods for deriving alternative fuels. On the other hand, due to the increase in production of plastic for packaging, the disposal of waste plastics has become an essential issue. In India there is a tremendous growth of consumption of plastics in the last decades. Consumption of plastic increased from 3 kg per capita to 13 kg per capita, but it is still very less as compared to the USA whose per capita consumption is 93 kg. Recycling of plastics is very less. On the other hand, due to fast depletion of fossil fuel, oil prices are increasing very rapidly. With this pyrolysis technique, we can solve the waste disposal problem, and on the other way, we get valuable petrol ileum products at economic price. On the basis of environmental issue, this is the best plastic recycling technique. While on economic level it cheaper than other as in other technique plastic has to sort as per their properties which make the process more expensive. It can be easily separated for metal and dust by floating technique process. It can be used for any type of waste irrespective of their chemical structure. Moreover it does not emit any harmful gases like H_2S, SO_2, HCl, etc. in the atmosphere as compared to thermal recovery or incineration.

Pyrolysis of waste plastic materials in the presence of a catalyst may be one of the useful solutions to this problem (Kasar et al., 2020). Plastics are long-chain polymers of hydrocarbons mainly derived from by-products of the petrochemical industry and are effectively a non-renewable fissile-based resource. In general, municipal solid waste consists of HDPE, LDPE, polypropylene (PP), polystyrene (PS), and polyethylene terephthalate (PET) (Shafferina et al., 2016). Fossil fuels also are of great concern today due to its declining resources. So, if depolymerized efficiently, they may produce smaller-size hydrocarbons which can be used as fuels. At a temperature above the ceiling temperature, the polymers generally decompose into smaller hydrocarbons. At the same time, some amount of foreign currency may also be saved. Addition of catalyst enhances the conversion and fuel quality. The addition of catalyst during pyrolysis significantly lowers pyrolysis temperatures and time. Also, there is an increase in conversion. Several investigators to recover gaseous as well as liquid fuel carried out thermal catalytic cracking of polymeric waste. Akpanudoh et al. (2005) carried out thermal degradation of polyethylene in the presence of two catalysts containing 20% and 40% ultra-stable zeolite in a fixed bed reactor and obtained a liquid fuel product having lighter fraction. Yanik et al. (2001) performed thermal degradation of polyvinyl chloride (PVC) containing polymer mixtures using red mud (a waste from alumina production) as the catalyst and flowing nitrogen, obtaining a product of fuel oil and hydrochloric acid. Miskoloczi et al. (2005) performed thermal cracking of waste polymer like polyethylene, polypropylene, and polystyrene and obtained liquid fuels and lubricant with low sulfur content. Areana and Mastellone (2000) cracked polyethylene and polyethylene terephthalate (PET) in a bubbling fluidized bed reactor made of quartz with operating variables (bed solids hold up, inert material size, fluidizing velocity, plastics feed rate) at a bed temperature range of 450–650 °C using sand as an inert material and obtained a fuel. Gersten et al. (1999) pyrolyzed waste rubber and plastics with oil share (25% by weight) at 500–900 °C (catalytically) and obtained a greater quantity of gas and an oil rich in valuable aromatic compounds. Walendziewski (2005) cracked waste polymers in a glass reactor at atmospheric pressure and in a temperature range of 320–420 °C and in autoclave under hydrogen pressure of 3–5 Mpa in the temperature range 380–440 °C using catalyst like acid silica-alumina, zeolites (HY, HZSM-5, modernize), as well as alkaline compounds like ZnO, CaO, and K_2O and obtained gaseous and liquid fuels whose composition varied on process condition and type of catalyst used; the solid product obtained was char. Williams and Williams (1997) carried out pyrolysis of plastic mixture at 500–700 °C and 9.8–88.8% of the feed being converted to a gaseous product, while the liquid product obtained was 18.44–57.11%. They also observed that the amount of aromatic compound in oil increased with increasing temperature. Bockhorn et al. (1999) measured the formal kinetic parameters of the degradation of polymers applying dynamic methods using thermogravimetry-mass spectrometry (TG-MS) and isothermal methods using a closed mixture in a fluidized bed reactor at temperature between 500 and 700 °C and obtained gases, oils, and char as products. Miskoloczi et al. (2004, 2005, 2006) carried out degradation of waste plastic (polyethylene and polystyrene) in a batch reactor within the temperature range 410–450 °C in the presence of catalyst (FCC, ZSM-5, and clinoptilolite) and obtained light liquid hydrocarbon (C_5–C_{28} range) and gaseous products (Shafferina et al., 2016). Gobin and Manos (2004) carried out thermal degradation of polyethylene semi-batch reactor over various micro-porous catalyst particles like zeolites and obtained a liquid fuel rich in light hydrocarbon similar to gasoline. Blazso (1999) applied pyrolysis-gas chromatography-mass spectrometry (Py-GCMs) for the fast online testing of solid catalyst activity in upgrading oil and/or gases obtained by pyrolysis of natural and synthetic polymers and found that intramolecular hydrogen transfer reactions prompted by Na-zeolite results in the isomerization of alkenes and alkadienes evolved by pyrolysis of polyethylene and polypropylene. Chang (1996) pyrolyzed waste tire at 400 °C to gas and liquid char and observed that the production of fuel oil is at maximum 350 °C and gas production rate was 30–53% of feed. Serrano et al. (2005) performed catalytic cracking of both LDPE and HDPE using three different samples of nanocrystalline HZSM-5 zeolite as catalyst at temperature of 340 °C for 2.0 h using a plastic/catalyst ratio of 100 and observed that the products obtained were light hydrocarbon (C_3–C_5) fractions with a considerable amount of olefins. Miskoloczi et al. (2005) performed pyrolysis of waste polymer like polystyrene and polycarbonate and obtained a diesel-like liquid fuel. Yoon et al. (1999) cracked polyethylene, polypropylene, polystyrene, polyethylene terephthalate, and acrylonitrile-butadiene-styrene with lubricating oil and obtained a fuel oil. Walendziewski (2005) performed cracking of polyethylene, polystyrene, and polypropylene in a continuous flow reactor at 420–480 °C with a feed ratio of 0.3–2.4 kg/h of polymer and a liquid product having 20–40% of gasoline fraction and 30–80% light gas oil, the solid product having same characteristics of brown coal. Zhibo et al. (1996) catalytically degraded polyethylene using silica-alumina and ZSM-5 zeolite to the obtained fuel oils. Some investigators have also investigated polymer modifications through catalytic hydrogenation. Karagoz et al. (2003) hydrogenated municipal waste plastic with vacuum gas oil over HZSM-5, DHC-8 (commercial silica-alumina catalyst), and cobalt-loaded active carbon catalyst at temperatures between 425 and 450 °C and obtained gaseous and liquid fuels, whose proportions were observed to be dependent on the type of catalyst used and operating temperature. Catalytic hydrogenation of polystyrene in solution in an organic solvent was studied by Roberts and Dawei (2003) using an agitator to produce other polymers like poly(vinyl cyclohexane) having higher glass transition temperature. They also hydrogenated poly(bisphenol-A carbonate), poly (ethylene terephthalate), polybutadiene, and copolymers thereof, polyisoprene, copolymers thereof and obtained different products. Xu et al. (2005) studied the phase equilibrium characteristics for hydrogenation of polystyrene using a heterogeneous catalyst suspended in a solvent swollen by supercritical CO_2. Rate of hydrogenation of polystyrene was found to be higher in this case than pure solvent. Lin et al. (2004) hydrogenated nitrile-butadiene rubber latex with diimide and observed that the hydrogenation efficiency varies with the type of catalyst and degree of hydrogenation. Hydrogenation efficiency is higher with boric acid compared to copper ion, silver ion, or ferrous ion. Feast et al. (2003) studied the highly polar polymers with low glass transition temperature via a combination of ring opening polymerization and hydrogenation resulting in a significant lowering of the glass transition temperature of the copolymers. The work of Luo and Curtis (1996) and Hovert and Ng (1999) has limited application as polymer-to-fuel conversion is very low. Most recently, Kremer et al., 2020 carried out catalytic cracking of mixed polymeric wastes followed by subsequent kinetic study of liquid products.

Blazso (1999) performed depolymerization of polymers, copolymers, and blends catalytically producing their monomers. Zhibo et al. (1996) catalytically degraded polystyrene into styrene monomer using BaO and K_2O. Shah et al. (2008) converted waste tyre rubber into hydrocarbon via base catalyst using catalytic pyrolysis technique. Lin and Yang (2008) reported the recycling via catalytic cracking which operates in mixing the polymer waste with fluid catalytic cracking (FCC) as a catalyst. Kaimal and Vijayabalan (2015) converted municipal disposal site's waste plastics into plastic oil with the yield of 80% by input weight. Ananthakumar et al. (2017) also studied the catalytic pyrolysis of waste plastics in the absence of atmosphere, and they got the different parameters like density (0.840 g/cm^3), caloric value (44.8 MJ/kg), flash point (50 °C), viscosity (2 mm^2/s), and cetane number (42) of the diesel fuel. Mishra et al. (2010) reported thermolysis of waste plastic to liquid fuel is a suitable method for plastic waste management. Sampath et al. (2011) tested waste plastic oil as a fuel in a diesel engine. The literatures are available using batch or continuous reactor (Passamonti & Sedran, 2012; Achyut et al., 2010; Goad & Ali, 2017; Thahir et al., 2019; Kasar et al., 2020; Kremer et al., 2020) for production of fuel from plastics.

After being discarded as trash due to its very large molecular structure, the success and dominance of plastics beginning in the early twentieth century led to environmental issues about its slow decomposition rate. One solution to this issue was met with broad efforts toward recycling toward the end of the century. Most recycling systems need a fairly pure feedstock containing only products made from a single form of polymer, such as high-density polyethylene (HDPE) commonly used to produce milk pouches, bottles of polyethylene terephthalate (PET), etc. Incineration creates a lot of unpleasant gases that pollute the atmosphere.

One of the useful solutions to this issue could be the pyrolysis of waste plastic materials in the presence of a catalyst. Plastics are primarily long-chain hydrocarbon polymers derived from petrochemical by-products and are, in turn, a non-renewable fissile resource. In general, HDPE, LDPE, polypropylene (PP), polyvinyl chlorides (PVC), polystyrene (PS), and polyethylene terephthalate are municipal solid waste (PET).

Today, fossil fuels are also of great concern because of their diminishing resources. Therefore, if efficiently depolymerized, smaller hydrocarbons that can be used as fuels can be generated. The polymers typically decompose into smaller hydrocarbons at a temperature above the ceiling temperature. A certain amount of foreign currency could also be saved at the same time. Catalyst integration increases conversion and fuel efficiency. The addition of a catalyst during pyrolysis greatly decreases the temperature of pyrolysis.

1.1 Importance of Plastic as Material

Over the last few decades, plastic materials play a vital role in enhancing the standard lives of human beings. It is used in various sectors such as packing, automotive, construction, healthcare, electronics, and others. The demand of plastics has been increased tremendously.

Since early times, human beings were using polymeric material though their exact existence was not known. In the early 1900s, first synthetic polymer, i.e., Bakelite, was prepared using phenol and formaldehyde. However, scientists were unaware of its true nature. Initially, scientists assume plastics are colloids, i.e., a material that is a molecular composite. In the 1920s, Herman Staudinger proved polymeric materials are giant molecules or macromolecules. Carothers in 1928 prepared polyesters and polyamides which are now known as nylon. In the 1950s, the work of Ziegler and Natta on anionic coordination that catalyzed polypropylene development, high-density linear polyethylene, and other polymers of stereo specificity was developed. Metals, ceramics, or polymers are the conventional materials which are commonly used. Among these conventional materials, polymers have many attractive benefits which make them key of innovation of many products in various sectors. Polymeric materials have many advantages such as low density, easy processing, corrosion resistant, water resistance, etc. Polymeric materials are mostly preferred very light weight component is required. Polymer materials are bad conductors of electricity, but with the addition of conductive filler, one can make it conductive. A comparative analysis of the properties of polymers against metals and ceramics was performed, with the results as shown in the following table (Table 1):

Table 1 Comparative study of polymer and metals

Materials		Specific gravity	Thermal conductivity (J-cm/°C cm^2 s)	Electrical resistivity (μΩ-cm)	Modulus (MPa)
Metals	Aluminum	2.7	2.2	2.9	70,000
	Brass	8.5	1.2	6.2	110,000
	Copper	8.9	4.0	1.7	110,000
	Steel (1040)	7.85	0.48	17.1	205,000
Ceramics	Alumina (Al_2O_3)	3.8	0.29	$>10^{14}$	350,000
	Concrete	2.4	0.01	–	14,000
	Borosilicate glass	2.4	0.01	$>10^{17}$	70,000
	MgO	3.6	–	10^5 (2000 °C)	205,000
Plastics	High-density polyethylene	0.96	0.0052	10^{14}–10^{18}	350–1250
	Polystyrene	1.05	0.0008	10^{18}	2800
	Polymethyl methacrylate	1.2	0.002	10^{16}	3500
	Nylon	1.15	0.0025	10^{14}	2800

1.2 Problems Due to Plastic

Plastic pollution includes the accumulation of environmental plastic materials that adversely impact ecosystems, the ecosystem of wildlife, or humans. Plastics which function as pollutants are classified according to size into micro-, meso-, or macro-debris. The importance of plastic contamination is associated with the inexpensiveness and resilience of plastics, which leads to high levels of plastics used by humans. However, degradation is slow. Lands, rivers, and oceans can be unfavorably affected by plastic pollution. Plastic pollution has tremendously affected the marine lives; they are affected by either entanglement or through direct ingestion of plastic wastes. Plastic chemicals also affect the biological function of the marine lives. Plastic contamination in human body causes destruction of the thyroid hormone axis or hormone levels.

1.3 Effects on Humans

During manufacturing of plastics, various chemical additives including plasticizers for phthalates and brominated flame retardants are being used which has serious consequences on human health such as carcinogenic and dermatitis or encourage endocrine disruption. Chemicals such as BPA and phthalates present in plastics have been detected in the human body. These harmful chemicals enter into the human body through the nose, mouth, or skin. While the degree of exposure differs based on age and geography, most people are exposed to all of these chemicals simultaneously. The overall daily exposure levels are below those considered unhealthy, but there is a need for further studies to be conducted on the effects of low-dose exposure on humans. A lot remain unclear about how badly these chemicals impact human's health. Toxic chemicals are only used in trace quantities in many plastics, but essential checks are also needed to ensure that the toxic elements contain inert material or polymer in the plastic.

1.4 Methods of Recycling of Plastics

Any operation concerning repossession, recovery, or reuse of materials or articles may be described as recycling. Any method of extracting value energy from waste at any time in the cradle-to-grave life cycle of a product may be included in the description in the form of energy or material. It may be mechanical, chemical, or thermal to recycle. The French word of valorization is now being used more commonly and includes waste energy as well as other ways of recycling. Recycling can be achieved primarily using the following three methods:

(A) Mechanical recycling.
(B) Chemical recycling.
(C) Thermal recovery (incineration).

1.4.1 Mechanical Recycling

Mechanical recycling with the retranslation of either process scrap (start-up, shut-down, rejects, trimmings, etc.) or post-consumer material so that it can be melted and re-formed. Sprues and runners and faulty moldings are very commonly regranulated next to an injection molding machine and inserted back into the feed material. Likewise, the extruders and blow molder edge trimmings are practically reprocessed at source. New articles can be generated from the waste directly. The study shows that many factories are reusing their process scrap, and a number are reprocessing it outside for their own use or selling it for others to eventually use.

1.4.2 Chemical Recycling

Chemical recycling includes pyrolytic, chemolytic, or hydro cracking to break down polymeric waste into simpler substances that are subsequently re-polymerized for heat application to manufacture virgin materials. Plastic materials are converted into smaller molecules (chemical intermediates) which are suitable for the use as a feedstock for the development of new petrochemicals and plastics, and may be liquids or gases or solids. It recovers clean plastics directly for reuse as plastics of high molecular weight in the manufacture of new plastic products. It can further be categorized as follows:

(a) Depolymerization for monomer recovery.
(b) Depolymerization for liquid fuel and gas.

1.4.3 Thermal Recovery or Incineration

The burning of waste for the production of heat for cooking, space heating, industrial processes, or the generation of electricity is thermal recovery. In order to further minimize the amount of material to be finally disposed of, obtained ash is utilized for the building and road construction sectors. For direct combustion, dry waste and dried sludge from waste water will be utilized as a feedstock. Solid waste can be incinerated in furnaces to generate process heat for supplying steam turbine generators on a larger scale. The size of power plants is often limited by local feedstock availability and is usually less than 25–40 MWe.

1.5 World Energy Scenario

Oil, coal, natural gas, hydropower, nuclear energy, renewable fuel waste, and other sources of energy are the world's main sources of energy. Animal products, biomass, and industrial waste contain combustible waste. The contribution of the various sources of energy to the overall supply of world supplies (Table 2).

Oil is the world's most significant and plentiful source of energy. It is perhaps the most heavily eaten. Crude oil prices, however, are very unpredictable, and production is driven by prices. Coal is the world's second most abundant energy source and is heavily used in power generation. In terms of consumption, coal ranks very low, but its consumption has risen in some regions. The poor quality of coal has been a problem for many developing nations. The energy source with the highest growth rates in recent years has been natural gas. For power generation projects, the high end-use efficiency of natural gas has made it a common option. Renewable power sources are gaining in popularity, where hydropower plays a significant role.

Research is underway to increase the contribution of other sources of energy, such as solar, wind, and geothermal. However, different countries' fuel prices and regulatory policies play an important role in the production of renewals.

1.6 Recycling Plastic

In India, nearly 60% of its plastics are recycled, compared to the world average of 22%. The caloric value of plastic waste is equivalent to that of petrol. India has one of the lowest plastic consumption per capita, and the production of plastic waste is also very poor, as seen from Table 3.

Table 2 Availability of conventional sources of energy

Oil	35.1%
Coal	23.5%
Natural gas	20.7%
Renewable combustible wastes	11.1%
Nuclear	6.8%
Hydro	2.3%
Other sources	0.5%

Table 3 Consumption and recycling of plastic

	Description	World	India
1	Per capita per year consumption of plastic (kg)	24	6–7
2	Recycling (%)	15–20	60
3	Plastic in solid waste (%)	7	9

2 Process

Thermolysis, in which the organic components of polymer waste are converted by heat into high-value refining goods such as naphtha, crude oil, diesel, and coal, is used to recycle feedstock. The use of high temperatures to cleave the bonds in the polymer backbone requires thermolysis. It can be split into three processes according to the decomposition process:

(a) Pyrolysis – when decomposition process is carried out in the absence of air, it is termed as pyrolysis.
(b) Hydrogenation – when it is performed in an atmosphere of hydrogen gas, it is referred to as hydrogenation (hydro cracking).
(c) Gasification – if the decomposition is done in the presence of controlled amount of oxygen, then it is known as gasification.

Pyrolysis is carried out in reducing atmosphere at temperature of 700–800 °C. On addition of catalysts, heating temperature comes down to 390–560 °C. Cleaned waste is charged into the reactor and heat supplied which convert plastic waste into pyrolytic oil and solid coke. According to process it can be divided into two types:

1. Thermal pyrolysis.
2. Catalytic pyrolysis.

2.1 *Thermal Pyrolysis*

Thermal pyrolysis is done by directly heating the grinded polymer waste at 700–800 °C in the absence of oxygen gas. During this heating polymer molecules degrade into smaller molecular weight molecules. These finally give a very wide range of mixture of hydrocarbons. Economically this method is not suitable for recycling due high energy requirement and higher residence time.

2.2 *Catalytic Pyrolysis*

Catalytic pyrolysis takes place in the presence of catalyst. In this, feedstock is mixed with catalyst and heated up to 400–500 °C. The polymer depolymerizes at this temperature and gives a mixture of liquid hydrocarbons ranging from C_5 to C_{30} and gaseous products C_2–C_4 and coke as the final product. This method is most economical and feasible. Catalysts used in pyrolysis decrease the temperature of degradation, and there is a significant decrease in reaction time and residence time. So many researchers have been done by scientists in order to decrease the operational temperature and residence time. Till now so many catalysts has been discovered.

2.3 *Catalysts Used for Pyrolysis*

Various catalysts have been used in the past for the pyrolysis process; some promising catalysts are:

1. Zeolite-based ZSM-5.
2. Amorphous silica-alumina.
3. Metal oxide such as CaO, ZnO, and MgO.
4. Copper chromite.
5. Spend FCC.

3 Mechanism of Pyrolysis

3.1 *Thermal Pyrolysis*

The free radical process is accompanied by thermal pyrolysis. Hydrocarbons occur first from hemolytic reaction to free radicals at high temperatures, and then free radicals crack at β-bonds to form smaller olefins and alkanes. Ethylene is the most stable olefin, while methane is the most stable alkane, so the weight percentage of ethylene in light olefins is very high by a free radical mechanism (Xiang-hai et al., 2010).

Initiation

$$—CH_2—CH_2—CH_2—CH_2— \xrightarrow{h\nu} —CH_2—CH^{\bullet}_2 + {}^{\bullet}CH_2—CH_2—$$

Propagation

$$—CH_2—CH^{\bullet}_2 + CH_3—CH_2—CH_2— \longrightarrow —CH_2—CH_3 + {}^{\bullet}CH_2—CH_2—CH_2—$$

Termination

$${}^{\bullet}CH_3 + CH_3—CH^{\bullet}_2 \longrightarrow CH_3—CH_2—CH_3$$

$${}^{\bullet}CH_3 + {}^{\bullet}CH_3 \longrightarrow CH_3—CH_3$$

3.2 *Catalytic Pyrolysis*

Thermal feedstock pyrolysis begins with the random formation of polymer radicals; then abstracting H^+ from the reactants to form anionic intermediates called carbanions initiates catalytic pyrolysis on a solid basis (Shah et al., 2008). The O^{2-} ion surface of abstract H^+ metal oxides and the metal cations stabilize the carbanions. The anionic species creations signify the presence of sites on the surface contributing an electron or electron pair. The additional base-catalyzed pyrolysis reaction is identical to thermal pyrolysis. The empirical β-rule states that at the C-C bond located β to the carbon atom having the unpaired electron, C-C bond scission occurs. The answer takes place in the following phases:

$$-CH_2-CH_2-CH_2-CH_2- \overset{-H^+}{\rightarrow} -CH_2-\bar{C}H-CH_2-CH_2-$$

$$CH_3-\bar{C}H-CH_2-CH_3 \rightarrow C^+H_3+\bar{C}H_2-CH_2-CH_3$$

$$\bar{C}H_2-CH_2-CH_3 \overset{2H^+}{\rightarrow} CH_3-CH_2-CH_3$$

$$C^+H_3 \rightarrow \bar{C}H_2+H^+$$

$$\bar{C}H_2 \rightarrow \text{Catalyst deposite}$$

Isomerization, addition, decomposition, and alkylation are typically the base-catalyzed reactions occurring over metal oxides.

4 Various Plants Working Presently

4.1 BASF Feedstock Recycling Plant

BASF has a plant in Ludwigshafen, Germany, having an annual capacity of 15,000 tn. It converts waste plastic to naphtha, aromatic compounds, and high boiling oils by pyrolysis. The intensive feedstock preparation for BASF process involves coarse shredding, drying, sieving, separation, iron removal, re-shredding, and compacting. These steps are very important in order to increase the bulk density, handling characteristic, and process efficiency of the plastic waste. Waste is converted to petrochemicals using a three-stage process:

(a) The plastic agglomerates are heated to about 300 °C in stirred kettles for melting.
(b) The molten plastic is heated further in gas-fired tabular furnaces to range 350°–480 °C for cleaving the polymer into smaller fragments.
(c) The oils formed in the second stage are then distilled into naphtha, aromatic compound, and high boiling oils.

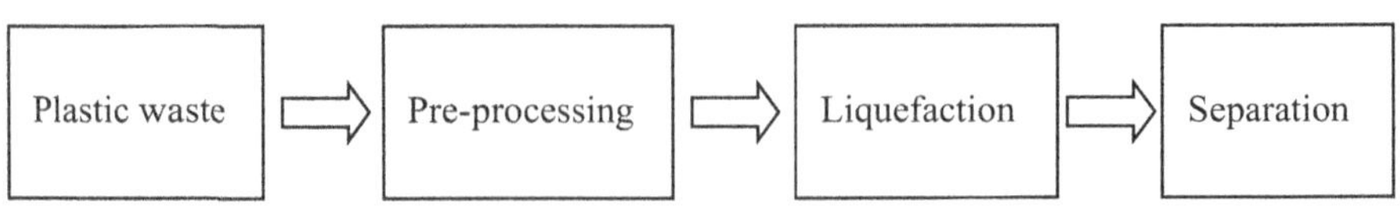

4.2 *VEBA Pyrolysis Plant*

Plastic waste is fed into the rotary kiln via a screw feeder and pyrolyzed using gas burners at a slight overpressure approx. 10 mbar in an oxygen-free environment. The residence time of the material feed can be varied by adjusting the inclination and RPM of kiln. The normal wall temperature of kiln is around 650 °C. It has been found that the throughput of rotary kiln could be doubled by raising temperature from 700 to 800 °C. The pyrolytic vapors are passed into a two-step condenser where in the first step they are cooled at around 200 °C by direct quenching with the condensed pyrolysis oil. Then it is cooled down to 35 °C for condensing the light hydrocarbons. This plant uses automotive shredder residue and vulcanized elastomers at a capacity of 12.5 tons/h.

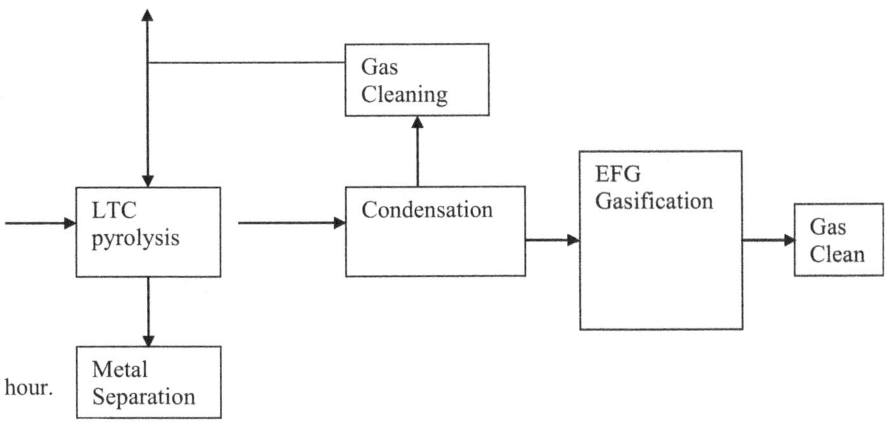

4.3 *Fuji Fixed-Bed Pyrolysis Plant*

A full-scale pyrolysis-catalytic cracking process for feedstock recovery of waste plastid is in use in Japan. The Fuji recycling industry is operating a 5000-tons-per-year plants for industrial plastic waste at Aiai City plant. The plastic waste is crushed into small pieces of less than 8 mm sq., then extruded at 300 °C, and mixed with the liquid product recycled from pyrolytic reactor. The low MW fragments are then cracked in a fixed-bed reactor using a zeolite-based catalyst to yield the following product:

(a) Oil – 80%.
(b) Gas – 15%.
(c) Residue – 5%.

4.4 Hamburg Fluidized-Bed Pyrolysis Plant

The pilot-scale fluidized-bed reactor at the University of Hamburg has an inner diameter of 450 mm. The plant maximum capacity is 40 kg/h. The reactor (Kaminsky, 2006) is heated indirectly by burning propane in a concentric combustion chamber. The plastics are fed into the reactor either through double flap gate or extruder. The pyrolyzed gas is preheated at 400 °C. The exhaust gases pass through a heat exchanger and separated from carbon residues and fine dust, using cyclone separator. The pyrolysate gases flow through the packed condensation columns where the product is condensed.

4.5 Author Tubular Pyrolysis Reactor

Polymer waste coming from the laboratory work is collected. Waste was grinded to increase its bulk density and its proper melting. Each material batch is prepared. Every batch constitutes of 1 kg of plastic and 1% ZnO mixed together. Every batch is fed into the reactor for 6–7 h at temperature 400 °C and 1 atm pressure. Condensed fuel is collected in bottle after cooling from the condenser. Catalytic pyrolysis is taking place in the pyrolyzer (reactor). It is a cylindrical vessel surrounded with band heater, which provides required heat to the reactor. Temperature is controlled by a thermostat. Temperature is kept constant at 400 °C. The feedstock is pyrolyzed in the reactor and converted into small fraction gaseous molecules. These gases were condensed in the condenser to a liquid state. Flow of gases is controlled by a valve. Cooling is done by circulating tap water through a condenser.

5 Summary

The plastic waste if used in a proper manner, i.e., whenever pyrolyzed, can provide cheap fuel oil as a substitute for diesel, also saving some foreign currency by importing less petroleum. The best results are obtained by blending this oil with diesel. In another manner also, it will be helpful in disposing of a huge amount of plastic wastes. Moreover, one will obtain valuable by-products apart from liquid fuel suitable for IC engine.

Acknowledgments The authors thanks Mr. Lakshya Dev, Mr. Mahboob Alam Siddiqui, Mr. Manish Kumar, and Mr. Kushagra Gangwar for their hardwork in finalizing this book chapter.

References

Achyut, K. P., Singh, R. K., & Mishra, D. K. (2010). Thermolysis of waste plastics to liquid fuel, a suitable method for plastic waste management and manufacture of value added products. *Renewable and Sustainable Energy Reviews, 14*, 233–248.

Akpanudoh, N. S., Gobin, K., & Manos, G. (2005). Catalytic degradation of plastic waste to liquid fuel over commercial cracking catalysts; effect of polymer to catalyst ratio/acidity content. *Journal of Macromolecular Catalyst A Chemical, 235*, 67–73.

Ananthakumar, S., Jayabal, S., & Thirumal, P. (2017). Investigation on performance, emission, and combustion characteristics of variable compression engine fuelled with diesel, waste plastics oil blends. *Journal of the Brazilian Society of Mechanical Sciences and Engineering, 39*(1), 19–28.

Areana, U., & Mastellone, M. L. (2000). Defluidization phenomena during the pyrolysis of two plastic wastes. *Chemical Engineering Science, 55*, 2849–2860.

Blazso, M. (1999). Recent trends in analytical and applied pyrolysis of polymers. *Journal of Analytical and Applied Pyrolysis, 50*, 77–101.

Bockhorn, H., Hornung, A., & Hornung, U. (1999). Mechanisms and kinetics of thermal decomposition of plastics from isothermal and dynamic measurements. *Journal of Analytical and Applied Pyrolysis, 50*, 77–101.

Chang, Y. M. (1996). On pyrolysis of waste tyre degradation rate and product yield. *Resources Conservation and Recycling, 17*, 125–131.

Feast, W. J., Gimeno, M., & Khosravi, E. (2003). Approaches to highly polar polymers with low glass transition temperature, 1. Fluorinated polymer via a combination of ring opening metathesis polymerization and hydrogenation. *Polymer, 44*, 6111–6121.

Gersten, J., Fainberg, V., Garber, A., Hetszoni, G., & Shindler, Y. (1999). Utilization of waste polymers through one stage low temperature pyrolysis with oil shell. *Fuel, 78*, 987–990.

Goad, M. A., & Ali, R. (2017). Thermal and catalytic cracking of plastic wastes into hydrocarbon fuels. *International Journal of Engineering and Information Systems, 1*, 56–61.

Gobin, K., & Manos, G. (2004). Polymer degradation to fuels over micro porous catalyst as a novel tertiary plastic recycling method. *Polymer Degradation and Stability, 83*, 267–279.

Hovert, N., & Ng, F. T. (1999). Tertiary polymer recycling; study of polyethylene thermolysis as a first step to synthetic diesel fuel. *Fuel, 78*, 459–470.

Kaimal, V. K., & Vijayabalan, P. (2015). A detailed study of combustion characteristics of a DI diesel engine using waste plastic oil and its blends. *Energy Conversion and Management, 105*, 951–956.

Karagoz, S., Karayildirium, S., Ucar Yuksel, M., & Yanik, J. (2003). Liquefaction of municipal waste plastic in VGO over acidic and non acidic catalyst. *Fuel, 82*, 412–423.

Kasar, P., Sharmab, D. K., & Ahmaruzzaman, M. (2020). Thermal and catalytic decomposition of waste plastics and its co-processing with petroleum residue through pyrolysis process. *Journal of Cleaner Production, 265*, 121639.

Kaminsky, W. (2006). The Hamburg fluidized-bed pyrolysis process to recycle polymer wastes and tires, chapter 17. In J. Schiers & W. Kaminsky (Eds.), *Feedstock recycling and pyrolysis of waste plastics* (p. 475). John Wiley & Sons.

Kremer, I., Tihomir, T., Zvonimir, K., Zlatan, H., Matko, E., & Daniel, R. S. (2020). Catalytic decomposition and kinetic study of mixed plastic waste. *Clean Technologies and Environmental Policy, 23*, 811. https://doi.org/10.1007/s10098-020-01930-y

Lin, X., Pan, Q., & Rempel, G. L. (2004). Hydrogenation of nitrile butadiene rubber latex with diemide. *Applied Chemistry A General, 276*, 123–128.

Lin, Y. H., & Yang, M. H. (2008). Chemical catalyzed recycling of polypropylene over a spent FCC catalyst and various commercial cracking catalysts using TGA. *Thermochimica Acta, 470*, 52–59.

Luo, M., & Curtis, C. W. (1996). Thermal and catalytic co processing of coal with model and commingled waste plastics. *Fuel Processing and Technology, 49*, 91–117.

Mishra, D. K., Singh, R. K., & Panda, A. K. (2010). Thermolysis of waste plastic to liquid fuel a suitable method for plastic waste management and manufacture of value added products -a world prospective. *Renew and Sustainable Energy, 14*, 233–284.
Miskoloczi, N., Bartha, L., & Deak, G. (2006). Thermal degradation of polyethylene and polystyrene from the packaging industries over different catalyst into fuel like feed stock. *Polymer Degradation and Stability, 91*, 517–526.
Miskoloczi, N., Bartha, L., Antal, F., & Dudas, C. (2005). Determination of sulfur in liquids obtained by thermal cracking of waste polymers and commercial fuels with different analytical method. *Talanta, 66*, 1264–1271.
Miskoloczi, N., Bartha, L., Deak, G., & Joves, B. (2004). Thermal degradation of municipal waste for production of fuel like hydrocarbon. *Polymer Degradation and Stability, 86*, 357–366.
Passamonti, F. J., & Sedran, U. (2012). Recycling of waste plastics into fuels. LDPE conversion in FCC. *Applied Catalyst B: Environment, 125*, 499–506.
Roberts G. and Dawei X (2003) Hydrogenation of polymers in the presence of non-reactive processing aid. US Patent No. 60/423,804.
Sampath, S., Nagarajan, G., & Mani, M. (2011). Characterisation and effect of using waste plastic oil and diesel fuel blends in compression ignition engine. *Energy, 36*, 212–219.
Serrano, D. P., Aguado, J., Escola, J. M., & Rodrigues, J. M. (2005). Influence of nano-crystalline HZSM-5 external surface on the catalytic cracking of polyolefin. *Journal of Analytical and Applied Pyrolysis, 74*, 353–360.
Shafferina, S., Faisal, A., Wan, D., & Mohamed, K. A. (2016). A review on pyrolysis of plastic wastes. *Energy Conversion and Management, 115*, 308–326.
Shah, J., Rasul, M. J., & Fazal, M. (2008). Catalytic pyrolysis of waste tyre rubber into hydrocarbon via base catalysts, Iran. *Journal of Cham Engineering, 27*, 2–10.
Thahir, R., Altway, A., Juliastuti, S. R., & Susianto. (2019). Production of liquid fuel from plastic waste using integrated pyrolysis method with refinery distillation bubble cap plate column. *Energy Report, 5*, 70–77.
Walendziewski, J. (2005). Continuous flow cracking of waste plastics. *Fuel Processing Technology, 86*, 1265–1278.
Williams, E. A., & Williams, P. T. (1997). Analysis of product derived from the fast pyrolysis of plastic waste. *Journal of Analytical and Applied Pyrolysis, 40*, 347–363.
Xu, D., Carbonael, R. G., Robert, G. W., & Kiserow, D. J. (2005). Phase equilibrium for the hydrogenation of polystyrene in CO_2 swollen solvents. *Journal of Supercritical Fluids, 34*, 1–9.
Xiang-hai, M., Jin-sen, G., Li, L., & Chun-ming, X. (2010). Advances in catalytic pyrolysis of hydrocarbons. *Petroleum Science and Technology, 22*, 1327–1341.
Yanik, J., Md, A., Uddin, I. K., & Sakita, Y. (2001). The catalytic effect of red mud on degradation of poly vinyl chloride containing polymixer into fuel oil. *Polymer Degradation and Stability, 73*, 335–346.
Yoon, W. L., Park, J. S., Jung, H., Lee, H. T., & Lee, D. K. (1999). Optimization of pyrolytic coprocessing of waste plastic and motor oil into fuel oils using statistical pentagonal experimental design. *Fuel, 78*, 809–813.
Zhibo, Z., Nishio, S., Morioka, J., Ucno, A., Ohkita, H., Tochihara, Y., Mizushima, T., & Kakuta, N. (1996). Thermal and chemical recycle of waste polymer. *Catalysis Today, 29*, 303–308.

Nanomaterials for Energy Storage Applications

K. M. G. C. Bandara, H. M. R. S. Herath, and Nadeesh M. Adassooriya

Abstract Electrical energy generation plays an important role in directly influencing the economic and social activities in a country. However, the amount of electricity that can be generated is relatively fixed over short periods of time, although demand for electricity fluctuates throughout the day. Therefore, developing technology to store electrical energy is extremely important to meet the demand in electricity specially when the sources of electricity generation include renewable ones such as solar and wind where power capacity varies with time. Also, energy storage is a vital requirement for portable/mobile electrical and electronic systems. In general, energy storage systems can be classified into batteries, fuel cells, and capacitors. Although fuel cells have very high energy densities, they are poor in terms of power density. However, batteries such as Li-ion show a good compromise between energy and power densities which have made it the preferred choice for portable electronics. In contrast, capacitors have very high power densities which can provide energy bursts of high current (at fixed voltage). The aim of the chapter is to focus on research studies which involve nanomaterials to improve the performance of aforementioned energy storage techniques, related challenges, and future prospect.

Keywords Clean energy · Electrolytes · Energy · Fuel cell · Nanomaterial · Storage · Fuels

1 Introduction

Energy can neither be destroyed nor created. Still, it can be transformed from one form to another and stored. Energy transformation is conversion of a certain form of energy to another form, while energy storage is harvesting the converted form of

K. M. G. C. Bandara · H. M. R. S. Herath · N. M. Adassooriya (✉)
Department of Chemical and Process Engineering, University of Peradeniya, Peradeniya, Sri Lanka
e-mail: nadeeshm@eng.pdn.ac.lk

D. B. Pal, J. M. Jha (eds.), *Sustainable and Clean Energy Production Technologies*, Clean Energy Production Technologies,
https://doi.org/10.1007/978-981-16-9135-5_4

energy for later use. Sustainable energy conversion and storage methods are becoming necessary due to rising energy demand worldwide. It is forecasted that the world's energy demand will be doubled by year 2050 (Armaroli & Balzani, 2007). So, development of high performance, efficient and low emission energy conversion, and storage devices has become a popular concern and is receiving extensive research.

Energy storage technologies that are present today can be sorted according to the medium or the method of implementation. Coal, oil, and natural gas are the commonly used fossil fuels whose energy is obtained mainly through combustion reactions. Energy is stored mechanically by hydropower, compressed air, flywheel, and hydraulic accumulator, while dielectric capacitors, electric double-layer capacitors, and superconducting magnetic devices store energy as electrical and electromagnetic energy. Besides, glycogen and starch also store energy in means of biological storage, and biofuels, hydrated salts, and hydrogen storages store energy chemically (Wang et al., 2019). On top of that, electrochemical energy storage which is the main concern of this chapter is achieved through fuel cells, batteries, and supercapacitors.

Electrochemical energy storage devices convert chemical energy to electrical energy through electrochemical redox reactions governed by Faraday's law. In an electrochemical system, one species removes electrons and undergoes oxidation, while the other species receives the electrons and reduces. The electrons given out by one electrode pass through the external circuit producing electricity. The oxidizing and reducing electrodes are placed in an electrolyte which is electronically isolated and ionically conductive. Along with these main components, an electrochemical system consists of a current collector, conductive additive, separator, and binder whose contribution is also important for the performance of the system. The energy storage process is merely the migration and storage of charges in the system. Electrochemical qualities of the materials used in the system play a major role defining the performance of the battery. Besides, size and structure of the materials have a great effect on the process of energy storage (Guo et al., 2008).

For the past few decades, nanoscale materials have been emerging as a promising solution for limitations in energy conversion and storage devices and have shown significant enhancements in power and efficiency (Abdalla et al., 2018). Nanoscale materials provide small particle size, less than 100 nm, and high specific surface area due to the porous structure. Owing to the small size of the particles, nanomaterials enable fast diffusion of ions within the particles. Moreover, high specific surface area and porous nature of nanomaterials provide a high area of contact for reactions or interactions to take place which is more advantageous for electrochemical systems (Wang et al., 2019). Accordingly, researchers have made great efforts to utilize the distinctive characteristics of nanomaterials in 0D, 1D, 2D, and 3D forms to develop high-performance energy storage devices. The present chapter is hence focus on the use of nanoscale materials for major components of electrochemical energy storage devices mainly fuel cells, batteries, and capacitors to enhance the electrochemical performance.

2 Energy Storage Devices

2.1 *Fuel Cells*

Fuel cell technology has gained a greater attraction of researchers during the twenty-first century as it is a clean source of energy and a substitute for limited fossil fuel resources on earth (Sun & Liao, 2020; Basri et al., 2010). Fuel cell is an efficient and environmentally friendly energy device which directly converts chemical energy to electrical energy. Water and heat are the only by-products (Zhong et al., 2008; Lade et al., 2017). It consists of an anode and a cathode connected with a load resistance and an electrolyte. A fuel, mostly hydrogen gas, is pumped on to the anode, and hydrogen splits into electrons and H^+ ions (protons). H^+ ions diffuse through the electrolyte toward the cathode, while electrons flow out of the anode to the outer circuit. Electrons moved through the load resistance, and H^+ ions ended up at the cathode. They combine with reduced O_2 at catalyst sites of the cathode and form water.

Fuel cells are categorized on the basis of electrolytes used (Suryawanshi et al., 2018). The main types of fuel cells are proton exchange or polymer electrolyte membrane fuel cell (PEMFCs), direct liquid fuel cells (DLFCs), solid oxide fuel cells (SOFCs), phosphoric acid fuel cells (PAFCs), molten carbonate fuel cells (MCFCs), and alkaline fuel cells (AFCs) (Abdalla et al., 2018). Fuel cells own many advantages like higher energy density, higher efficiency, easy handling of liquids, possibility of applying in micro-fuel cells, and very low environmental impact with negligible emissions compared to conventional energy storage devices (Guo & Wang, 2011; Suryawanshi et al., 2018). Yet electrocatalyst materials are a major obstacle for fuel cell technology (Basri et al., 2010). Electrocatalysts influence the speed of reactions at anode and cathode substantially. Although many materials have been synthesized as electrode materials, parameters such as cost and durability are still challenging. Therefore, research is being made to develop materials at nanoscale to overcome these problems and improve the fuel cell activity by increasing the potential of sluggish anodic and cathodic reactions (Wang et al., 2016a, 2016b; Abdalla et al., 2018).

2.1.1 Proton Exchange/Polymer Electrolyte Membrane Fuel Cells (PEMFCs)

Out of the fuel cell types, proton exchange or polymer electrolyte membrane fuel cell (PEMFC) is the commonly used fuel cell type and considered as the potential next-generation power source. Hydrogen is mainly used as a fuel in PEMFC. Higher power density, low operating temperature, absence of moving parts, rapid startup, and low emissions are some advantages of PEMFCs (Suryawanshi et al., 2018). Due to its numerous advantages, this type of fuel cell has broad applications in fields such as transportation, chip-integrated micro-devices, portable and stationary power systems, automobiles, and space shuttles to power grids for buildings and factories (Suryawanshi et al., 2018; Lykhach et al., 2017; Zhong et al., 2008).

In a PEMFC oxidation reaction takes place at the anode, and reduction reaction occurs at the cathode as follows (Choudhury et al., 2020; Suryawanshi et al., 2018):

Reaction at the anode:

$$H_2 \rightarrow 2H^+ + 2e^- \tag{1}$$

Reaction at the cathode:

$$\frac{1}{2}O_2 + 2H^+ 2e^- \rightarrow H_2O \tag{2}$$

Complex heat and water management and costly catalysts are found to be restrictions to PEMFC. Therefore, development of effective electrocatalysts and other components of fuel cells have become a greater concern. Most commonly used catalyst in PEMFC is Pt, and global availability of Pt is limited while it is very expensive (Suryawanshi et al., 2018). To overcome these problems, noble metal is partially replaced by non-noble metal. Due to improved specific surface area, nanosized catalysts can reduce the platinum amount at the cathode. This reduction amount is more than 1–0.4 mg/cm^2 approximately. Nanocatalysts are also durable and stable (Dhathathreyan et al., 2017).

Noble Nanomaterial Catalysts

Pt and Pt-based materials are the fundamental and most adequate catalysts for fuel cells (Guo & Wang, 2011). Due to its abovementioned restrictions for commercialization, many researches aim in synthesizing advanced materials to minimize Pt loading amount while increasing performance. In achieving this goal, synthesizing nanomaterials is a greater consideration. There are several methods of synthesizing nanoparticles such as wet impregnation, colloidal synthesis, chemical reduction, electroless deposition, sonochemical approaches, solution plasma sputtering, etc. (Suryawanshi et al., 2018).

Carbon-supported Pt and Pt alloys are made by incorporating metals such as Ni, Co, Mn, Cu, etc. (Shao et al., 2007; Suryawanshi et al., 2018; Suryawanshi et al., 2018). $Pt_{30}Ni_{10}/C$ and $Pt_{30}Co_{10}/C$ nanocatalysts can be made via a sonochemical method which possesses advantages over conventional methods such as decreased process time, increased heat and mass transfer rates, and improved surface-to-volume ratio (Bang & Suslick, 2010). These Pt alloy nanocatalysts are more effective than the Pt/C electrocatalyst. Treating nanocatalysts with HNO_3 will enhance the electrochemical properties. HNO_3 treating will increase the rate of oxygen reduction reaction (ORR) by increasing electrochemically active surface area (EASA) (Suryawanshi et al., 2018). The encapsulation of Pt-Co bi-metallic nanoparticle PANI (polyaniline) improves the electrochemical activity by increasing current density. PANI is one of the best conducting polymers where the metal nanoparticles can be uniformly distributed throughout the support (Kakkar et al., 2020).

Shape controlling of nanocrystals acquires a significant part in fuel cell activity. Studies reveal that 1D Pt-based catalysts (Pt_3Co, PtMn, PtCu) functioned as nanorods, nanotubes, and nanowires are more effective than the zero-dimensional Pt nanoparticles as their aggregation is comparatively less. Also, nanocubes which have the ability to expose facet on all six faces exhibit noticeable ORR performance. Among different shapes Pt with concave surfaces shows a two to four times higher ORR activity in comparison to Pt nanocubes, octahedrons, and conventional Pt/C (Guo et al., 2013a).

Studies have been done regarding shape-controlled synthesis of Pd as an electrocatalyst which is also a noble metal. According to the findings, Pd nanocubes show about ten times improved cathodic reaction than Pd octahedrons (Dhathathreyan et al., 2017).

Other than the shape, mesoporous and hollow structure affects positively in enhancing ORR activity, current density, and EASA. Nanoporous alloys such as PtNi, Pt_3Ni, PtNiAl, PtCu, and PtRuNi are reported. Furthermore, the smaller the pore, the higher the current density (Zhou et al., 2013b). Researchers have also synthesized AuPt alloy nanoparticles with different compositions to investigate the increased performance compared to conventional AuPt alloy (Zhong et al., 2008).

Carbon and Non-carbon Nanomaterials as Catalyst Support

There are several carbon nanomaterials developed to operate as catalyst supports for Pt and Pt-based alloys in cathode. Single-walled nanotubes (SWNT), multi-walled carbon nanotubes (MWCNT), carbon nanocoil (CNC), carbon nanofibers (CNFs), carbon nanohorns (CNH), ordered mesoporous carbon (OMC), and graphene are some examples. N-, B-, P-, and S-doped carbon nanomaterials are also available as Pt support materials. Among these, Pt supported on an N-doped carbon nanocage has shown a noticeable activity in ORR. In recent years graphene and doped graphene have also been developed as catalyst supports for Pt, Pd, and Pt alloys due to their improved properties (Jafri et al., 2010).

Oxides, nitrides, and conducting polymers have been developed as non-carbon electrocatalyst supports in the recent past. However further development of these supports have been weakened due to the low corrosion resistance for oxygen reduction reaction (Dhathathreyan et al., 2017).

Non-noble Metal-Free Nanocatalysts

Heteroatom-doped graphene are the most considered metal-free nanocatalysts for ORR. B-, P-, I-, and S-doped graphene are few examples. Other than these 3D N-CNt/graphene, nitrogen-doped few layers of graphene/MWCNT (FLG/MWCNT) are some hybrid structures. These nanocatalysts absorb a reduced energy from anions in ORR. But these nanocatalysts are not commercialized in practical scenarios as they need further developments. Though these materials are cost-effective, it is restricted by the ohmic resistance of the cell (Dai et al., 2015; Dhathathreyan et al., 2017).

Nanostructured Catalyst Layers

Nanostructured catalyst layer is a significant development in PEMFC under wet conditions. These layers can increase the efficiency while reducing the catalyst loading. Higher performance can be obtained even with a low Pt loading of about 0.026 mg/cm^2. PtCoMn and PtNi alloys exhibit a higher performance than Pt (Dhathathreyan et al., 2017; Debe, 2013).

2.1.2 Solid Oxide Fuel Cells (SOFCs)

Solid oxide fuel cell (SOFC) is the most reliable type of fuel cell consisting of two porous electrodes detached by a dense electrolyte such as yttria-stabilized zirconia (YSZ) or gadolinium doped ceria (GDC), Y-doped $BaZrO_3$ (BZY) (Abdalla et al., 2018; Frattini et al., 2015). It is highly efficient, and natural gas also can be used as a fuel. Short lifespan, high heat-up due to operating temperature, and extensive startup are some drawbacks of the cell (Suryawanshi et al., 2018). SOFCs are mostly used in medium- to large-scale power stations. Overall performance of the SOFC depends on the improvements in the surface properties, interface between materials, and atmosphere. The function inside the cell which converts chemical energy to electrical energy is by means of H^+ and O^- ions which are as follows (Abdalla et al., 2018):

Anodic reaction:

$$H_2 + O_2 \rightarrow H_2O + 2e^- \tag{3}$$

$$CO + O_2 \rightarrow CO_2 + 2e^- \tag{4}$$

Cathodic reaction:

$$\frac{1}{2}O_2 + 2e^- \rightarrow O^{2-} \tag{5}$$

Here the fuel oxidation reaction takes place at the anode. Many researches have been done to develop suitable materials in nanoscale in order to lower the operating temperature while retaining stability and required conductivity, avoid deposition of carbon on anode when operating with hydrocarbon fuels, tolerate sulfur, and introduce pores and contact points to enhance performance (Dhathathreyan et al., 2017).

Nanomaterials for Anode

Anode is the main part of SOFC components. It consumes about 95% of the total materials used in fuel cells and should be catalytic active for H_2 that splits H_2 into protons. Porous, organized structure and fine particle structure will greatly influence the performance enhancement of a fuel cell. Other than these features, anode material should be highly conductive and thermally stable as most of the operations are done at higher temperatures. Also, the material used in fabricating the anode should be chemically compatible with other fuel cell materials.

Among the several methods of incorporating nanoparticles in anode matrix thermally driven diffusion processes, sonoelectrochemical methods, electrodeposition, and electroless deposition are generally used. Ni cermet is the ordinarily used material in anode in SOFCs that operate under higher temperatures. Incorporating nanoscale Ni particles will increase the power density of fuel cells. Sulfur tolerance and hydrocarbon fuels without carbon deposition can be achieved by adding Pd nanoparticles.

There is an influence of the morphology of the nanoparticle toward the fuel cell performance in increasing the power density. Results have been obtained that a porous flower-like structure of CeO_2 in Ru-CeO_2 anode could increase the power density. Pt, Co, Ni, and Pd nanoparticles can be coated in a vertically oriented morphology with a high porosity using pulsed laser deposition method to improve fuel cell activity. Studies have been done to prove that the nanorod structure of NI-YSZ can increase the surface area by three times than the conventional Ni anodes. Also, hierarchically ordered nanocomposites like NiO-$Ce_{0.8}Gd_{0.2}O_{1.9}$ exhibit reduced activation energy and polarization resistance over conventional Ni-based anode materials (Dhathathreyan et al., 2017). In addition to improved electrochemical properties, nanosized particles show better sinterability (Sun & Liao, 2020).

Nanostructured Electrolytes

The electrolyte of a SOFC is a highly dense media which possesses a high ionic conductivity and chemical and mechanical compatibility with other materials. Making the electrolyte as a thin layer without losing its characteristic properties to reduce the resistance inside the cell during the electrochemical reactions is challenging. As a result, single-phase electrolytes have been dominated by composite electrolytes with a higher number of interface regions to provide higher conduction. Furthermore, to reduce the operating temperature, nanocomposite electrolytes have become the researchers' focus.

Some of the commonly used single-phase electrolytes are yttria-stabilized zirconia (YSZ), Sc_2O_3-doped ZrO_2 (ScSZ), and gadolinium-doped ceria (GDC). Nanostructuring of electrolytes which can be obtained via pulse layer deposition method results in a 3D thin-film architecture allowing the electrolyte to reach a peak value of power density at lowered temperatures of about 450 $^{\circ}$C. NiO-YSZ, samarium-doped ceria (SDC)-$(Li/Na)_2CO_3$), and SDC/Na_2CO_3 are few examples of versatile nanocomposite electrolytes (Frattini et al., 2015).

Nanomaterials for Cathode

Cathode is the second major component in SOFC (Abdalla et al., 2018). Large particle size and low porosity result in the low performance of a fuel cell. According to research studies, incorporating nanomaterials in cathodes operating at higher temperatures is not feasible. Furthermore, nanostructures can be supported on micro-sized cathodes (Dhathathreyan et al., 2017). Most commonly used 0D nanomaterials are gadolinia-doped ceria (GDC), $Sm_{0.5}Sr_{0.5}CoO_{3-\delta}$(SSC), yttria-doped ceria, and cobalt oxide (Co_3O_4). These nanomaterials display increased electronic and decreased ionic conductivities raising the overall fuel cell activity. Studies have reported about Pd and Pt nanomaterials in cathodes. But Pd nanoparticles infiltrated on porous YSZ scaffold block the oxygen diffusion path while increasing the polarization losses. Nanoparticle-incorporated $La_{0.8}Sr_{0.2}MnO_{3-\delta}$ (LSM) is a structurally stable, high electronic conducting porous nanocomposite cathode which shows power densities greater than the pure LSM cathode. Cathode performance can be increased by using 1D nanotubes of perovskites through enlarged surface area (De Larramendi et al., 2011). $La_{0.7}5Sr_{0.25}Cr_{0.5}Mn_{0.5}O_{3-\delta}$ (LSCM) is another captivating material which exhibits both oxidation and reduction reactions. Nanopowder of LSCM shows enhanced fuel cell performance (Chanquía et al., 2014).

Cathode materials with proton conduction have become the latest attraction in SOFC research. The use of nanomaterials in protonic SOFCs positively influences fuel cell cathode performance by increasing triple-phase boundaries (TPB) which connect electrolyte and cathode. Sr-doped $LaMnO_3$ nanoparticles are commonly used in this type of cells for impressive performance (Sun & Liao, 2020).

2.1.3 Direct Liquid Fuel Cells (DLFCs)

Direct liquid fuel cell (DLFC) is an additional, significant type of fuel cell used as an energy production device since several years. This type of fuel cells generates energy from various types of fuels which include methanol, ethanol, formic acid, dimethyl ether, ethylene glycol, etc. Hence the fuel cells are categorized accordingly as direct methanol fuel cell (DMFC), direct ethanol fuel cell (DEFC), direct formic acid fuel cell (DFAFC), direct ethylene glycol fuel cell (DEGFC), and direct dimethyl ether fuel cell (DDEFC) (Kumaran et al., 2018). Stable operation, higher efficiency, higher power density, lower emissions, easiness in fuel refilling, and reduced operating temperature are some factors that draw attraction toward direct liquid fuel cells as a power source in movable and transport utilizations (Pu et al., 2015). Pt is the widely used electrocatalyst, but due to its drawbacks such as high sensitivity toward CO and high cost, it has initiated to explore new materials with minimum noble metal loading and higher electrocatalytic performance (Wang et al., 2007).

Direct Methanol Fuel Cells (DMFCs)

Direct methanol fuel cell (DMFC) consumes methanol as the fuel. It is a proton exchange fuel cell. The chemical reactions happening inside the fuel cell presence of an acidic electrolyte are as follows (Yola et al., 2016):

Anodic reaction:

$$CH_3OH + H_2O \rightarrow CO_2 + 6H^+ + 6e^- \tag{6}$$

Cathodic reaction:

$$\frac{3}{2}O_2 + 6H^+ + 6e^- \rightarrow 3H_2O \tag{7}$$

Slow methanol oxidation at the anode is one of the major challenges in a DMFC. The main reason behind this is the self-poisoning of the surface by CO. Thus, anode catalysts should be a material withstand CO poisoning in order to enhance efficiency of the cell (Jusys et al., 2002). As a result, developing nanostructured materials with improved effective surface area has become a research trend.

Particle agglomeration is a common problem encountered when synthesizing nanoparticles. This can be minimized by using surfactants and polymer holders. But, introducing surfactants and polymer holders can result in internal resistance of the fuel cell. Pt-decorated carbon nanotubes developed via atomic layer deposition method are used in DMFC electrodes to overcome this problem (Wang et al., 2016a, 2016b). Three-dimensionally connected, porous Pd nanoflowers are novel electrocatalysts for Pt-free DMFCs. In comparison to Pd nanoparticles, Pd nanoflowers exhibit an improved stability and performance (Yin et al., 2009). Studies have recorded on electrocatalysts developed by integrating mono-metallic and bi-metallic nanoparticles like rod silver (rAgNPs), rod gold (rAuNPs), rod silver-platinum (rAg-Pt NPs), and rod gold-platinum (rAu-Pt NPs) with 2-aminoethanethiol functionalized graphene oxide (AETGO). rAu-Pt NPs/AETGO has the highest specific area and peak current for methanol oxidation reaction among all these nanomaterials under same Pt amount (Yola et al., 2016).

Using carbon nanostructures over carbon black as catalyst supports has led to significant changes in DMFCs. Out of the many carbon nanostructures, carbon nanotubes are the widely used nanostructure for fuel cells operating at lower temperatures. High crystalline arrangement of these nanostructures delivers increased conductivity, porosity, and specific surface area that will result in increased diffusivity. Therefore, carbon nanostructures such as carbon nanotubes, carbon nanofibers, carbon nanohorns, carbon nanocoils, etc. positively influence in enhancing catalytic activity as catalyst supports in fuel cells (Basri et al., 2010).

Direct Ethanol Fuel Cell (DEFC)

Although methanol owns many positive properties as a fuel in fuel cells, still there are some problems. Ethanol is a distinguished substitute to methanol. Nontoxicity, easy production, and high energy density are some factors of ethanol as a fuel to use over methanol (Du et al., 2018). Compared to Pt, Pd and Pd-based catalysts are more efficient for ethanol oxidation. Hence the development of Pd and Pd-based nanomaterials started. Initially, the research work was done on Pt-Pd bi-metallic nanoparticles. Pt_3Pd_3/MWCNT (multi-walled carbon nanotube) electrocatalyst is an example that shows enhanced peak current for ethanol oxidation reaction compared to Pt/MWCNTs (Ding et al., 2013). Pd-NiO nanoparticles incorporated Pd-NiO/MWCNT nanocomposite supported on graphene oxide support (Pd-NiO/MWCNT/rGO) offer improved catalytic activity greater than Pd-NiO/MWCNT/rGO and Pd/C electrocatalysts. It also exhibits a comparatively higher electrochemical stability (Rajesh et al., 2018).

2.2 *Batteries*

A battery can be defined as an energy storage device which is operated by electrochemical reactions. Batteries are available mainly in two types: primary cell which is not rechargeable and can be used only one time and secondary cell which is rechargeable. Despite the complexity of the arrangement of the batteries, they are made up with two key components: electrodes and electrolytes (Armand & Tarascon, 2008). Electrode can be either the anode (negative electrode) or the cathode (positive electrode). The two electrodes are separated and immersed in an electrolyte which is ionically conductive (Wang et al., 2019). The redox reactions occurring near the electrodes help transform the chemical energy to electrical energy to get the external work done (Wang et al., 2019). Redox reaction is a transfer of electrons among two species where one species reduces and one species oxidizes. The electrochemical reactions occurring in a primary cell is not reversible, and after all the chemicals being used up, the cell becomes useless. Zinc-carbon and alkaline cells are the common types of primary cells (Wang et al., 2019). Secondary cells are known for recharge ability as the electrochemical reactions occurring in a secondary cell are reversible and cycled for many times before its service life ends (Wang et al., 2019). The typical rechargeable batteries include lead-acid, nickel-cadmium (NiCd), nickel-metal hydride (NiMH), and lithium-ion cells (Wang et al., 2019).

A battery operates by exchanging electrons between two electrodes by chemical reactions taking place in the electrochemical system. The number of charges transferred and the potential difference between the electrodes mainly decide the output of the battery. Therefore, the energy given out is decided by the kinetic and thermodynamic characteristics of the materials used for the construction of electrochemical system (Wang et al., 2019). The use of nanomaterial in electrochemical systems has come to attention due to characteristics like high surface-area-to-volume ratio, better transport kinetics, and nanopore confinement. In the case of batteries, these properties allow to introduce new active storage sites, reduce ion transport distances, and create conductive channels for electrons, decrease rate of specific surface current, and enhance stability of the system (Wang et al., 2019). The variation of these characteristics with the use of different types of nanomaterial in the electrochemical system will be discussed for some common battery types.

The use of nanomaterials in battery electrochemical system has become popular in the scientific community as it is beneficial in many ways. The electrochemical reactions that are not possible with micrometer-sized particles can easily take place with nanosized particles. The rate of insertion and removal of ions increases due to the reduced dimensions. Electron transport within the particles also speeds up due to the reduced dimensions. High specific surface area of nanomaterials facilitates better contact between electrodes and electrolyte.

2.2.1 Li-Ion Batteries

Lithium-ion battery consists of an anode, a cathode, and an electrolyte. The anode is usually graphitic carbon, and the cathode is lithium transition metal oxide (Wang et al., 2019). During the charging phase, lithium ions (Li^+) are extracted from the cathode and migrate through the organic electrolyte to the anode to form LiC_6. At the same time, an electron from the transition metal is removed and passed through the external circuit producing electric current. During the discharging phase, Li^+ ions are drawn out from the anode and intercalate back into the cathode (Wang et al., 2019). The reversible cell reaction is given as Reaction 8. The performance of the Li-ion battery with the use of nanostructured materials for the electrodes is illustrated in next sections:

$$LiCoO_2 + 3C \leftrightarrow \tfrac{1}{2}LiC_6 + Li_{1/2}CoO_2 \tag{8}$$

Nanomaterial for Li-Ion Cathode

The cathode of the Li-ion battery should be the Li source of the system, and it can be intercalated with a nanomaterial to enhance the performance by gaining a high energy density (Wang et al., 2019).

Nanostructured vanadium oxides offer great advantages for Li-ion intercalation applications (Wang et al., 2019). V_2O_5 is a highly abundant high-valent oxide of vanadium with a high theoretical capacity (294 mAh g^{-1}) (Wang et al., 2019). Modified V_2O_5 nanostructures could enhance the electrochemical activity and shorten the ion diffusion distance while eliminating the limitations of conventional V_2O_5 such as low conductivity, low diffusion coefficient, and poor structural stability (Cao et al., 2005; Tan et al., 2015; Yue & Liang, 2017). V_6O_{13} also offers high energy density and theoretical specific capacity up to 417 mAh g^{-1}. The developed V_6O_{13} nanosheets and 3D nanotextiles further improve the electrochemical performance of V_6O_{13} (Zou et al., 2014; Ding et al., 2015). $Li_3V_2(PO_4)_3/C$ nanocomposite developed by coating graphene on monoclinic lithium vanadium phosphate ($Li_3V_2(PO_4)_3$) also offers high rate performance and cycling stability (Liu et al., 2011).

Cathode made up with nanostructured transition metal oxides of Li has proven elevated energy density and specific capacity in Li-ion batteries (Wang et al., 2019). Lithium cobalt oxide ($LiCoO_2$) offers a high theoretical capacity of 274 mAh g^{-1} (1.0 Li/Co) with a better rate capability and most importantly facile and scalable production capability (Kalluri et al., 2017). Coating $LiCoO_2$ with nanoparticle coatings such as $AlPO_4$, $FePO_4$, and $CePO_4$ has resulted in high intercalation capacities (Lee et al., 2004). Due to the cost and safety issues aroused with the $LiCoO_2$ cathode, lithium nickel oxide ($LiNiO_2$) came into play (Wang et al., 2019). $LiNiO_2$ resulted in enhanced electrical and ionic conductivity and thermal and structural stability when doped with nano-CO^{3+}, Al^{3+}, and Mn^{4+} (Delmas et al., 1993; Park et al., 2001; Guilmard et al., 2003). $LiMnO_2$ nanorods have also shown improved cyclability and storage capacity (Xiao et al., 2010). All the above-discussed nanostructures improve the battery performance either by shortening the Li^+ ion migration path enhancing the rate capability or accommodating large strain increasing the cycling ability (Wang et al., 2019).

Nanomaterial for Li-Ion Anode

Anode of a Li-ion battery is usually a graphite electrode due to the structural stability, chemical stability, desirable electrical conductivity, and low cost (Wang et al., 2019). Li^+ ions intercalate between planes of graphite and occupy interstitial sites. In graphite, the Li^+ intercalations are limited because an inserted ion does not allow another ion to occupy the directly adjacent site (De Las Casas & Li, 2012). It directly affects the energy storage density of the battery. Therefore, many research works have done to develop nanostructured materials, especially carbonaceous materials with enhanced power density and energy.

Zero-dimensional nanoparticles; one-dimensional nanowires, nanotubes, and nanosprings; and two-dimensional thin films are widely studied as anode materials for Li-ion batteries. Carbon nanotubes have shown high rigidity, low density, high tensile strength, and higher capacity compared to graphite as nanotubes allow Li^+ to diffuse into individual nanotubes and also interstitial sites between layers of carbon nanotubes (De Las Casas & Li, 2012). Carbon nanotubes show twice the lithium storage compared with graphite (Bruce et al., 2008).

Metal electrodes are also used as anodes in Li-ion batteries, where Li is stored via alloy formation. However, due to the expansion of the anode during the operation, they are not recommended for rechargeable batteries. This problem has been attempted to solve by developing nanoparticles, nanotubes, and thin films. Metals, specially silicon and metal oxides, are combined with carbon nanotubes to form nanocomposites with much higher capacities (De Las Casas & Li, 2012).

Nanomaterial for Electrolyte

Electrolytes are usually in liquid phase, amorphous polymer phase, and crystalline polymer phase. Addition of nanomaterial for these phases could enhance the properties of the electrolyte. Addition of compounds such as SiO_2, Al_2O_3, and ZrO_2 in powdered form can efficiently enhance the conductivity of the electrolyte by a factor of 6 (Bhattacharyya & Maier, 2004).

2.2.2 Li-Metal Batteries

In contrast to Li-ion batteries, Li-metal batteries employ lithium metal as anode material. Li-metal batteries are usually known for providing extremely high energy densities (Wang et al., 2019). Li-S and Li-air are the common types of Li-metal batteries that have very high theoretical specific energies of 3500 and 2600 Wh kg^{-1}, while 387 Wh kg^{-1} for Li-ion (Wang et al., 2019).

But the abovementioned theoretical energy densities are yet to achieve. As sulfur is an electrical insulator and air is in gas phase, the electrochemical performance of the battery could be reduced leading to limited energy densities. As Li-ion batteries involve intercalation reactions for operation, Li-metal batteries perform conversion reactions that can cause structural changes in electrodes leading to reduced battery life (Wang et al., 2019). Therefore, to improve the Li-metal battery performance, efforts have made to incorporate nanomaterials with suitable nanostructures for electrodes by changing the battery material design and structure.

Nanomaterials for Li-Metal Anode

Although Li-metal anode leads to high theoretical capacities, Li-metal anode construction has become problematic on account of complications like dendrites and intrinsic generation when plating Li, unstable nature of Li-metal in electrolytes, electrical detachment of Li by repetitive plating due to formation of dendrites, and large changes in electrode volume during Li plating (Wang et al., 2019).

Nanostructured host materials are used to provide high effective electroactive area for Li anodes. Copper is proposed to use as a host in the form of three-dimensional current collector to accommodate Li-metal anode, as it provides high electroactive area to lower the local current density, and a porous volume preserved to accommodate the anode and lessen the volume change (Yang et al., 2015). Porous carbon which is inexpensive and light in weight compared to Cu is also considered as a host material for Li-metal anode. A 3D structure formed by packing 2D graphite flakes which is called massive artificial graphite owns many edge planes and empty spaces, and Li-metal can be accommodated inside individual particles of the 3D structure. It lessens the formation of dendrites and any possible chemical reactions with the electrolyte (Sun et al., 2016). Graphene network is also widely studied as a host material due to properties like high conductivity and mechanical strength

(Cheng et al., 2015; Zhang et al., 2016). Li-metal anode with synergistic effects and dendrite-free Li deposition is obtained by a 3D graphene on a Ni scaffold as it provides high porosity and surface area while acting as an artificial protection area (Xie et al., 2016). Large pore volume is desirable to have in a host in order to sufficiently accommodate the Li-metal anode. Therefore, interconnected carbon nanofibers having large free spaces compared to graphene and graphite are also in consideration as host materials for Li-metal anode.

Nanomaterials for Li-Air Cathode

Having Li-metal as the anode, Li-air battery uses air (usually oxygen and referred as Li-O_2) as the cathode infused in open porous conductive electrode. Other than aqueous electrolytes, it also uses organic electrolytes in some cases (Wang et al., 2019). Depending on the type of the electrolyte, either aqueous or non-aqueous, the electrochemical reaction occurring in the system varies. Reactions 9 and 10 show the electrochemical reaction taking place in a Li-air battery with an aqueous electrolyte and a non-aqueous electrolyte, respectively (Choi & Aurbach, 2016):

$$2Li^+ + 2e^- + \frac{1}{2}O_2 + H_2O \leftrightarrow 2LiOH \tag{9}$$

$$2Li^+ + 2e^- + O_2 \leftrightarrow Li_2O_2 \tag{10}$$

Cathode material of a Li-O_2 battery should have properties like good conductivity, electrochemical stability, good specific surface area, and low reactivity with the electrolyte (Wang et al., 2013). Nanostructured carbon materials have shown great benefits as Li-O_2 cathode. Carbon provides a porous structure to accommodate discharge ions and provide pathways for diffusion of oxygen. It also acts as an electrocatalyst to reduce oxygen (Wang et al., 2019). Carbon nanotubes and carbon nanofibers exhibit excellent chemical stability and conductivities and high tensile strength, while graphene nanosheets exhibit excellent electrical conductivity, superior mechanical flexibility, high specific surface area, and diffusion channels for electrolyte and O_2 (Xiao et al., 2011). Hierarchical carbon architectures like graphene and 3D-ordered mesoporous/microporous carbon sphere arrays found to be fitting well as cathode materials. The mesoporous structures boost the immersion of electrolyte and ease the transfer of electrons and ion diffusion (Guo et al., 2013a, 2013b; Zhao et al., 2016).

In Li-air batteries, catalysts are used to increase the performance of the system by enhancing oxygen reduction reaction during discharge and oxygen evolving reaction during charge. Catalyst nanomaterials which are also called nanocatalysts are developed to further improve the performance of the system. Catalyst in the Li-air system is expected to have superior catalytic activity, good conductivity, high porosity and specific surface area, and low cost (Wang et al., 2013). N-doped carbon materials proved to have good catalytic activity as it enhance the kinetics of O_2 reduction and discharge capacity (Ding et al., 2014). Noble metals and transition metal oxides in nanocrystal, nanosheet, nanocubes, and 3D form structures also have shown good catalytic activity in Li-air electrochemical system (Zhang et al., 2014; Liao et al., 2015; Su et al., 2015; Liu et al., 2016).

2.3 Capacitors

Electrostatic capacitors are the generally known capacitors that consist of two conductive electrodes in the form of metal plates or surfaces separated by a dielectric material and are also known as supercapacitors or ultra-capacitors. This type of capacitors are used in high-speed circuits, but the capacitance and energy density are much lower (Zhao et al., 2011). In electrochemical capacitors (ECs), ions are reversibly adsorbed onto porous electrodes to form an electric double layer at the electrode-electrolyte interface and store the electrical energy (Zhao et al., 2011). ECs have higher energy densities compared to electrostatic capacitors and longer life cycle than conventional Li-ion batteries (Huang et al., 2020). ECs with higher energy and power densities are becoming useful for applications like electrical vehicles and next-generation electronic systems. As these properties are dependent on surface area and crystalline properties of the electrode material (Kumar et al., 2018; Sun et al., 2019), electrochemical performance of the capacitor can be improved by incorporating nanomaterials for the main components of the capacitor.

In electrochemical capacitors, charge is stored either as surface charge storage or bulk charge storage. According to these charge storing mechanisms, capacitors are categorized into electric double-layer capacitors, pseudocapacitors, and hybrid capacitors. In electric double-layer capacitors, two layers of electrical charges are formed between electrode and the electrolyte. Apart from that, the operation of the electric double-layer capacitor is similar to electrostatic capacitor. This type of capacitors is ideal with electrodes having properties like rapid charge and discharge rates, good conductivity, large potential window, and high specific surface area (Choi & Yoon, 2015). In pseudocapacitors, fast and fully reversible Faraday's charge transfer occurs at the electrode-electrolyte interface (Abbas et al., 2019). Therefore, pseudocapacitors have a performance similar to a battery, and they often exhibit higher capacitance than electric double-layer capacitors. The capacitance of a pseudocapacitor can be enhanced by considering few factors: fast charge and discharge rates, large surface area for redox reactions, wide potential window, and increasing conductivity by doping conductive polymers (Choi & Yoon, 2015). Hybrid capacitor is a combination of electric double-layer capacitor and pseudocapacitor. The charge storage also occurs as a combination of electrostatic adsorption-desorption and a reversible redox reaction at the electrode (Choi & Yoon, 2015).

The performance of an electrochemical capacitor is mainly measured using energy and power density expressed as a quantity per unit weight. The latest research on electrochemical capacitors is mostly on developing high-performance capacitors by improving the materials used in the capacitor electrochemical system. The use of nanomaterials for electrodes in electric double-layer capacitors and pseudocapacitors is discussed in this section.

2.3.1 Electric Double-Layer Capacitors (EDLCs)

Electric double-layer capacitor (EDLC) is the main type of supercapacitors which has a surface charge storage (Abbas et al., 2019). Capacitance of EDLC generates from a physical phenomenon involving separation of charge at electrode-electrolyte interface (Abbas et al., 2019). Carbon-based materials are the most commonly used electrode materials in EDLC due to its merits such as good electronic conductivity, high chemical endurance, open porosity, and vast surface area (Huang et al., 2020; Yu et al., 2013). Pore size distribution, architecture, and specific surface area of an electrode greatly influence the performance of a capacitor. While EDCL acquires a high power and longer cycle life, its energy density is comparatively low. Therefore, materials fulfilling the mentioned properties have become a research attraction. In this scenario nanomaterials play an important role as the micrometer-sized large pores limit the power density via sluggish ion transport kinetics (Huang et al., 2020).

Porous carbon materials play a major role in EDLC electrodes as they provide numerous structural benefits for electrochemical energy storage. Many studies have been done to investigate nanoporous materials in order to enhance the capacitor performance over the conventional materials. Hierarchical nanoporous structures can develop charge storage and transfer with the increase of active sites. The nanopore acts as active sites, while the material functions as the routes that carry electrons and points where ion interactions take place. N-doped carbon nanosheets (HPNC) is a 2D porous material which exhibits an improved specific surface area along with enhanced energy density, capacitance, and cyclic stability (Peng et al., 2018).

2D Ggraphene is another promising nanomaterial which shows excellent properties for EDLCs. It owns the theoretically highest surface area compared to other porous nanomaterials (Qin et al., 2014). Although graphene displays fascinating properties due to its atom-thick 2D structure, forming irreversible agglomerates blocking the graphitic structure is a drawback. However many studies have recorded that synthesizing hybrid materials uses carbon nanotubes (CNT) or carbon nanofibers (CNF) along with graphene (Chen et al., 2013). Electrospinning is a best synthesizing method for these types of hybrid materials (Krishnan et al., 2014), (Choi & Yoon, 2015). Nanofibers of graphene nanoribbon (GNR)/carbon composite are examples of hybrid nanomaterials which possess enlarged surface areas and good electrochemical activities. Recent studies also demonstrate a ternary hybrid nanostructure comprising CNFs, manganese oxide (MnO), and graphene. These materials give a 3D nanohybrid architecture which enhances the ion diffusion.

Newer types of carbon nanomaterials have been developed by incorporating various polymers. As the morphology and the composition of a polymer are easily controllable, carbon nanomaterials with controlled structures and properties can be easily synthesized with the introduction of polymer materials. Cellulose nanofiber

(CeNF)/carbon nanohybrid aerogel electrode material made with the use of the most abundant natural polymer cellulose displays outstanding specific surface area and electrolyte absorption properties (Zheng et al., 2015).

Outstretched research works have revealed 3D nanomaterials with well-defined structures. Three-dimensionally ordered mesoporous (3DOM) carbon exhibits an improved capacitance over the ordinary carbon electrodes (Kumaran et al., 2018).

2.3.2 Pseudocapacitors (PCs)

In contrast to EDLCs, pseudocapacitors (PCs) have speedy reversible redox reactions that occur at the surface of electroactive materials during the charge/discharge process (Yu et al., 2013), (Choi & Yoon, 2015). Even though PCs show higher capacitance than the EDLCs, the charge/discharge rate is slow. Transition metal oxides and electronically conductive redox polymers are major types of conventional materials used in electrodes (Naoi & Simon, 2008).

Increasing redox reactions and conductivity by doping conducting polymer, improving charging and discharging rate, increasing surface area encountered in reactions, and enhancing potential are the main factors to be considered when developing capacitance of the PCs (Nejati et al., 2014). Most of the time, a PC's redox reactions are confined to electrode material surfaces. Hence, developing nanostructured materials with enlarged effective surface area has become a research attraction.

When considering conducting polymers, size and morphology affect the performance of the PC. The instability of polymers at nanoscale is a challenge. However, there are several novel synthesizing methods of nanostructured polymer materials. Polyaniline (PANI) is one of the commonly used polymers for electrode materials in PCs. Studies have revealed that the aspect ratio of PANI nanoparticles has a great impact on the capacitor activity. The interparticle resistance decreases in the order of nanospheres, nanorods, and nanofibers. As a result PANI nanofibers show the highest discharge capacitance (Park et al., 2012). It is understood that there is a major influence for the performance of PCs. Multidimensional nanostructures with improved surface-area-to-volume ratio such as poly(3,4-ethylenedioxythiophene) (PEDOT) and multidimensional polypyrrole (PPy) nanotubes also exhibit excellent charge transport properties while enhancing capacitance (Choi & Yoon, 2015). Enhanced capacitance can be obtained by incorporating ultrathin polythiophene with porous nanostructures of anodized aluminum oxide (AAO), activated carbon, and TiO_2 via oxidative chemical vapor deposition (OCVD) technique (Nejati et al., 2014). Unique surface structures such as nanonodules synthesized using ultrathin poly(methyl methacrylate) (PMMA) are beneficial as PCs (Choi & Yoon, 2015). In order to reduce the interparticle aggregation, conductive polymer coatings can be done on metal nanoparticles. Ag-PPy core-shell nanoparticle is one such example (Chang et al., 2012).

Transition metal (TMO) oxides are also promising candidates for pseudocapacitor electrodes. Ruthenium oxides (RuO_2), iron oxide (Fe_3O_4), manganese oxide (MnOx), and nickel oxide (NiO) are few typical electrode materials. Various studies have been done regarding different morphologies of TMOs such as nanoparticles (zero dimensional), nanowires, and nanorods (one-dimensional), nanosheets (two-dimensional), and nanospheres (three-dimensional) in order to assess the increment in electrode performance.

Two-dimensional nanosheet of nickel-cobalt oxide (NCO) having a hollow structure which promotes penetration of electrolyte through improved active sites is a material that exhibits high specific capacitance (Peng et al., 2018). As a remedy for the poor conductivity of metal oxides, nanocomposites consisting of metal oxide and conductive materials have been developed (Choi & Yoon, 2015). CoO-based PPy nanowire is a nanocomposite electrode material which displays a higher capacitance and a good rate capability (Zhou et al., 2013a). Hierarchical nanostructures also display excellent specific capacitance values through maximized specific surface area. Substantial research has been done on this and 1D hierarchical nanostructured $NiCo_2O_4$ nanosheets is a one of the example which exhibits an extremely high capacitance and durability as a PC electrode (Liang et al., 2014). RuO_2 and MnO_2 combined with carbon nanomaterials and metal nanostructures exihibit higher conductivities and identified them as widely used materials. However their practical applications are limited due to the higher cost (Yu et al., 2013).

2.3.3 Dielectric Capacitors

Dielectric capacitors have gained high popularity due to comparatively higher power and lower energy densities (Nelson & Hu, 2005). Although dielectric capacitor does not fall into the category of electrochemical capacitors, it is a strong energy storage technique that should not be missed. Incorporating nanomaterials in dielectric capacitors also has become promising due to their enhanced dielectric properties. Organic and inorganic nanocomposites have gained a great research attraction due to the significant upgraded properties compared to the conventional dielectric materials (Madusanka et al., 2017). Increased interface between polymer and nanofiller in a nanocomposite positively increases properties (Fan & Bai, 2015). However, particle aggregation is a common challenge encountered in developing dielectric nanocomposites. Surface modification of nanoparticles will promote homogeneous dispersion (Niu & Wang, 2019).

TiO_2 owns an improved dielectric constant and a bandgap. Therefore, TiO_2 is widely used as a dielectric material. Many researches have been done on nanocomposites that comprise TiO_2 nanoparticles and different types of polymers. Polystyrene, epoxy resin, cross-linked poly-4-vinylphenol, cyanoethylated cellulose (CRS), and polypyrrole are some polymer materials which are blended with TiO_2 nanofiller to fabricate dielectric nanocomposites. TiO_2 nanoparticles offer enhanced dielectric properties including ultrahigh dielectric constants (Madusanka et al., 2016). Hexagonal boron nitride nanosheets (BNNS) are emerged as multifunctional fillers in polymer dielectrics which exhibit a wide bandgap, higher dielectric strength, improved thermal conductivity, and enhanced capacitive performance

(Li et al., 2020). 1D and 2D carbon nanomaterials also display excellent dielectric properties when incorporated in nanocomposites of dielectric capacitor (Dang et al., 2016).

3 Challenges and Future Prospects

Nanostructured materials own a significant part in enhancing performance in energy generation and storage devices. With novel technologies different types of nanomaterials in different morphologies have been developed to uplift the electrochemical properties of energy storage devices. Despite these distinguished advances, there are some challenges that can be encountered at different stages of incorporating nanostructured materials in different components of energy storage devices.

Although nanoparticles exhibit excellent performance, effective control of the particle size (to obtain maximum electrochemically active sites), morphology (to understand growth mechanisms to effectively customize geometries of complex structures), high-index facets (to increase catalytic activity), and composition and architecture (to enhance activity and stability) during synthesizing is a great challenge (Guo & Wang, 2011; Chen & Ostrom, 2015). Increase in inter-surface area between electrolyte and electrode due to uncontrolled conditions can direct to noticeable side reactions with electrolyte which will be an obstacle to maintain interparticle contact (Bruce et al., 2008). Particle agglomeration and non-uniform distribution are also encountered problems. Developing surfactants to minimize agglomeration should be considered.

Noble materials such as Pt are promising materials for electrode materials. However, the high cost of these noble materials is a restriction for their utilization in energy storage devices (Suryawanshi et al., 2018). Development of nanostructured composites and hybrid materials incorporating non-noble materials has become a recent research trend, and advanced studies should be done to synthesize nanomaterials while safeguarding the stability and enhanced activity (Ghosh et al., 2010; Zhou et al., 2013b). Also, introducing more user-friendly techniques and scalable production methods for the practical, scaled-up applications and commercialized applications through future research work would be a great benefit. Also, research areas should expand to recover the gap in reproducibility of these devices. The improvements in modern quantum chemistry and computer science can be integrated in developing nanomaterials for different components in energy storage devices (Shao et al., 2007).

4 Conclusion

The demand for energy in the current world is increasing rapidly. Limited availability of most widely used fossil fuels has paved the way for alternative energy generation and storage methods. Fuel cells, batteries, and capacitors are promising energy storage devices where the performance improvement of devices has become a hot research area. Incorporating nanostructured materials in the components (anode, cathode, and electrolyte) of the energy storage devices offers remarkable improvements in the electrochemical activity comparatively to conventional materials. Increased specific surface area, particle interactions, and active sites are some of the key factors led toward the enhancement of cathodic and anodic reactions. Controllable synthesis of nanoparticles is very important to secure properties. Morphology and architecture are also significant as well as the particle size. Replacing noble materials by non-noble materials by introducing nanocomposite or nanohybrid materials can reduce the material cost. Nanocomposites also exhibit improved electrochemical properties compared to single nanomaterials due to their merged properties. Future research work can be executed to further improvement of nanomaterials and synthesis processes to expand in commercial scale, securing the stability and enhancing performance.

References

Abbas, Q., et al. (2019). Heteroatom doped high porosity carbon nanomaterials as electrodes for energy storage in electrochemical capacitors: A review. *Journal of Science: Advanced Materials and Devices, 4*(3), 341–352. https://doi.org/10.1016/j.jsamd.2019.07.007

Abdalla, A. M., et al. (2018). Nanomaterials for solid oxide fuel cells: A review. *Renewable and Sustainable Energy Reviews, 82*(August 2017), 353–368. https://doi.org/10.1016/j.rser.2017.09.046

Armand, M., & Tarascon, J. M. (2008). Building better batteries. *Nature, 451*, 652–657.

Armaroli, N., & Balzani, V. (2007). The future of energy supply: Challenges and opportunities. *Angewandte Chemie – International Edition, 46*(1–2), 52–66. https://doi.org/10.1002/anie.200602373

Bang, J. H., & Suslick, K. S. (2010). Applications of ultrasound to the synthesis of nanostructured materials. *Advanced Materials, 22*(10), 1039–1059. https://doi.org/10.1002/adma.200904093

Basri, S., et al. (2010). Nanocatalyst for direct methanol fuel cell (DMFC). *International Journal of Hydrogen Energy, 35*(15), 7957–7970. https://doi.org/10.1016/j.ijhydene.2010.05.111

Bhattacharyya, B. A. J., & Maier, J. (2004). Second phase effects on the conductivity of non-aqueous salt solutions. *Soggy Sand Electrolytes, 9*, 811–814. https://doi.org/10.1002/adma.200306210

Bruce, P. G., Scrosati, B., & Tarascon, J. M. (2008). Nanomaterials for rechargeable lithium batteries. *Angewandte Chemie – International Edition, 47*(16), 2930–2946. https://doi.org/10.1002/anie.200702505

Cao, A.-M., Hu, J.-S., Liang, H.-P., & Wan, L.-J. (2005). Self-assembled vanadium pentoxide (V2O5) hollow microspheres from nanorods and their application in lithium-ion batteries. *Angewandte Chemie International Edition, 44*, 4391–4395.

Chang, M., et al. (2012). Imparting chemical stability in nanoparticulate silver via a conjugated polymer casing approach. *ACS Applied Materials and Interfaces, 4*(8), 4357–4365. https://doi.org/10.1021/am3009967

Chanquía, C. M., et al. (2014). A bottom-up building process of nanostructured La0.75Sr 0.25Cr0.5Mn0.5O3-δ electrodes for symmetrical-solid oxide fuel cell: Synthesis, characterization and electrocatalytic testing. *Journal of Power Sources, 245*, 377–388. https://doi.org/10.1016/j.jpowsour.2013.06.124

Chen, A., & Ostrom, C. (2015). Palladium-based nanomaterials: Synthesis and electrochemical applications. *Chemical Reviews, 115*(21), 11999–12044. https://doi.org/10.1021/acs.chemrev.5b00324

Chen, X., et al. (2013). Novel electric double-layer capacitor with a coaxial fiber structure. *Advanced Materials, 25*(44), 6436–6441. https://doi.org/10.1002/adma.201301519

Cheng, X., et al. (2015.) 'Xin-Bing Cheng, Hong-Jie Peng, Jia-Qi Huang, Rui Zhang, Chen-Zi Zhao, and Qiang Zhang*). *Beijing key Laboratory of Green Chemical Reaction Engineering and Technology, department of chemical engineering*. Tsinghua University.

Choi, H., & Yoon, H. (2015). Nanostructured electrode materials for electrochemical capacitor applications. *Nanomaterials, 5*(2), 906–936. https://doi.org/10.3390/nano5020906

Choi, J. W., & Aurbach, D. (2016). Promise and reality of post-lithium-ion batteries with high energy densities. *Nature Reviews Materials, 1*. https://doi.org/10.1038/natrevmats.2016.13

Choudhury, R. R., Sahoo, S. K., & Gohil, J. M. (2020). Potential of bioinspired cellulose nanomaterials and nanocomposite membranes thereof for water treatment and fuel cell applications. *Cellulose, 27*(12), 6719–6746. https://doi.org/10.1007/s10570-020-03253-z

Dang, Z. M., Zheng, M. S., & Zha, J. W. (2016). 1D/2D carbon nanomaterial-polymer dielectric composites with high permittivity for power energy storage applications. *Small, 12*(13), 1688–1701. https://doi.org/10.1002/smll.201503193

Debe, M. K. (2013). Tutorial on the fundamental characteristics and practical properties of nanostructured thin film (NSTF) catalysts. *Journal of the Electrochemical Society, 160*(6), F522–F534. https://doi.org/10.1149/2.049306jes

De Las Casas, C., & Li, W. (2012). A review of application of carbon nanotubes for lithium ion battery anode material. *Journal of Power Sources, 208*, 74–85. https://doi.org/10.1016/j.jpowsour.2012.02.013

De Larramendi, I. R., et al. (2011). La0.6Sr0.2Ca0.2Fe0.8Ni 0.2O3 thin films obtained by pulsed laser ablation: Effect of the substrate on the electrochemical behavior. *Solid State Ionics, 192*(1), 584–590. https://doi.org/10.1016/j.ssi.2010.02.012

Delmas, C., Saadoune, I., & Rougier, A. (1993). The cycling properties of the LixNi1−yCoyO2 electrode. *Journal of Power Sources, 44*, 595–602.

Dhathathreyan, K. S., Rajalakshmi, N., & Balaji, R. (2017). Nanomaterials for fuel cell technology. *Nanotechnology for Energy Sustainability*, 569–596. https://doi.org/10.1002/9783527696109.ch24

Dai, L., et al. (2015). Metal-free catalysts for oxygen reduction reaction. *Chemical Reviews, 115*(11), 4823–4892. https://doi.org/10.1021/cr5003563

Ding, K., et al. (2013). Electrocatalytic activity of multi-walled carbon nanotubes-supported PtxPdy catalysts prepared by a pyrolysis process toward ethanol oxidation reaction. *Electrochimica Acta*. Elsevier Ltd, *100*, 147–156. https://doi.org/10.1016/j.electacta.2013.02.130

Ding, N., et al. (2014). Influence of carbon pore size on the discharge capacity of Li-O2 batteries. *Journal of Materials Chemistry A, 2*(31), 12433–12441. https://doi.org/10.1039/c4ta01745e

Ding, Y. L., Wen, Y., Wu, C., van Aken, P. A., Maier, J., & Yu, Y. (2015). 3D V6O13 nanotextiles assembled from interconnected nanogrooves as cathode materials for high-energy lithium ion batteries. *Nano Letters, 15*, 1388–1394.

Du, H., et al. (2018). Carbon nanomaterials in direct liquid fuel cells. *Chemical Record, 18*(9), 1365–1372. https://doi.org/10.1002/tcr.201800008

Fan, B., & Bai, J. (2015). Composites of hybrids BaTiO3/carbon nanotubes/polyvinylidene fluoride with high dielectric properties. *Journal of Physics D: Applied Physics*. IOP Publishing, *48*(45), 455303. https://doi.org/10.1088/0022-3727/48/45/455303

Frattini, F. M., *et al.* (2015) 'Frontiers and recent developments in nanomaterials for fuel cell applications: An overview of performance and perspectives', *European Fuel Cell Technology & Applications Conference-Piero Lunghi Conference*.

Ghosh, T., et al. (2010). Pt alloy and intermetallic phases with V, Cr, Mn, Ni, and cu: Synthesis as nanomaterials and possible applications as fuel cell catalysts. *Chemistry of Materials, 22*(7), 2190–2202. https://doi.org/10.1021/cm9018474

Guilmard, M., Croguennec, L., & Delmas, C. (2003). Thermal stability of lithium nickel oxide derivatives'. Part II: LixNi0.70Co0.15Al0.15O2 and LixNi0.90Mn0.10O2 (X = 0.50 and 0.30). Comparison with LixNi1.02O2 and LixNi0.89Al0.16O2, *chemistry of materials, 15*,4484–4493. Ion battery anode material. *Journal of Power Sources, 208*, 74–85.

Guo, S., et al. (2013a). FePt and CoPt nanowires as efficient catalysts for the oxygen reduction reaction. *Angewandte Chemie, 125*(12), 3549–3552. https://doi.org/10.1002/ange.201209871

Guo, S., & Wang, E. (2011). Noble metal nanomaterials: Controllable synthesis and application in fuel cells and analytical sensors. *Nano Today, 6*(3), 240–264. https://doi.org/10.1016/j.nantod.2011.04.007

Guo, Y.-G., Hu, J.-S., & Wan, L.-J. (2008). Nanostructured materials for electrochemical energy conversion and storage devices. *Advanced Materials, 20*(15), 2878–2887. https://doi.org/10.1002/adma.200800627

Guo, Z. *et al.* (2013b). 'Ordered Hierarchical Mesoporous/ Macroporous Carbon: A High Performance Catalyst for Rechargeable Li-O2 Batteries', pp. 1–5. doi: https://doi.org/10.1002/adma201302459.

Huang, T., et al. (2020). Recent advances of two-dimensional nanomaterials for electrochemical capacitors. *ChemSusChem, 13*(6), 1093–1113. https://doi.org/10.1002/cssc.201903260

Jafri, R. I., et al. (2010). Nanostructured Pt dispersed on graphene-multiwalled carbon nanotube hybrid nanomaterials as Electrocatalyst for PEMFC. *Journal of the Electrochemical Society, 157*(6), B874. https://doi.org/10.1149/1.3374353

Jusys, Z., et al. (2002). Activity of PtRuMeOx (Me = W, Mo or V) catalysts towards methanol oxidation and their characterization. *Journal of Power Sources, 105*(2), 297–304. https://doi.org/10.1016/S0378-7753(01)00955-7

Kakkar, S. *et al.* (2020). *Nanotechnology for energy and environmental engineering*. Available at: http://link.springer.com/10.1007/978-3-030-337742_2%0A http://link.springer.com/10.1007/978-3-030-33774-2.

Kalluri, S., et al. (2017). Surface engineering strategies of layered LiCoO2 cathode material to realize high-energy and high-voltage Li-ion cells. *Advanced Energy Materials, 7*, 1. https://doi.org/10.1002/aenm.201601507

Krishnan, D., et al. (2014). Graphene oxide assisted hydrothermal carbonization of carbon hydrates. *ACS Nano, 8*(1), 449–457. https://doi.org/10.1021/nn404805p

Kumar, K. S., et al. (2018). Recent advances in two-dimensional nanomaterials for supercapacitor electrode applications. *ACS Energy Letters, 3*(2), 482–495. https://doi.org/10.1021/acsenergylett.7b01169

Kumaran, R., et al. (2018). The morphology-dependent electrocatalytic activities of spinel-cobalt oxide nanomaterials for direct hydrazine fuel cell application. *New Journal of Chemistry*. Royal Society of Chemistry, *42*(15), 13087–13095. https://doi.org/10.1039/c8nj01622d

Lade, H., et al. (2017). Sulfonated poly(arylene ether sulfone) nanocomposite electrolyte membrane for fuel cell applications: A review. *International Journal of Hydrogen Energy, 42*(2), 1063–1074. https://doi.org/10.1016/j.ijhydene.2016.10.038

Lee, J. G., Kim, B., Cho, J., Kim, Y. W., & Park, B. (2004). Effect of AlPO4-nanoparticle coating concentration on high-cutoff-voltage electrochemical performances in LiCoO2. *Journal of the Electrochemical Society, 151*, A801–A805.

Liang, J., et al. (2014). Jin Liang, Zhaoyang Fan, Sheng Chen, Shujiang Ding, and Guang Yang). *Chemistry of Materials, 26*, 4354–4360.

Liao, K., et al. (2015). An oxygen cathode with stable full discharge-charge capability based on 2D conducting oxide. *Energy and Environmental Science, 8*(7), 1992–1997. https://doi.org/10.1039/c5ee01451d

Li, H., et al. (2020). Recent progress in polymer dielectrics containing boron nitride nanosheets for high energy density capacitors. *High Voltage, 5*(4), 365–376. https://doi.org/10.1049/hve.2020.0076

Liu, H., Gao, P., Fang, J., & Yang, G. (2011). Li3V2(PO4)3/graphene nanocomposites as cathode material for lithium ion batteries. *Chemical Communications, 47*, 9110–9112.

Liu, L., et al. (2016). Self-assembled 3D foam-like NiCo2O4 as efficient catalyst for lithium oxygen batteries. *Small, 12*(5), 602–611. https://doi.org/10.1002/smll.201502924

Lykhach, Y., et al. (2017). Oxide-based nanomaterials for fuel cell catalysis: The interplay between supported single Pt atoms and particles. *Catalysis Science and Technology, 7*(19), 4315–4345. https://doi.org/10.1039/c7cy00710h

Madusanka, N., et al. (2016). Nanocomposites of TiO2/cyanoethylated cellulose with ultra high dielectric constants. *Nanotechnology*. IOP Publishing, *27*, 19. https://doi.org/10.1088/0957-4484/27/19/195402

Madusanka, N., Shivareddy, S. G., Eddleston, M. D., Hiralal, P., Oliver, R. A., & Amaratunga, G. A. J. (2017). Dielectric behaviour of montmorillonite/cyanoethylated cellulose nanocomposites. *Carbohydrate Polymers, 172*, 315–321. https://doi.org/10.1016/j.carbpol.2017.05.057

Naoi, K., & Simon, P. (2008). New materials and new confgurations for advanced electrochemical capacitors. *Electrochemical Society Interface, 17*(1), 34–37. https://doi.org/10.1149/2.f04081if

Nejati, S., et al. (2014). Enhanced charge storage of ultrathin polythiophene films within porous nanostructures. *ACS Nano, 8*(6), 5413–5422. https://doi.org/10.1021/nn500007c

Nelson, J. K., & Hu, Y. (2005). Nanocomposite dielectrics—properties and implications. *Journal of Physics D: Applied Physics, 38*(2), 213–222. https://doi.org/10.1088/0022-3727/38/2/005

Niu, Y., & Wang, H. (2019). Dielectric nanomaterials for power energy storage: Surface modification and characterization. *ACS Applied Nano Materials*. American Chemical Society, *2*(2), 627–642. https://doi.org/10.1021/acsanm.8b01846

Park, H. W., et al. (2012). Anisotropic growth control of polyaniline nanostructures and their morphology-dependent electrochemical characteristics. *ACS Nano, 6*(9), 7624–7633. https://doi.org/10.1021/nn3033425

Park, S. H., Park, K. S., Sun, Y. K., Nahm, K. S., Lee, Y. S., & Yoshio, M. (2001). Structural and electrochemical characterization of lithium excess and Al-doped nickel oxides synthesized by the sol-gel method. *Electrochimica Acta, 46*, 1215–1222.

Peng, L., et al. (2018). Holey 2D nanomaterials for electrochemical energy storage. *Advanced Energy Materials, 8*(9), 1–19. https://doi.org/10.1002/aenm.201702179

Pu, L., et al. (2015). High performance platinum nanorod assemblies based double-layered cathode for passive direct methanol fuel cells. *Journal of Power Sources*. Elsevier B.V, *276*, 95–101. https://doi.org/10.1016/j.jpowsour.2014.11.100

Qin, Z., et al. (2014). Effect of wrinkles on the surface area of graphene: Toward the design of nanoelectronics. *Nano Letters, 14*(11), 6520–6525. https://doi.org/10.1021/nl503097u

Rajesh, D., et al. (2018). Pd-NiO decorated multiwalled carbon nanotubes supported on reduced graphene oxide as an efficient electrocatalyst for ethanol oxidation in alkaline medium. *Applied Surface Science, 442*, 787–796. https://doi.org/10.1016/j.apsusc.2018.02.174

Shao, Y., Yin, G., & Gao, Y. (2007). Understanding and approaches for the durability issues of Pt-based catalysts for PEM fuel cell. *Journal of Power Sources, 171*(2), 558–566. https://doi.org/10.1016/j.jpowsour.2007.07.004

Su, D., Dou, S., & Wang, G. (2015). Gold nanocrystals with variable index facets as highly effective cathode catalysts for lithium-oxygen batteries. *NPG Asia Materials, 7*(1), e155. https://doi.org/10.1038/am.2014.130

Sun, Y., et al. (2016). Graphite-encapsulated Li-metal hybrid anodes for high-capacity Li batteries. *Chem, 1*(2), 287–297. https://doi.org/10.1016/j.chempr.2016.07.009

Sun, Z., et al. (2019). Bifunctional iron disulfide nanoellipsoids for high energy density supercapacitor and electrocatalytic oxygen evolution applications. *Inorganic Chemistry Frontiers, 6*(3), 659–670. https://doi.org/10.1039/c8qi01230j

Sun, Z. and Liao, T. (2020). *Springer Series in Materials Science 297 Responsive Nanomaterials for Sustainable Applications*.

Suryawanshi, P. L., Polagani, R. K., & Sonawane, S. H. (2018). Synthesis of nanomaterials for fuel cell applications. *Nanomaterials for Green Energy*, 205–226. https://doi.org/10.1016/B978-0-12-813731-4.00007-2

Tan, H. T., Rui, X., Sun, W., Yan, Q., & Lim, T. M. (2015). Vanadium-based nanostructure materials for secondary lithium battery applications. *Nanoscale, 7*, 14595–14607.

Wang, C. H., et al. (2007). High performance of low electrocatalysts loading on CNT directly grown on carbon cloth for DMFC. *Journal of Power Sources, 171*(1), 55–62. https://doi.org/10.1016/j.jpowsour.2006.12.028

Wang, J., Li, Y., & Sun, X. (2013). Challenges and opportunities of nanostructured materials for aprotic rechargeable lithium-air batteries. *Nano Energy, 2*(4), 443–467. https://doi.org/10.1016/j.nanoen.2012.11.014

Wang, P. F., Niu, Y. Bin and Guo, Y. G. (2019) *Nanostructures and nanomaterials for sodium batteries, Nanostructures and Nanomaterials for Batteries: Principles and Applications*. https://doi.org/10.1007/978-981-13-6233-0_6.

Wang, W., et al. (2016a). Tuning nanowires and nanotubes for efficient fuel-cell electrocatalysis. *Advanced Materials, 28*(46), 10117–10141. https://doi.org/10.1002/adma.201601909

Wang, Y., et al. (2016b). Enhanced methanol oxidation with annealed atomic layer deposited platinum nanoparticles on carbon nanotubes. *Journal of the Electrochemical Society, 163*(2), F1–F10. https://doi.org/10.1149/2.1001514jes

Xiao, J., et al. (2011). Nl203332E.Pdf. *Nano Letters, 11*, 5071–5078.

Xiao, X., Wang, L., Wang, D., He, X., Peng, Q., & Li, Y. (2010). Hydrothermal synthesis of orthorhombic LiMnO2 nano-particles and LiMnO2 nanorods and comparison of their electrochemical performances. *Nano Research, 2*, 923–930.

Xie, K., et al. (2016). Toward dendrite-free lithium deposition via structural and interfacial synergistic effects of 3D graphene@Ni scaffold. *ACS Applied Materials and Interfaces, 8*(39), 26091–26097. https://doi.org/10.1021/acsami.6b09031

Yang, C. P., et al. (2015). Accommodating lithium into 3D current collectors with a submicron skeleton towards long-life lithium metal anodes. *Nature Communications, 6*(May), 1–9. https://doi.org/10.1038/ncomms9058

Yin, Z., et al. (2009). Porous palladium nanoflowers that have enhanced methanol electro-oxidation activity. *Journal of Physical Chemistry C, 113*(3), 1001–1005. https://doi.org/10.1021/jp807456j

Yola, M. L., et al. (2016). Direct-methanol fuel cell based on functionalized graphene oxide with mono-metallic and bi-metallic nanoparticles: Electrochemical performances of nanomaterials for methanol oxidation. *Electroanalysis, 28*(3), 570–579. https://doi.org/10.1002/elan.201500381

Yu, G., et al. (2013). Hybrid nanostructured materials for high-performance electrochemical capacitors. *Nano Energy, 2*(2), 213–234. https://doi.org/10.1016/j.nanoen.2012.10.006

Yue, Y., & Liang, H. (2017). Micro- and nano-structured vanadium pentoxide (V2O5) for electrodes of lithium-ion batteries. *Advanced Energy Materials, 7*, 1602545.

Zhang, J., et al. (2014). Porous perovskite LaNiO3 nanocubes as cathode catalysts for Li-O2 batteries with low charge potential. *Scientific Reports, 4*, 2–7. https://doi.org/10.1038/srep06005

Zhang, R., et al. (2016). Conductive nanostructured scaffolds render low local current density to inhibit lithium dendrite growth. *Advanced Materials, 28*(11), 2155–2162. https://doi.org/10.1002/adma.201504117

Zhao, G., et al. (2016). A graphitic foam framework with hierarchical pore structure as self-supported electrodes of Li-O2 batteries and Li ion batteries. *Journal of Materials Chemistry A, 4*(4), 1399–1407. https://doi.org/10.1039/c5ta09033d

Zhao, X., et al. (2011). The role of nanomaterials in redox-based supercapacitors for next generation energy storage devices. *Nanoscale, 3*(3), 839–855. https://doi.org/10.1039/c0nr00594k

Zheng, Q., et al. (2015). Cellulose nanofibril/reduced graphene oxide/carbon nanotube hybrid aerogels for highly flexible and all-solid-state supercapacitors. *ACS Applied Materials and Interfaces, 7*(5), 3263–3271. https://doi.org/10.1021/am507999s

Zhong, C. J., et al. (2008). Fuel cell technology: Nano-engineered multimetallic catalysts. *Energy and Environmental Science, 1*(4), 454–466. https://doi.org/10.1039/b810734n

Zhou, C., et al. (2013a). Construction of high-capacitance 3D CoO@Polypyrrole nanowire array electrode for aqueous asymmetric supercapacitor. *Nano Letters, 13*(5), 2078–2085. https://doi.org/10.1021/nl400378j

Zhou, X., et al. (2013b). A review of hollow Pt-based nanocatalysts applied in proton exchange membrane fuel cells. *Journal of Power Sources, 232*, 310–322. https://doi.org/10.1016/j.jpowsour.2013.01.062

Zou, Z., et al. (2014). V6O13 nanosheets synthesized from ethanol-aqueous solutions as high energy cathode material for lithium-ion batteries. *Electrochimica Acta, 135*, 175–180. https://doi.org/10.1016/j.electacta.2014.04.163

Biomass Energy Conversion Using Thermochemical and Biochemical Technologies

S. Aishwarya, G. Sruthi, M. N. Aditya, K. Sivagami, and Samarshi Chakraborty

Abstract Increasing oil prices, energy security and rising greenhouse gas concentrations necessitate the need for the development of biomass-based energy conversion technologies. Extensive lab-scale scientific studies have been performed on the conversion of various biomass feedstocks to energy in the form of electricity, heat and fuels/chemicals in solid, liquid and gases. The current review gives an overview of different thermal and biochemical technologies (incineration, gasification, pyrolysis, liquefaction, carbonization, cogeneration, anaerobic digestion, ethanol fermentation, transesterification, etc.) for bioenergy generation. The advantages and limitations of primary conversion technologies like thermochemical and biochemical conversion techniques have also been discussed here. Different type of biomass-based feedstocks, their pretreatment requirements, are reviewed. The health and environmental hazards of biomass energy conversion plants and their advantages over conventional fuel sources are discussed in detail. The review also discusses air pollution hazards related to different pollutants and diverse preventive measures taken to mitigate the problem. The current study also discusses the potential and limitations of biomass energy generation projects considering process economy, technological limitation, socio-political aspect and the role of government policies from the Indian context. The extended discussion on the aforementioned criteria further reaffirms the necessity of the current review.

Keywords Biomass · Bioenergy · Thermochemical-biochemical conversion · Biogas · Economical aspects · India

---

S. Aishwarya · G. Sruthi · M. N. Aditya · K. Sivagami · S. Chakraborty (✉)
Colloids and Polymer & Mass Transfer Research Group, School of Chemical Engineering, Vellore Institute of Technology, Vellore, Tamil Nadu, India
e-mail: samarshi.chakraborty@vit.ac.in

D. B. Pal, J. M. Jha (eds.), *Sustainable and Clean Energy Production Technologies*, Clean Energy Production Technologies,
https://doi.org/10.1007/978-981-16-9135-5_5

Abbreviations

BOD	Biological oxygen demand
CDM	Clean Development Mechanism
CPCB	Central Pollution Control Board
ESP	Electrostatic precipitators
GHG	Greenhouse gas
IPP	Independent power producer
LCA	Life cycle analysis
MSW	Municipal solid waste
MNRE	Ministry of New and Renewable Energy
MW	Megawatt
NNBOMP	New National Biogas and Organic Manure Programme
NGCC	Natural Gas Combined Cycle
TWh	Terawatt hour

1 Introduction

Across the world, several million tonnes of solid and liquid waste are produced each year. These wastes are of different types depending on their region of generation. Their accumulation over a long period leads to various harmful effects both to nature and mankind, such as air, land and water pollution; the spread of infectious diseases; etc. (Wang & You, 2021, Zhou et al., 2014). Further, the sources of energy (fossil fuel) on which modern human civilization is dependent are shrinking by every passing year (Abowei et al., 2009). The overuse of fossil resources has caused serious resource shortage and environmental problems. The degradation of fossil fuels has instilled a great amount of interest in the field of biofuels by researchers and technocrats. The key focus of converting biomass to energy is owed to the quick and efficient disposal of a variety of wastes to generate energy from an abundant and low-cost resource. It can be viewed as one of the biggest solutions to limit our dependency on fossil fuels (Moh & Manaf, 2014) and also can be used as a sustainable waste management strategy that can be implemented worldwide. Different types of wastes require diverse methods of conversion and disposal, which have to be analysed carefully and implemented accordingly.

The ways of waste disposal can be categorized broadly into two types: (a) thermochemical, and (b) biochemical methods. The above classification can be further subdivided based on the type of feed, components, agents needed for processing and the final product (Sivan, 2011; Jadhao et al., 2017). Some processes focus on converting the wastes into easy disposable forms, whereas few processes focus on obtaining useful products. These can be processed or used directly as energy sources which help in preserving the environment (Erdinc & Uzunoglu, 2012). Biomass can be considered as a potential resource for the conversion of

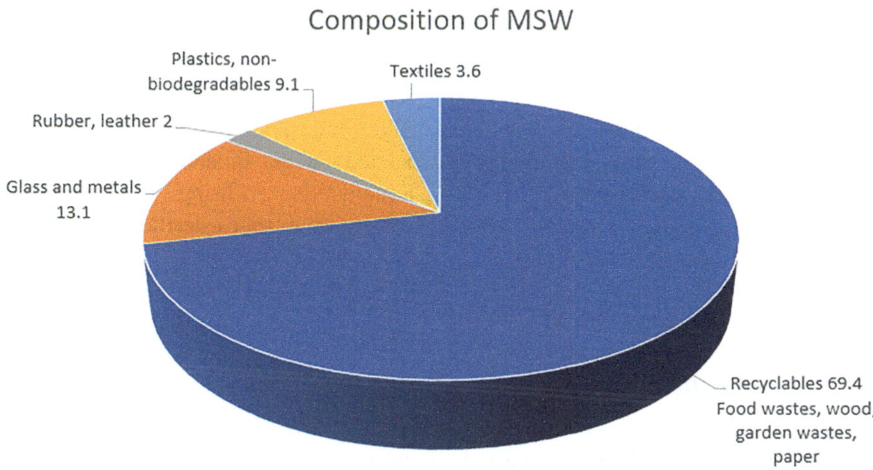

Fig. 1 Generic composition of municipal solid waste (Adopted from Speight, 2015)

waste material into biofuel which gives rise to many value-added products, i.e. bio-ethanol, biodiesel and biogas.

The feed source for thermochemical and biochemical processes are the municipality solid wastes (MSW) which comprise of paper, cardboard waste, food wastes (fruit and vegetable peels), leftover food wastes, agricultural waste and garden wastes, wood, cloth, textiles (i.e. fibre, jute, cotton), leather, metals, glass, ceramics, cement, plastic, rubber, sewage and sludge, ash residue, electrical appliances and hazardous wastes (Edjabou et al., 2015). Figure 1 depicts the generic composition of municipal solid waste. Waste disposal methods such as thermochemical and biochemical processes use a variety of strategies for energy conversion i.e. application of heat, chemical reactions, and microbes to break down the wastes and release hydrocarbon and biogas. The current chapter sheds light on various thermal and biochemical processes like incineration, gasification, pyrolysis, liquefaction, co-firing, carbonization, cogeneration, anaerobic digestion, ethanol fermentation and transesterification for bioenergy generation which are discussed below. The chapter also provides insight into the advantages and limitations of these processes which is essential for selecting the correct conversion strategy for different biomass-based feedstock. The air pollution hazards associated with these processes and different preventive measures necessary to mitigate such problems have also been discussed. Detailed discussion on the health and environmental hazards of biomass energy conversion plants and their advantages over conventional fossil fuels has been presented in this chapter. The current chapter discusses the potential and limitations of biomass energy generating plants in India and provides a comparative analysis of global biomass energy generation plants with the Indian scenario.

2 Introduction to Thermochemical and Biochemical Conversion Technologies

The following section provides a detailed insight into various thermochemical and biochemical conversion techniques associated with bioenergy generation, i.e. incineration, gasification, pyrolysis, liquefaction, co-firing, carbonization, cogeneration, anaerobic digestion, ethanol fermentation, transesterification, photobiological hydrogen production, etc.

2.1 *Incineration*

Incineration is a widely used method of converting waste to energy. A waste incineration plant generally comprises of following units, namely, (a) rotary kiln, (b) combustion chamber, (c) post-combustion chamber, (d) boilers for cooling of exhaust gases, (e) catalytic or non-catalytic reduction chambers (for NO_x reduction), (f) acid gas neutralization and absorption section, (g) filter used for removal of particulate matter and (h) scrubber fitted for removal of heavy metals (Jiao et al., 2016). The rotary kiln operates at a high operating temperature of 1000°C where the solid waste is burnt. The exhaust gases released from the plant depend on the composition of the wastes, based on which conditions inside the plant are modified and samples are monitored regularly (de Titto & Savino, 2019). The ash generated after incineration is collected from the combustion chamber which is then quenched in a spray chamber using water as the quenching agent. The bottom ash usually contains zinc, lead, iron and magnesium in the following amount 0.06%, 0.02%, 0.017% and 0.079%, respectively (Shelef, 1995). These levels will increase depending on the amount of pharmaceutical contaminants present in the waste feed (Hjelmar et al., 2010). Post-quenching, the water is usually contaminated with heavy metals from the wastes which need to be treated before release. According to studies, the analysis of the level of contamination in water is done using an atomic absorption spectrophotometer with acetylene flame. The levels of mercury in the wastewater were found to be in the range of 170–230 mg/L, depending on the type of waste feed.

New advancements in the incineration process include the use of a fluidized bed for combustion with a new multi-fuel boiler. The feed is segregated based on its composition (organic and inorganic wastes and physical states), and size is reduced to increase the surface area, thus benefitting quicker combustion and lower power consumption. This makes the process energy efficient, and combustion time is also reduced (Ji et al., 2016). Fluidized beds have a simple and compact design and can be utilized for a wide variety of wastes. They also operate in uniform temperature ranges and reduce the emissions of NO_x and SO_x, hereby reducing the environmental and health hazards. The efficiency of reaction is increased with reduced corrosion of the equipment. Removal of the ash produced from combustion also becomes easier. However, this method has drawbacks such as clinker formation, clogging of the bed due to carbon deposition and high emission of SO_2 ($SO_2 > 1000$ ppm) which needs to be monitored. The wastes that the incineration plant receives as feed may be classified as "National Research Council" (2000) (Rao et al., 2016):

- Non-hazardous municipal solid waste (MSW) containing dry waste like plastics, paper, cardboard, cloth, rubber, leather, wood and glass
- Hazardous solid wastes comprising of dry wastes: metals, batteries, chemical powders and fluorescent bulbs
- Hazardous liquid wastes with wet wastes: oil, grease, pesticides and waste sludge from households
- Medical and other pharmaceutical wastes which include dry and wet waste: drugs, needles, tissues, cotton and fluids (blood, plasma, urine, etc.)

These wastes pose a severe hazard to the environment and human health if it is not contained or disposed of appropriately. Care should be taken during their storage in bins or drums to prevent leakages. Segregation of waste plays a major role in mitigating the adverse environmental impact of waste incineration which is done after waste collection and is sorted according to size and nature (Cheng & Yuanan, 2010, Lam et al., 2010).

Beneficial attributes of incineration lie in its quick and easy waste removal capacity which not only helps in the rapid reduction of waste but also helps in overall waste management. The process also helps in curbing the odour released from the waste, thereby reducing air pollution, and saves the space occupied by the waste in landfill sites, which otherwise would have served to be breeding grounds for disease-causing growth of vectors and other microorganisms (Pandiyan et al., 2019). Incinerators can be operated under different weather conditions and also reduce the risk of groundwater contamination.

Other than its beneficial attributes, the negative aspects of incineration lie in the health hazards and environmental pollution associated with the emissions from incinerators. The bottom ash produced by the incinerator is also difficult to dispose of.

2.2 *Gasification*

Gasification refers to the process which utilizes reaction between fuel and gasification agent for converting biomass or fossil fuels (hydrocarbon-based feedstock) into useful gases, i.e. producer gas and syngas which majorly contain N_2, CO, H_2, CO_2 and some hydrocarbons. A small amount of H_2S, NH_3 and tar may also be present along with additional chemicals (Zhang et al., 2019). The process is exothermic in nature with temperatures around 800°C. The syngas obtained from gasification have high calorific value and serves as a good fuel. Improvements to the initial process have led to the design of different types of gasification units based on three factors:

(a) type of gasification agents like air, steam and pure oxygen; (b) reactor designs; and (c) operational costs.

The performance of a gasifier depends majorly on the type of gasifier and its design characteristics (Laciak et al., 2016, Puig-Arnavat et al., 2010). Based on the position of air inlet and product gas outlet, gasifier can also be classified into three types: i.e. downward draft, upward draft and cross-draft gasifiers (Mahmoudi, 2015). In case of upward draft gasifier, the hot air/oxygen/steam enters the gasification unit from the bottom of the ash bed (through the grate section) of the gasifier and moves up the unit to leave from the top, whereas the biomass feed moves in downward direction (Virmond et al., 2013). It creates a countercurrent flow within the gasifier as biomass solids go down and hot air used for combustion travels upward. As the air/steam comes in contact with heated ash bed, majority of the oxygen gets consumed due to partial combustion of char. The hot gases produced by combustion helps in char gasification and biomass pyrolysis. However, the main demerit of this process is observed higher up the gasifier unit where due to low temperature and lack of oxygen, higher quantity of tar is produced which makes this process completely unsuited to deal with highly volatile feedstock. This type of gasifier is rather suited for high ash and moisture containing biomass feedstock. The schematic diagram of updraft gasifier unit is depicted in Fig. 2. On the contrary to the updraft gasifier, a downward draft gasifier works on a co-current operating mode as biomass feed, and

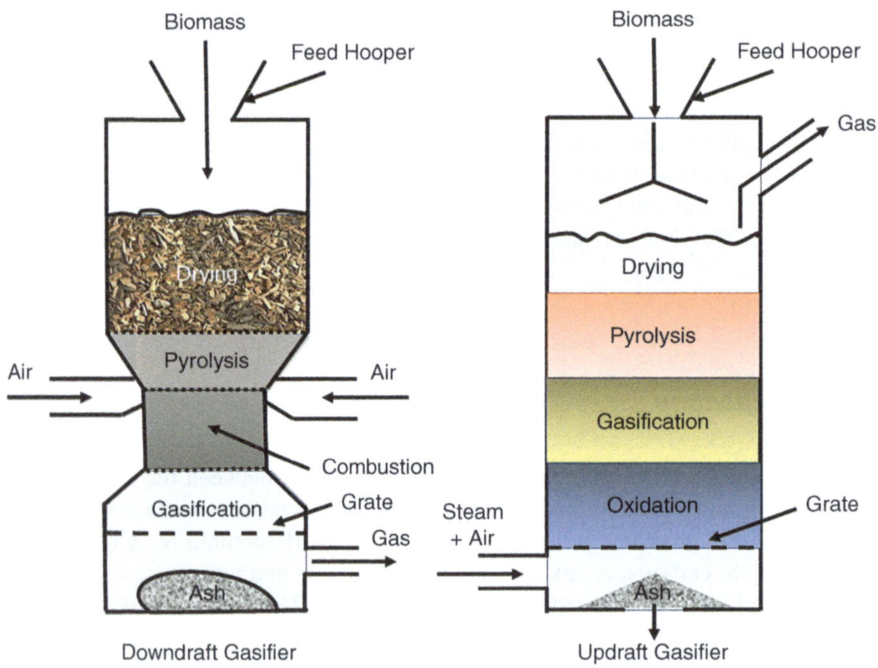

Fig. 2 Schematic representation of downdraft and upward gasifier (Basu, 2010)

air enters the reactor from the top and flows/travels in downward direction. The product gases produced by the gasification process also exit from the bottom of the reactor which can be clearly seen through the schematic diagram (see Fig. 2). The tar produced by this process is significantly less compared to updraft gasifier due to high-temperature cracking of tar into several other gas products.

In case of cross-draft gasifier, the biomass (fuel) is fed from the top of the reactor, whereas air used for combustion is injected from the one side, and product gas is extracted from opposite side of the gasifier. As the air enters in the gasifier unit through the air nozzles, a portion of the char produced by pyrolysis burns in the area close to air inlet, thereby creating a high-temperature region. The residual char is then gasified to form carbon monoxide gas. The heat generated in the combustion zone is utilized for drying and pyrolysis of biomass feed within the reactor. This type of gasification process is well suited for small-scale biomass gasification unit. However, such gasifiers have faster response and startup time, smaller reaction area and lower thermal capacity compared updraft and downdraft gasifiers. This type of gasifier is not compatible to handle biomass feed containing high ash and tar content. These types of gasifier falls under the classification fixed-bed type of gasifier. A schematic diagram of downdraft and updraft gasifier units has been depicted in Fig. 3. In addition to be above type, fluidized-bed gasifier and entrained-flow gasifier are other types of widely used gasifier units. Each type of gasifier works with specific feed input which in turn gives additional by-products as output. These by-products can be further processed to produce value-added

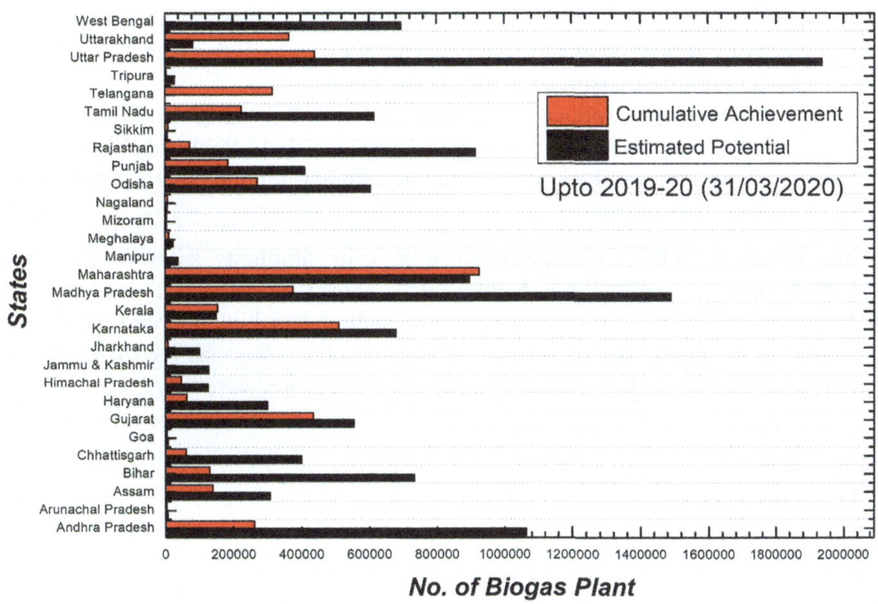

Fig. 3 Comparative depiction of estimated potential and state-wise achievements for family-type/small biogas plants from 1981–1982 to 2019–2020

products. This makes the process more economic and also minimizes environmental damage. The widely used types include (Hofbauer & Materazzi, 2019, and Phillips, Jeffrey, *The gas turbine handbook* 1 (2006)):

- ***Fixed-bed type of gasifier*** – It has a comparatively small power generation capacity and is used for handling large/coarse particles (around 51 mm or greater). The temperature of produced syngas using such gasifier lies in the range of 450–650°C. This process needs a higher amount of gasification agents and produces a higher amount of particulate matter which is considered to be environmentally unfriendly. The gasifier also produces ash as a by-product which also needs to be disposed of as slag.
- ***Fluidized-bed type of gasifier*** – This type of gasifier has a relatively medium power generation capacity. Similar to the fixed-bed gasifier, this system can also handle large and coarse particles (usually around 6–10 mm), as feed input and syngas produced in this gasifier are in the temperature range of 1000–1200°C. Some beds are capable of handling larger particles (around 15–25 mm). The particulate matter released is low as compared to its fixed-bed counterpart. This unit is suitable for feed that has a low ash fusion temperature. Drawbacks of this gasifier are bed clogging, agglomeration and non-uniform velocity distribution within the gasifier.
- ***Entrained-flow type of gasifier*** – It has a moderately large power generation capacity compared to fixed- and fluidized-bed gasifier. These gasifiers are capable of handling fine feed particles (0.15 mm and lower-sized feed) with low ash content and produce high-temperature syngas (1200°C). Similar to the fixed-bed gasifiers, the particulate matter release is high. The process involves high consumption of gasification agents like oxygen. Drawbacks of these gasifiers include high consumption of gasification agents (oxygen).

  In general, syngas produced from the above gasifier has high methane content. Common by-products of this gasification process include chemicals like tar, oils, phenols along with product gases containing higher amounts of H_2 and CO and other lighter hydrocarbons.

The benefits of the gasification process lie in its simplicity, effective design at minimum expenses and ability to handle a wide variety of wastes. Other advantages of this process lie in its low dust emission feature and high thermal efficiency. However, to maintain and improve the efficiency of the process, feed size specifications and quantity of tar produced needs to be monitored carefully.

2.3 Pyrolysis

Pyrolysis is an efficient method of waste disposal and treatment which involves the decomposition of organic matter via a thermochemical route. The feed input to pyrolysis plant includes residues from the forest, agriculture, textile industry, plastics and pulping industry (Chen et al., 2014). The advantages of the pyrolysis process over other thermochemical processes lie in its low operating temperature and relatively less air pollution hazard (Czajczyńska et al., 2017). Pyrolysis plants compared to incineration plants are more flexible based on their ability to generate a plethora of products and by-products. The operating parameters, i.e. temperature, pressure, additional agents, feed loading, heating time or heating rate, are varied to obtain the desired products (Basu, 2018). An increase in pressure in the reactor leads to an increase in the yield of charcoal. This effect is more prominent at higher pressures. The operating pressure of the process is maintained around 1–6 bar (absolute pressure) to ensure slower rate of pyrolysis, and increasing the pressure results in decrease of biomass residence time.

A brief overview of different pyrolysis processes has been discussed below (Foston & Ragauskas, 2012; Zaman et al., 2017; Jahirul et al., 2012):

- ***Fast pyrolysis*** – The feed for the fast pyrolysis process comprises forest and agricultural residue. The operating temperature of such pyrolysis was kept at 500 °C with a low residence time of 2 s and pressure of 20 bar. The process involves drying to feed to remove moisture, rearrangement of molecules and breaking of the polysaccharide chains. Bio-oil is obtained as a product along with a minimum amount of tar and gas (Zhang et al., 2013).
- ***Flash drum pyrolysis*** – The feed for the flash drum-type process includes forest, agricultural and garden wastes. The flash drum pyrolysis process is performed at a temperature of 650 °C with a very low residence time of less than 1 second. For fast pyrolysis of the cellulose present in feed, pressure has been kept around 50 bar. The process involves three main steps: (a) heating of feed, (b) breaking of the polysaccharide chains and (c) release of free radicles. The main product obtained from the pyrolysis process is bio-oil. In addition to bio-oil, by-products, i.e. acetaldehyde and gases like CO and CO_2, is also obtained.
- ***Ultra-rapid pyrolysis*** – The feed for this process is biomass. Temperature is maintained around 1000 °C with a very low residence time of fewer than 0.5 seconds, and pressure is maintained at 27 bar. The process involves heating of feed for rapid dehydration followed by chain fission. A mixture of products is obtained which includes gases like CO and CO_2 along with carboxylic acids and ketones (C=O and COOH).
- ***Vacuum-type pyrolysis*** – The feed includes forest, agricultural and garden wastes (Archer et al., 2005). The pyrolysis process is performed at a temperature of 400 °C, and the reactant residence time varies between 2 and 30 s. The operating pressure is sustained between 40 and 50 bar during the process. The process involves the following steps: (a) feed heating for dehydration, which is then followed by (b) molecular rearrangement and breaking of polysaccharide chains. Bio-oil is obtained along with tar and anhydrides as products and by-products of the process.
- ***Hydro-pyrolysis*** – The feed needed for this process is mainly plant residues. The pyrolysis temperature is maintained at 500 °C with a residence time of fewer than 10 seconds, and operating pressure is kept between 20 and 40 bar within the reactor. For higher temperature around 1000 °C, pressure is maintained at 70 bar.

Hydro-pyrolysis involves heating of the feed and breaking of polysaccharide chains with the release of free radicles. Bio-oil is the main product of this process. Additional components like aldehydes are also obtained.

- ***Conventional pyrolysis*** – Feed needed for this process is forest and agricultural residue (Foong et al., 2020). The temperature for the process is kept around 600 °C with a residence time between 5 and 30 min and pressure of 5–15 bar. The process involves low heating of feed followed by condensation of unsaturated components. Biochar residue is obtained with oil and gas.

Pyrolysis process helps in combating problems such as greenhouse gas emission and leaching of wastes from the landfill sites which tend to contaminate the groundwater. The complexity of the pyrolysis process and high operational and investment costs are some of the barriers which need to be addressed before its wide-scale implementation.

2.4 Co-firing

Co-firing is the process that involves synchronous combustion of fuels like biomass and fossil fuels, in the same boiler. This process helps in the proper usage of renewable fuels like fossil fuels and reduces energy consumption costs and waste disposal costs to a great extent (Seepana et al., 2018; Boardman et al., 2014). Reactor designs are modified based on the requirements and the feed input given to the specific unit (Agbor et al., 2014). Higher moisture content in the biomass feed increases the time for the reaction and hereby increases energy consumption. Corrosive agents like sulphur or chlorine in the feed pose a threat to the reactor and hence should be segregated before the reaction. A feed of uniform size is preferred. Different feed sizes result in a non-uniform reaction, overloading of the reactor and reduced yield. Reactor clogging and risk of fire persist with long-term storage of the feed. Factors such as boiler design, biomass supply, coal prices, national and international market conditions and the quantity of waste generated majorly affect the performance and economics of the co-firing process. Coal-fired boilers are most suitable for this process (Madanayake et al., 2017; Lüschen & Madlener, 2010). Co-firing reactors or boilers are feed-specific and provide a maximum output when used with the feed it is designed for. The process helps in reducing sulphur and carbon dioxide emissions, thereby contributing to decreasing levels of greenhouse gases in the atmosphere. The level of emission of mercury is also reduced as a result of co-firing.

The boiler design, feed type, handling and storage determine the efficiency of the co-firing process (Roni et al., 2017; Verma et al., 2017). Different boilers used for co-firing process based on their feed input are listed below:

- ***Stoker boiler*** – The feed for this type of boilers comprises wood residues, plant wastes, peat and straw of larger sizes.

- ***Fluidized-bed boiler*** – Wood waste, forest products, herbs and agricultural wastes of larger sizes are suitable as feed for fluidized-bed type of boiler.
- ***Cyclone boiler*** – The feed comprises small-sized particles with low ash content.
- ***Pulverized coal boiler*** – The feed comprises fine-sized materials like sawdust, coal fines, powdered wastes and wood shavings.

2.5 Liquefaction

Liquefaction is the process by which the municipal solid wastes (MSW) are thermally depolymerized to bio-crude (Jena & Das, 2011). The MSW wastes include organic and inorganic components. Separation of feed is done by hydrothermal treatment (Katakojwala et al., 2020). Temperatures between 300 and 400 °C are usually maintained during hydrothermal treatment of the wastes and polymers present in the feed (Demirbas, 2008). The reactions are carried out usually in batch or semi-batch reactors, and it can be concluded that the oil yield increases in the latter case (Gollakota et al., 2018). The efficiency of the liquefaction reactions is proportional to the operating temperatures and pressures of the reactors in the absence of a catalyst. The process is sensitive to the moisture content in the feed. The segregation of solids from liquid waste is critical before commencing the process (Yang & Yang, 2019). Changes in operating temperature dictate the pressure requirement for liquefaction reactions. Higher pressure is applied to shorten the reactant residence time. The process ensures better conversion of the feed, and the liquefied products obtained are easy to store and transport and can be used as fuels due to their high calorific values (Berstad et al., 2010). The high operating temperature also helps in reducing reactant residence time within the liquefaction reactor, resulting in high conversion rates (Wang et al., 2018). The moisture content of the waste to be liquefied plays a critical role in determining the quality and quantity of the oil product and also dictates the reaction time needed for the process. Reactors have different specifications of feed and operating parameters that need to be maintained for efficient processing (Castello et al., 2018). The different types of reactors used for liquefaction are (Anastasakis et al., 2018):

- ***Tubular reactor*** – The feed contains biodegradable waste like wood, bagasse, beet pulp, vegetable wastes and sludge from sewage. The operating temperature and pressure within the reactor is maintained in the range of 300–350 °C and 120–180 bar, respectively. Reactants are kept inside the reactor for a longer time (few hours). However, when the operating temperature increases to 380–420 °C, then the reactant residence time drops to 15–20 min.
- ***Combined reactors*** – The feed contains wet wastes, paper, plastics and agriculture residues. The temperature and pressure in combined reactors is maintained around 300–420 °C and 200 bar, respectively. The residence time is kept between 5 and 10 min in the presence of solvent and 7–30 min with the process performed in a sand bath.

- ***Stirred reactor*** – The feed input includes wood and agricultural wastes, partially treated sewage, algae and microorganic wastes. The operating temperature and pressure of after treatment of food and organic wastes between 300 °C and 4000 °C and 80 bar and 100 bar, respectively. The reactant residence time is 30–50 min which is significantly higher than the combined reactors.
- ***Fixed-bed reactor with loop*** – The feed comprises paper and pulp waste from industries, cardboard and few chemicals like chlorine, lime, enzymes and starch. The operating temperature in fixed-bed reactors is maintained between 300 and 360 °C, whereas the operating pressures is kept around 100–250 bar. Residence time for the process is about 5–15 minutes depending on the use of a catalyst (Balagurumurthy et al., 2015; Muhtadin et al., 2020).

2.6 Carbonization

Biomass carbonization is another method of waste valorization that is used to convert organic biodegradable matter and wastes into charcoal which is also known as biochar. This process is less time-consuming and easy to perform. The feed for this process usually includes waste from a plethora of sources like humans, animals, food, forest and agriculture. Biochar has many nutrients, acts as a good natural fertilizer and can also be used as a fuel source. This not only helps in waste reduction but also pollution control (Tradler et al., 2018). The time required for the carbonization process is less but can vary depending on the feed composition and quality. The carbonization process has benefits of lower consumption of energy, simple and flexible design of reactors and faster conversion of feed to char (Lin et al., 2017; Libra et al., 2011).

The compositional analysis of flue gases produced from this process helps us determine the reaction time needed (Lu et al., 2012). Based on the type of feed used, the product obtained varies to a certain extent. When dealing with agricultural wastes, it requires segregation, drying and binding, and the feed is then carbonized at a temperature of 300–450 °C for a longer residence time. The product of this process is char having high organic residue. Wood and forest wastes are needed to be pre-processed (cutting and reshaping) for better char formation. These wastes are carbonized at a temperature of 350–550 °C for a long residence time with the formation of char. Paper wastes were segregated, dried and treated at a temperature of 300–350 °C. Char with a small amount of chemical residue is obtained. Food wastes and organic feed with high carbon content need drying at 300–500 °C with a long residence time. Char with organic residue is obtained as a product after treatment of food and organic wastes (Lohri et al., 2015). Generally, fixed-bed reactors made of stainless steel are used for the carbonization process which is capable of withstanding high temperatures and pressures and is also corrosion-resistant. This ensures the longevity of the reactors with minimum maintenance and reduces operational costs. Carbonization processes produce emission with less or almost no smoke and higher steam production which can further be utilized for other purposes. The biochar yield is also high, and the entire process is easy to control and maintain.

2.7 Advantages and Disadvantages of Thermochemical Processes

Thermochemical processes are used for biomass feed of any composition/type. It involves the use of various chemical agents, gases, heat and fuel sources and reactors for the completion of the process. The yield and conversion efficiency is higher as there is a complete breakdown of the feed by heat and catalyst. Many intermediate high-value products can be obtained by fractional separation from the process (Carneiro et al., 2017). The effect of ambient and atmospheric temperature is minimal as the reactions operate at a higher range of temperature (Verma et al., 2012). However, the pollution hazard (except anaerobic digestion) of such processes is significantly higher than the biochemical processes (Primo & García, 2013). On the other hand, thermochemical processes are much more energy-intensive, and additional chemical agents and catalysts further aggravate the operation cost.

3 Introduction to Biochemical Processes

The exponential increase in population, urbanization and modern industrialization increased the energy requirement. However, the improvement of fuel sources is not at the same pace as the expanding utilization. The significant energy demand is supplied from conventional fossil fuel sources like coal, oil, gaseous petrol and so on. The fast exhaustion of petroleum derivatives and disintegration of the climate has prompted the utilization of non-customary/renewable fuel sources. Different biochemical cycles are utilized for the transformation of biomass into fuel for energy utilization. The accompanying conversation centers around various biochemical techniques like aerobic and anaerobic digestion, ethanol ageing, and photobiological hydrogen production. There are a plethora of factors that need to be considered for choosing to pick a transformation cycle. These factors include the amount of the biomass, type of the biomass and its properties, item necessities, plant location and other task explicit elements. The fuels extracted from biomass are generally in the form of solid, liquid and gaseous states. Fuel in liquid and gaseous states can be utilized for transportation purposes, whereas solid fuels can be utilized for heat generation. (Hermansen et al., 2017; Menon & Rao 2012; Barik, 2018). (Chen & Wang, 2016).

3.1 Anaerobic Digestion

It is a method where microorganisms separate biomass into fuel without oxygen. It starts with the bacterial hydrolysis of the biomass. Insoluble natural materials, for example, starches, are separated into sugars and amino acids. These substances are then changed over into carbon dioxide, hydrogen, smelling salts and natural acids by acidogenic microbes. On the off chance that acetogenic microscopic organisms are included, they convert sugar and amino acids into acidic corrosive alongside hydrogen, alkali and carbon dioxide. These items are additionally changed over into methane and carbon dioxide by methanogens (Tabatabaei et al. (2010); "Anaerobic absorption reference sheet", 2011). There are usually four stages involved in this process which are hydrolysis, acidogenesis, acetogenesis and methanogenesis. The entire process can be explained in terms of organic matter such as when glucose is decomposed into carbon dioxide and methane in the absence of air. The stoichiometric reaction can be expressed as

$$C_6H_{12}O_6 \rightarrow 3CO_2 + 3CH_4$$

Anaerobic digesters can be classified into a group based on several factors, i.e. interaction mode, temperature conditions (mesophilic and thermophilic) and solid content (high and low). In thermophilic conditions, microorganisms require more heat energy compared to mesophilic organisms. Thermophiles produce more methane content in a lesser time. (Goodrich & Schmidt, 2002). (Ziemiński & Frąc, 2012, Kardos et al., 2011).

It has different applications considering present realities utilized for mechanical gushing and wastewater sewage treatment. Because of the expansion in strong waste in agricultural nations, there has been an increment in the utilization of anaerobic processing. It is utilized either to handle the source separator portions from the natural waste or combine it with mechanical arranging administrators to deal with the remaining blended organic waste (Archer et al., 2005). As this process produces methane as a by-product, the release of methane into the atmosphere may cause significant damage to the environment and contribute significantly to the global warming process. Nations that gather household waste utilize anaerobic digestion to decrease the transport of natural waste that must be burnt from a landfill. This likewise diminishes the weight of fossil fuel by-products from the gathering vehicles (De Mes et al., 2003; Reay et al., 2018); Anderson et al., 2016).

3.2 Aerobic Digestion

It is the digestion of decomposing materials in the presence of oxygen. The process gives out products like carbon dioxide, ammonia water and heat. Aerobic digestion requires the right type of feedstock and conditions for producing good product quality. This can include factors such as moisture content of around 60–70% and a desired carbon-nitrogen ratio of 30:1. Feedstock that usually provides a high content of carbon is wood, paper, sewage waste and food waste. There are many advantages of aerobic digestion which include lower biological oxygen demand (BOD) under supernatant conditions and odourless and stable products. The capital cost required for this process is relatively low. The residue formed at the end is usually in liquid form due to the poor dewatering process. One of the main disadvantages of aerobic production is that the process is adversely affected by the climate conditions around the plant.

3.3 Ethanol Fermentation

Ethanol fermentation is a process that converts sugars such as glucose, sucrose and fructose into alcohol using yeast in the absence of oxygen (Kachrimanidou et al., 2020). About 4.5 billion gallons of ethanol are sold per annum as indicated by US reports. As a rule, ethanol maturation is prepared out with *Saccharomyces cerevisiae* as the yeast specialist because of its low pH and high ethanol resistance (Poe et al., 2020). According to Gay Lussac's equation, the stoichiometric equation for ethanol fermentation reaction is

$$C_6H_{12}O_6 \rightarrow 2CH_3CH_2OH + 2CO_2$$

The literature review has shown different percentage yields in the ethanol fermentation cycle experiments. In a trial, fermentation of sugar by *Saccharomyces cerevisiae* BY4742 was led to the creation of ethanol in a clump (Karavalakis et al., 2012). It was then discovered that the thermotolerant capacity of *S. cerevisiae* to age glucose at high temperatures was like the optima for saccharification. Various researchers also reported that the process parameters like temperature, substrate fixation and pH on ethanol maturation and changed the yield. When the temperature was expanded to 45 °C, high cell development and higher ethanol formation rate were observed, and this was hindered at 50 °C. The cell development rate and ethanol creation rate were in the range of 30–45 °C. At 30 °C, the highest sugar conversion was attained after an incubation period of 72 h. The sugar conversion varies with the glucose concentration, and the highest conversion of 59.9% is achieved at glucose concentration of 40 kg/m^3. At pH 5.0, the ethanol production rate was 410 g kg^{-1} h^{-1} of suspended solids with a production rate of 61.93%. At pH 4.0, the ethanol creation rate was 310 g kg^{-1} h^{-1} of SS. When the pH was below 4.0, acidic corrosion develops and above 5.0 which creates butyric acid corrosion in the process (Lin et al., 2012).

3.4 Photobiological Hydrogen Production

The microorganisms along with sunlight convert water and biomass into hydrogen in photobiological hydrogen production. Microorganisms such as green microalgae or cyanobacteria utilise sunlight to break down water into oxygen and hydrogen ions. The hydrogen ions are then combined to form hydrogen gas (Akkerman et al., 2002; Sakurai et al., 2013 Ghirardi et al., 2009; McKinlay & Harwood, 2010). The stoichiometric reaction of photobiological hydrogen production is

$$N_2 + 8H^+ + 8e^- + 16\ ATP \Rightarrow 2NH_3 + H_2 + 16\ ADP + 16\ Pi$$

Some microorganisms use sunlight only as a means to break down the organic matter so that organic matter can decompose and release out hydrogen on its own. This is also known as photo-fermentative hydrogen production (McKinlay & Harwood, 2010). This is a newly formed method to create fuel, and in the long term, it may provide economical hydrogen production from sunlight with almost low to zero net emissions of carbon. There are many types of microorganisms such as *Chlamydomonas reinhardtii* CC-124, *Anabaena azollae* which can fix N_2 from the atmosphere and *Rhodobacter capsulatus* with acetate as substrate with pH less than 8 which can produce hydrogen. Green algae contain chlorophyll which is used for photosynthesis. The bacteria produce hydrogen by biophotolysis indirectly and photo-fermentation directly without the presence of oxygen. Cyanobacteria can produce hydrogen by nitrogenase or hydrogenase. The process occurs without the presence of oxygen and in the absence of light by the decomposition of glucose, whereas purple non-sulphur bacteria decompose nitrogen and oxygen using oxygenic photosynthesis to produce hydrogen (Sharma & Arya, 2019).

3.4.1 The Advantages and Disadvantages of Photobiological Hydrogen Production

Since the microorganisms utilize solar energy to change biomass into hydrogen, the use of sustainable power sources prompts the generation of unadulterated and clean hydrogen without delivering poisonous gases and liquid by-products into the environment. Organisms contributing to green growth like cyanobacteria and photosynthetic microorganisms are plentiful in nature. Their presence in nature does not cause any damage to the environment. The effectiveness of change of biomass to hydrogen production in the presence of daylight is very high, approximately 10–16%. Creating energy by utilizing daylight is an environmentally safe and sustainable alternative compared to energy from petroleum derivatives (Kruse et al., 2005; Prince & Kheshgi, 2005). Hydrogen movement gets inactivated effectively if oxygen is available in the microorganism and is generally surpassed by the simultaneous creation of hydrogen and oxygen in green algae species. There can be a decline in hydrogen production during the addition of hydrogenase (enzymes) in cyanobacteria and photosynthetic microbes. The specific pathway for hydrogen generation by microorganisms remains to be properly understood by researchers. Similarly, the research could not find or develop engineered or naturally occurring microorganisms that can digest waste effectively to increase the hydrogen production rate. It is also to be pointed out that photosynthetic microbes can't satisfy the public demand for hydrogen production (Dasgupta et al., 2010).

3.5 Transesterification

Transesterification is a process in which alkyl groups of vegetables/plant oil are converted into a methyl group by an acid or a base catalyst. The attacking area gets converted into a tetrahedral intermediate by nucleophilic attack by an alkoxide ion which further decomposes to form our product and glycerol (Patel et al., 2017; Speight, 2019). The stoichiometric reaction of transesterification is as follows:

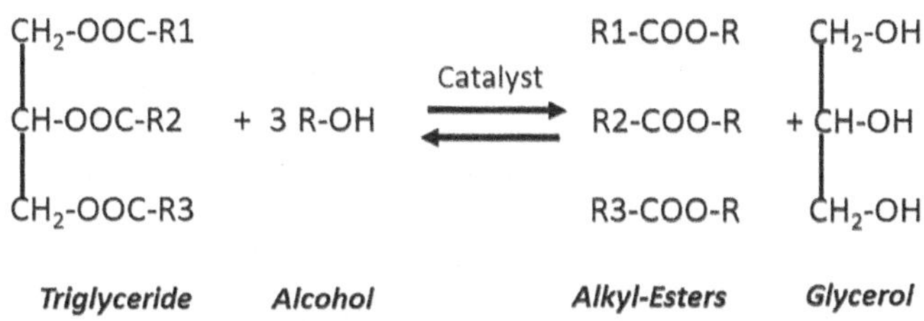

The feeds used for transesterification are methanol and ethanol. Feed selection depends on two factors, mainly (a) easy availability and (b) low cost of production. Transesterification can be of two types: catalytic and non-catalytic transesterification. **Catalytic esterification** is a process that takes place with the help of a catalyst. One of the main problems is that the rate of reaction is slow. The separation of oil or alcohol from impurities in the mixture is very difficult (Bello et al., 2012; Kandasamy et al., 2020; Bhatia, 2014). **Non-catalytic transesterification** is a process that utilizes ethanol at supercritical conditions for biodiesel production (da Silva & Vladimir Oliveira, 2014). The supercritical technique has a major benefit over conventional techniques because of its fast reaction rate, feedstock flexibility and increase in the product yield, and it is also environmentally friendly. However, this method has few limitations such as very high operating temperature and pressure and high feed requirement. This results in high energy costs and poor product quality (Chundawat et al., 2011). In supercritical conditions, alcohol acts as both a reactant and catalyst (Boer and Bahri (2011); Deshpande et al., 2010; Abbaszaadeh et al. (2012)). Temperature plays a major role in ester production. The esterification process is performed between the temperature range of 25 and 60 °C. Increasing the temperature beyond the optimal point leads to denaturing of the products, deactivation of enzyme and decline of catalyst activity (Fillion et al., 2005; Freedman et al., 1984). The thermal stability of an enzyme can be improved by its method of immobilization and external carrier (Köse et al., 2002; Iso et al., 2001).

3.6 Advantages and Disadvantages of Biochemical Processes

It is considered a more cleaner and efficient conversion technology when compared to thermochemical processes. It can be turned into various products such as hydrogen, biogas, organic acids, ethanol, butanol, etc. which are considered as the replacement of petroleum-based products (Goodrich & Schmidt, 2002). It is one of the few techniques that provide an environmentally friendly direction for obtaining energy and fuel from MSW. Anaerobic digestion is used to reduce the solid waste quantity as well as improves energy recovery. It needs low temperatures for processing with high selectivity of products generated. It can be used for a wide range of substrates with high moisture content and impurities. Carbon dioxide and sulphur emission is less than 0.1% by weight which reduces the chance of acid rain. Unlike the thermochemical processes, it involves the use of microbes, enzymes or chemicals. Pre-treatment is essential for biochemical energy conversion as it controls the product quality. The product generation is limited due to constraints of biological conversion. For a higher yield of products, additional microbes/enzymes need to be added, unlike thermochemical processes which result in higher feedstock cost compared to biochemical processes. Biochemical processes are very sensitive to temperature and produce secondary wastes which completely utilize the waste or biomass. It has higher intrinsic stability and higher reaction time compared to the thermochemical process (Garhyan & Elnashaie, 2004). Since the processes cannot take place under sterile conditions, large-scale industrial reactors have to utilize enzymes or organisms that are robust enough to withstand any contamination. The conversion of organic material into energy can lead to loss of biodiversity and transformation of the ecosystem in the region of biochemical energy conversion plants (Speight, 2019; Teter et al., 2014; Balagurumurthy et al., 2015). However, the biochemical processes cannot be used for all types of biomass. The yield and conversion efficiency is limited due to biochemical conversion efficiency. It can be increased by higher capital investments, bigger reactors and proper maintenance of reaction conditions. The effect of ambient temperature is highest as the reaction operates at moderate ranges of temperature and even small changes in operating conditions may affect the microbial activities. Complete utilization is not attained in biochemical processes as it forms sludge as residue. Different thermochemical and biochemical processes, feedstocks, process conditions and products obtained are provided in Table 1.

Table 1 Review of different thermochemical and biochemical processes to convert biomass waste to energy

Process name	Feed input	Products and by-products	Temperature (°C)	Process specifications	References
Incineration	Plastics, paper, cardboard, cloth, rubber, leather, wood, glass, metals, batteries, chemical powders and fluorescent bulbs Wet waste like oil, grease, pesticides, waste sludge from households, medical drugs, needles, tissues, cotton, fluids (blood, plasma, urine, etc.)	Gases like CO, CO_2 and water vapour. Ash and char are left behind. Impurities and hazardous substances in feed produce gases like SO_x, NO_x and organic residue	600–750 °C for dry waste 800–1200 °C for pharmaceutical waste Less than 600 °C for heat-sensitive materials including rubber, leather and polymer	Very low residence time should be ensured. Proper segregation and storage of the wastes	Tian et al. (2012), Gautam et al. (2010)
Gasification	Coal fluxes containing coal particles and water slurry, biomass, petroleum residue, wastes	Syngas (high methane content, sometimes may contain H_2 and CO_2), tar, oils, phenol, char particles and liquids	450–650°C for fixed-bed type 1000–1200 °C for entrained- and fluidized-bed gasifier with fine feed as input	Proper maintenance of temperature with a continuous supply of water for cooling Clogging of the bed should be prevented Feed size should be maintained	Szwaja et al. (2019)
Pyrolysis	Dry and semi-wet organic waste is easy to handle.	Bio-oil with tar and gases like CO, CO_2, H_2S, NH_3, SO_x. Additional components like char, acetaldehyde and anhydrides are released along with compounds C=O and COOH	400 °C to 500 °C for moderate heating. 500°C to 650°C for the flash drum process Up to 1000 °C for ultra-rapid pyrolysis	Proper maintenance of retention time and temperature in the reactor	Chen et al. (2014), Czajczyńska et al. (2017)
Liquefaction	Biodegradable waste like paper and pulp waste, wood, bagasse, beet pulp, vegetable wastes, sludge from	Bio-oil with a mixture of hydrocarbons, volatile components, organic compounds like acids, alcohols,	300–350 °C for tubular and combined reactors 400–420 °C for stirred and fixed-bed type of reactors	15–20 minutes of residence and reaction time is maintained which may	Berstad et al. (2010) Theegala and

(continued)

Table 1 (continued)

Process name	Feed input	Products and by-products	Temperature (°C)	Process specifications	References
	sewage Wet wastes, paper, plastics, agriculture residues, partially treated sewage, algae, microorganic wastes	aldehydes, anhydrides, ethers, esters, ketones, furans and phenols		increase up to 30 min Proper segregation of feed	Midgett (2012)
Carbonization	Agricultural wastes, wood and forest wastes, paper wastes, food wastes, organic feed with high carbon content	Char or charcoal with organic and chemical residue	300–450 °C and sometimes 550 °C for hard plant waste 300–350 °C for lighter waste like paper	Segregation, drying and binding of the feed Reshaping is done for a better reaction	Lin et al. (2017), Libra et al. (2011) Lohri et al. (2015)
Co-firing	Wood residues, forest and plant, herb and agricultural wastes, peat, straw, sawdust, small coal particles, powdered wastes with low ash content	Combustion residue of the wastes with the lower release of waste gases	1000 °C to 1200 °C for hard waste 600–800 °C for heat-sensitive materials	Segregation, drying and binding of the feed Feed is separated based on size and then fed to different reactors	Tumuluru et al. (2012)
Anaerobic digestion	Biodegradable wastes like paper and pulp waste, wood residues, forest and plant, herb and agricultural wastes, bagasse, beet pulp, vegetable wastes, sludge from sewage	Biogas comprising of CH_4, CO, CO_2, H_2, N_2, H_2S, O_2	35°C to 37°C. Major temperature variations influence the product and affect enzyme activity	The temperature inside the digester should not vary drastically. Variation is less than 0.6°C every day. No impurities in the digestor	Duan et al. (2012) Donoso-Bravo et al. (2011)
Ethanol fermentation	Biomass like crop waste, wood, forest and agricultural waste, bagasse, with rich glucose content	Ethanol, CO_2, CO	32–35 °C	Temperature and pH maintenance with regular monitoring. No impurities in the reactor	Rivera et al. (2017)
	Biodegradable wastes like paper and pulp waste, wood	Hydrogen with or without impurities		Temperature maintenance with an adequate supply of	Aghbashlo et al. (2016)

Photobiological hydrogen production	residues, forest and plant, herb and agricultural wastes, bagasse, beet pulp, vegetable wastes, sludge from sewage		30–33 °C. Major variations in temperature influence the product	light, with regular monitoring. No impurities in the reactor	
Transesterification	Fatty acids and oils like cooking oil from households, restaurants and food outlets	Biodiesel and glycerol and fatty acid methyl ester (FAME) biodiesel on using methanol for the process	25–60 °C. Varies based on the oil used	Maintenance of purity in the oil used and minimum variations in temperature conditions. No impurities in the reactor	de Luna et al. (2017)

4 Potential of Biomass Energy Generating Plants in India

The green development mechanism of the Kyoto Protocol provides countries to look into the issue of reducing fossil fuel consumption. India has invested more than $9251 million per annum in biomass projects to increase electricity and power generation to about 5000 million units per annum. By 2022, the Ministry of New and Renewable Energy (MNRE) has decided to increase the biogas power generation by 10 GW. However, there are many obstacles on the way. Firstly, there is only one biogas energy project has been registered under the Clean Development Mechanism (CDM) in India. According to researchers, almost 74 million tonnes of agricultural wastes can be used as feedstock for biogas production which is not utilized properly. If these wastes are properly utilized, biogas projects could reach up to 31 GW per annum which ultimately could produce 67 TWh more electricity every year (Hiloidhari & Baruah, 2011). According to 2016 studies, the following states that possess the highest power generation capacity using biogas are Maharashtra, Uttar Pradesh (UP) and Karnataka with a cumulative sum of more than 1 GW of biomass power. The otential states with a lesser power generation capacity are Punjab and Bihar. Uttar Pradesh is one of the leading producers of biogas power in India. A case study in UP showed the possibility of using biomass to replace fossil fuel usage in UP (Hiloidhari et al., 2019).

5 Current Installed Capacities of Biomass Energy Conversion Plants in India

Recent statistics from the Ministry of New and Renewable Energy (MNRE) by the Government of India reported that around 32% of the complete essential energy asset of the nation can be obtained from biomass and it can satisfy the energy need for 70% of the country's population (Archer et al., 2005; Energy Statistics, C.S. Office, 2014; MNRE, 2015). The currently accessible measure of biomass in India is approximately 500 million metric tonnes. Every year the capacity of biomass utilization from horticulture and forestry residue is rapidly growing. The power generation potential by utilizing agricultural and agro-industrial waste is around 18,000 MW. With a bagasse cogeneration facility in sugar mills, additional 8000 MW of power can be generated. Considering both the sources, the cumulative potential of biomass-based energy generation is 26,000 MW. As per the MNRE data, there are 550 biomass-independent power producers, and bagasse cogeneration-based power plants are currently installed (up to December 2020) in different states and union territory which has a cumulative power generation capacity of 9373 MW. The biomass and bagasse power plants installed so far are mainly located in the following states: Punjab, Maharashtra, West Bengal, Uttar Pradesh, Tamil Nadu, Karnataka, Chhattisgarh and Andhra Pradesh. These include power generated from bagasse cogeneration plant in 360 sugar mills which has a capacity of 7547 MW.

India has more than 540 sugar mills spread across the country so there is still room for further improvement.

In rural India, biogas plants have immense potential. As per MNRE report 2020, to fully utilize India's biomass potential, 1.23 crore biogas plants can be installed all over India. As per the latest MNRE report, only 50 lakh biogas plants are currently in operation with less than 50% of the overall potential including all states and union territory. The overall summary of India's current biogas generation plants has been depicted in Fig. 3. Under the New National Biogas and Organic Manure Programme (NNBOMP), the target was to install 60,000 new small biogas plants for the financial year of 2020–2021 among which only 8483 units could be installed due to the current pandemic situation.

As per the latest MNRE report, as of December 2020, the cumulative biogas generation capacity of India is 7,43,508 m^3 day^{-1}, whereas bio-compressed natural gas and compressed biogas production capacity is 97,199 kg day^{-1}. The overall power generation capacity of waste to energy projects in India is 291.34 MW.

6 Limitations and Challenges of Biomass Energy in India

Biomass is the fourth biggest fuel source in the world and the first in the Indian context (Capellán-Pérez et al., 2014). This energy is wasted because of inefficient technology, improper planning and financial limitations. The high rate of population growth and the rising prices of commercial energy sources like electrical, gas and petroleum products increase the need for biomass energy. However, biomass energy generation processes have their limitations such as the problem of maintaining the same feed type throughout the year, high labour requirement, less process flexibility and rapid increasing price of most biomass-based feedstock for energy generation. Such a price increase needs to be regulated by the government to ensure the economic feasibility of biomass-based energy generation. Despite having an enormous biomass reserve, the lack of planning to utilize agricultural wastes and the lack of investments hinder the potential of the biomass-based energy sector in India. As Indian agricultural lands are highly defragmented, it increases biomass production cost and thereby increases the cost of bioenergy production. High transportation costs and lack of proper infrastructure prevent the Indian states such as Andhra Pradesh, Maharashtra and Karnataka from fully utilizing their bioenergy generation potential. However, the sugarcane industry and sugar plants supply the majority of biomass feedstock in India (Delivand et al., 2011).

One of the major bottlenecks of biogas plant installation for rural India lies in its high capital requirement for installation and maintenance. The capital requirement for installation ($348) and operational cost are significantly higher than the affordability of a standard rural Indian family ($150) which acts as the hindrance for its widespread rural implementation. The operational cost per capita varies between $ 50 and 60, and for a family of five individuals, the total cost will reach up to $300 which is costlier than the cost of energy consumption ($220) in the average rural household (Jana et al., 2018; Bansal et al., 2013). This makes the implementation of a biogas plant in rural Indian households economically impractical.

7 Lack of Proper Financial Investment and Economic Limitation

The rapid exploitation and depletion of conventional energy sources like coal and petroleum have encouraged the utilization of biomass energy for satisfying the energy needs (Ghimire, 2013). In conjunction with the energy needs, biomass utilization as energy sources give an option to countries to diversify their energy resources also helps significantly in sustainable waste management in rural and urban parts of the country (Williams et al., 2017). Biomass-based feedstock for energy conversion includes biodegradable agriculture, forest and municipal waste (Williams et al., 2016). Based on the composition, size and moisture content of the biomass, a plethora of energy conversion or valorization methods can be implemented such as carbonization, liquefaction, pyrolysis or biochemical conversion as discussed earlier. The efficient use of biomass as a source of energy would help in mitigating the environmental damage caused by the burning of fossil fuels which has risen over the past decade at an alarming rate (Yousuf et al., 2016).

As discussed in the previous section, the main barriers which prevent the rapid growth of biomass energy production in India lie in financial and social factors (Mittal et al., 2018). A large portion of biomass (60–70%) obtained from agricultural waste have variable biomass composition based on soil properties, moisture, atmospheric variation due to season change and growth state of the crop (Liu et al., 2017) which have a significant impact on the quality of products produced from these biomass resources. Further, the land required for biomass treatment and energy production could alternatively be used as agricultural land, which the farmers would benefit from, and it is necessary to satisfy the increasing demand for food in a densely populated country like India (Liu et al., 2017). The fragmented nature of agricultural land (absence of landowner having large landmass for agriculture) is another major obstacle in the implementation of biomass-based energy production plants in rural India.

In rural India, establishing thermochemical and biochemical treatment facilities for biomass conversion requires resources like land, initial capital investment, maintenance cost, water, power and air supply. These resources cannot be afforded by the farmers in the agricultural land (Roopnarain & Adeleke, 2017) without continuous financial support from the government. In addition to operational costs, transportation, storage and distribution of both raw material and finished products lead to high additional expenditure. Another major reason behind the lack of financial investments is owed to the ownership of lands. It is often observed that the majority of farmers are not the real owners of the land and do not have the authority to make decisions regarding the use of land (Rupf et al., 2015). The current price of biomass energy in the market is not adequate to generate enough revenue to compensate for the expenses incurred in its production. The first step towards mitigating these problems is to create stable demand for biomass energy which is lacking significantly in India.

8 Conflict of Public Interest

The true owners of agricultural lands lack the initial capital and are focused on maximizing income from crop sales, so they are hesitant to invest in the small-scale use of land for biomass energy production, which does not guarantee large financial benefits. The issue of land acquisition for setting up biomass generation plants in rural setup is a key concern for the investors. For the establishment of plants, residents must be evacuated from the area, and a portion of agricultural lands must be cleared to accommodate instruments and plant utilities. Additional resources like water and energy required to operate biomass units could also be used to benefit crop production, harvesting, irrigation and transportation, illustrating the disparity in public opinion (Mwirigi et al., 2014). The location of the plant should be such that the necessities like water, power, heat, air, manpower and a continuous steady supply of biomass can be met, which is difficult to achieve. The degree to which biomass is used is also influenced by various superstitions and social stigma associated with its use. When it comes to the acceptance of biogas plants in rural areas, there are many barriers. Owners of biogas plants are usually apprehensive about using night soil because of its texture and feed ingredients. The absence of awareness about the innovation of biogas plants can also be one of the primary explanations behind the smaller number of installations of biogas technology in rural areas. Since individuals don't know about the way that they can utilize biogas plant assets, a lot of them opt for natural fuels instead (Blenkinsopp et al., 2013; Raha et al., 2014).

9 Environmental and Technological Difficulties of Biogas Production in Indian Scenario

Lack of resources and feedstock can lead to a low output performance and yield of biogas energy alongside the development of toxic by-products. This makes the biogas plant hard to work with and increases the maintenance cost of the plant (Rupf et al., 2015). For a biogas plant to work efficiently, the right proportions of water and organic wastes are required. Scarcity of water is a typical obstruction that numerous families face in rural India. This causes inefficient working of a biogas plant especially in deserted areas like Rajasthan (Samar & Meena, 2016). In such places, the temperature has a vital factor in regard to the rate of biogas production.

During cold seasons, the pace of biogas energy is lesser than in summer. Additionally, the hydraulic retention time of the plant increases to almost 120–150 days at a temperature below 15°C (Milieu & Shukla, 2007). The shortfall of skilled/unskilled labour to fix any technical issues that occur in a biogas plant also acts as a technological barrier in rural regions (Kaniyamparambil & Krantzberg, 2011). In metropolitan areas, without advanced waste segregation infrastructure, it is very difficult to isolate natural waste from non-organic wastes which diminish the quality of the feedstock provided to the biogas. Improper segregation of impurities such as dust and inert material remains in the feedstock which increases the generation cost and the complexity of the entire process. Poor collection and organization of the wastes from households also increase supply chain difficulty to biogas plants (Planning Commission, 2012).

10 Comparative Analysis of Global Biomass Energy Generation with Asian/Indian Scenario

According to the global bioenergy statistics provided by the World Bioenergy Association, biomass provided 637 TWh of electricity globally in 2018. Solid biomass sources accounted for 66% of total bioenergy produced, with municipal and industrial waste accounting for 19%. The overall growth of biopower generation was 8% between 2000 and 2018, owing primarily to the increased use of solid biomass for biopower generation. With 243 TWh of output in 2018, Asia accounted for 38% followed by Europe with 35% of global biopower generation. European countries mainly use municipal waste for biopower generation which accounts for 63% of total production, whereas Asia expertises in the utilization of industrial waste for energy production, accounting for 79% of overall production. Europe dominates in the biogas generation which accounts for 72% of global production. The use of biomass energy by each continent during year 2018 has been illustrated in Fig. 4. In the global context, the prominent countries that have significant biomass-based energy generation facilities include China, Japan, Germany, the USA, the UK and India. Figure 5 explains the country-wise capacity of biomass energy generation plants. India has the third highest biomass-based energy generation capacity after China and the USA.

11 Treatment and Removal of Major Air Pollutants and Health Hazards

Various pollutants are generated as a by-product of biomass to energy conversion processes. These pollutants need to be separated from the desired product and disposed of efficiently. The methods of pollutant separation alter based on the type

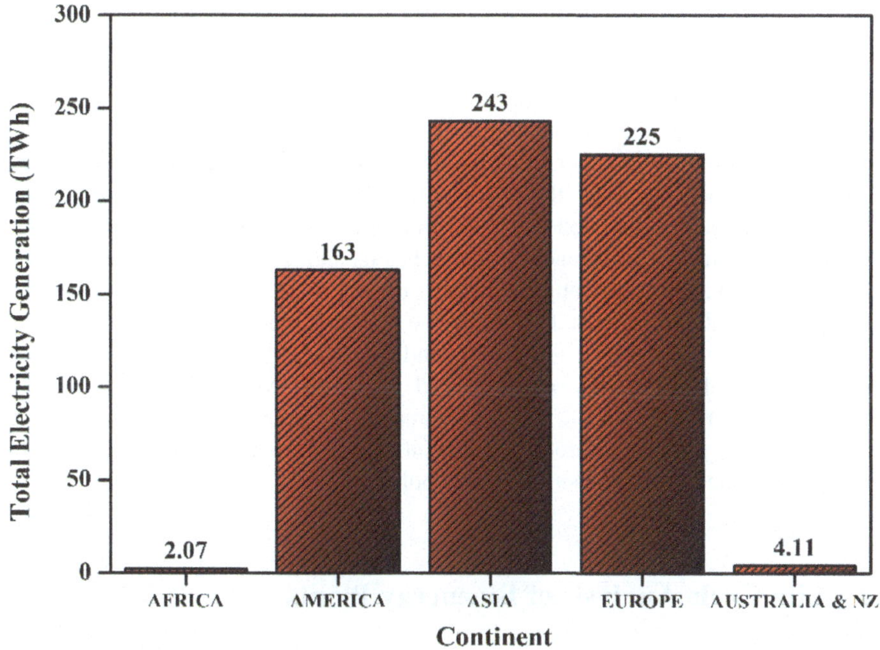

Fig. 4 Continent-wise power generation by biomass energy plant during 2018

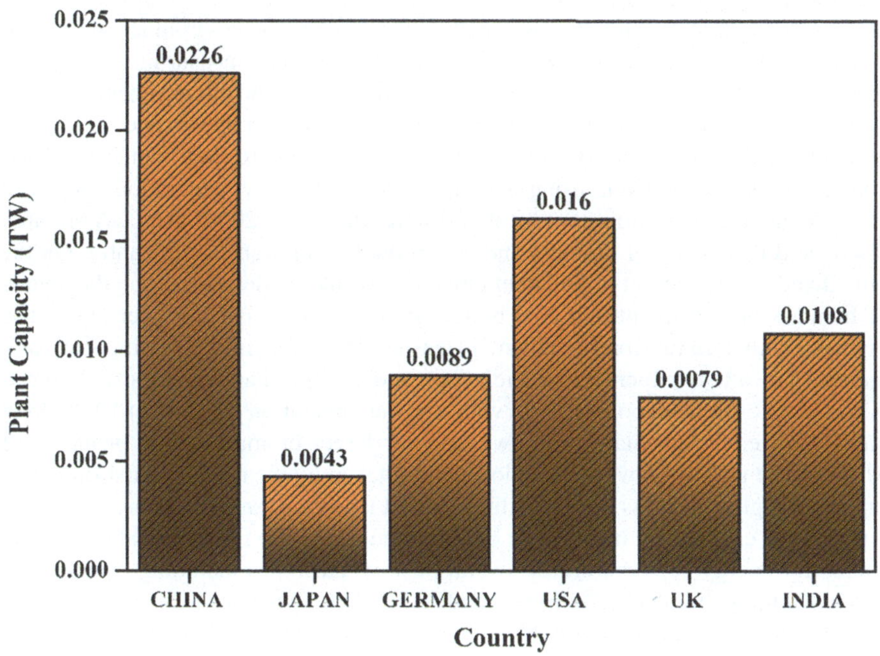

Fig. 5 Biogas plant capacities in different countries

of energy conversion process and composition of feedstock which involves additional costs and equipment usage. The major pollutants include SO_x, NO_x, CO_2, CO, CH_4, chlorides, bromides, fluorides, ash, soot, char residue, hazardous metal residue, oils, organic wastes comprising of microbes, wastewater and biological residue (Sharifian et al., 2021). Removal of the pollutants ensures mitigation of severe environmental damage and increases the efficiency of the process (Liu et al., 2020). This section discusses the major pollutants released from each process and many health hazards associated with such emission or discharge. Different materials are released from the plants, each of which has its side effects. Some lead to short-term illnesses like cold and cough; breathing disorder; asthma; pulmonary disorders (Karavalakis et al., 2012); irritation in the eye, throat and nose; and headaches (Pena & Rollins, 2017). In the long term, they tend to cause chronic diseases like cancer and cardiovascular disorder on prolonged exposure (Manisalidis et al., 2020, Mannucci & Franchini, 2017). Table 2 summarizes pollutants produced in different wastes to energy generation process, pollutant treatment method and different health hazards associated with exposure to those pollutants.

12 Life Cycle Analysis of Bioenergy Plants

It is critical to weigh on the benefits and drawbacks of bioenergy production from a life cycle assessment (LCA) standpoint. LCA aids in the streamlining of the manufacturing process by recommending the best solutions for reducing overall environmental effects. The LCA is a method for defining the environmental loads of a final product, process or operation by recognizing and measuring energy and materials used, along with waste discharges, quantifying the environmental effects of these wastes and evaluating the potential for eco-friendly advances over the product's entire life cycle (Canter, 1996). As compared to coal and natural gas combined cycle (NGCC), using bioenergy for electricity generation has been shown to reduce GHG emissions by 40–80% (Rahimi et al., 2020). An LCA research discovered that co-firing biomass and coal reduces emissions significantly. On the other hand, the reduction in pollutant emissions would be determined by the degree of biomass pre-treatment and coal boiler efficiency, so it is recommended to use biomass with minimal pre-treatment to ensure reduction in emissions. Emission patterns are often influenced by the scope and design characteristics of biomass power plants. Biomass power plants with a power generation capacity of 10–25 MW have better environmental results, whereas for larger biomass power plants (>25 MW), the benefits of power generation are overtaken by the negative environmental impact of higher biomass transport distance and land constraints. The biogas reactor type plays a critical role in emission control. The use of biomass as an alternate fuel source for energy conversion has several positive attributes including decreases in global warming, human health hazard and eco-toxicity, whereas the negative aspects include increased risk of eutrophication and respiratory illness (C.P.C. BOARD, 2013; Almutairi et al., 2018; Bacenetti, 2020; Wu et al., 2018). Results of LCA studies have shown that, if biogas is used sustainably, it has several environmental

Table 2 Different flue gas treatment processes with different wastes to energy conversion processes

Process name	Pollutants	Treatment methods	Health effects	References
Incineration	SO_x, NO_x, ash, chlorides and bromides, CO_2, CO, hazardous metals	***Spray scrubbers, absorbers and lime injection*** (dry) are used for SO_x, NO_x and chlorides. ***Electrostatic precipitators*** are used for fine particles and hazardous metals. Activated charcoal for particulate matter and metal emission	Respiratory and cardiovascular diseases, asthma, irritation in the eyes and nose, bronchial infection, haemoglobin poisoning, neurological disorder, cancer, mortality ***Permissible limits:*** SO_x (50–80 μg/m^3), NO_x (40–80 μg/m^3), ash (70 μg/m^3), chlorides and bromides (<1 ppm), CO_2 and CO (2–4 μg/m^3), metals (50 μg/m^3–0.1 mg/m^3)	National Research Council report (2000), Kurella et al. (2017)
Gasification	CO, CO_2, CH_4, H_2S, particulate matter, oils, phenols	***Wet scrubbers for gases*** and ***activated charcoal*** for oil and phenols. ***Candle filters*** for oils and ***ESP*** for particulate matter	Respiratory and cardiovascular diseases, asthma, bronchitis, nausea, system disorders ***Permissible limits:*** CO_2 and CO (2–4 μg/m^3), particulate matter (60–100 μg/m^3), hydrocarbons (500–1000 ppm), oils and grease (10 mg/l) in effluent water	Kurella et al. (2017)
Pyrolysis	CO, CO_2, oils, organic compounds, char residue	***Wet scrubbers*** for gases and ***activated charcoal*** for oil and organic degradation. Char residue can be converted to charcoal	Respiratory and cardiovascular diseases, asthma, bronchitis, digestive disorders ***Permissible limits:*** CO_2 and CO (2–4 μg/m^3), char residue	Ortuño et al. (2014)

(continued)

Table 2 (continued)

Process name	Pollutants	Treatment methods	Health effects	References
			(50–70 μg/m^3), organics (10–1000 ppm), oils and grease (10 mg/l) in effluent water	
Liquefaction	Organic substances, CO, CO_2, SO_x, NO_x	***Spray and wet scrubbers and absorbers*** for gases. Organic components are left for self-degradation	Ill functioning of body systems, digestive disorders, headaches, impaired vision, cancer, cardiac disorders ***Permissible limits:*** SO_x (50–80 μg/m^3), NO_x (40–80 μg/m^3), organics (10–1000 ppm), CO_2 and CO (2–4 μg/m^3)	Gollakota et al. (2018), Yuan et al. (2020)
Carbonization	CO, CO_2, ash	***Wet scrubbers, filters and absorbers*** for gases and ash (ESP)	Breathing disorders, sensory infection, haemoglobin poisoning, impaired vision, dizziness, skin irritation ***Permissible limits:*** CO_2 and CO (2–4 μg/m^3), ash (70 μg/m^3)	Voskian and Alan Hatton (2018), Sharifian et al. (2021), Yuan et al. (2020)
Co-firing	SO_x, NO_x, ash, CO_2, CO	***Spray scrubbers and absorbers*** used for acid gases. The use of low sulphur and low ash coal for the process. The use of fluidized-bed process	Respiratory and cardiovascular diseases, asthma, irritation in the eyes and nose, bronchial infection ***Permissible limits:*** SO_x (50–80 μg/m^3), NO_x (40–80 μg/m^3), CO_2 and CO (2–4 μg/m^3), ash (70 μg/m^3)	Rahimi et al. (2020)
Anaerobic digestion	SO_x, NO_x, CO_2, CO, CH_4	Iron wool ***CO_2 scrubber*** device used for CO_2 removal and	Breathing disorders, sensory infection, haemoglobin	Mamun et al. (2017)

(continued)

Table 2 (continued)

Process name	Pollutants	Treatment methods	Health effects	References
		selective membrane filtration for CO_2 and CH_4, ***scrubbers*** for acid gas removal	poisoning, nausea, impaired vision, dizziness, skin irritation ***Permissible limits:*** SO_x (50–80 μg/m^3), NO_x (40–80 μg/m^3), hydrocarbons (500–1000 ppm), CO_2 and CO (2–4 μg/m^3)	
Ethanol fermentation	CO_2, CO, CH_4	Oxidation of CH_4 to CO_2. ***Scrubbers and filter beds*** for CO_2. Natural methods of afforestation and sequestration for CO_2 capture	Respiratory and cardiovascular diseases, asthma, nausea, damaged vision, vomiting, sensory defects ***Permissible limits:*** hydrocarbons (500–1000 ppm), CO_2 and CO (2–4 μg/m^3)	Liu et al. (2017), Marchuk et al. (2019)
Photobiological hydrogen production	CO, CO_2, NO_x, SO_x, C_xH_x, soot, ash	***Electrostatic precipitators*** are used for fine particles, soot and ash. ***Spray scrubbers and absorbers*** used for acid gases. ***Catalytic converters*** for C_xH_x	Breathing disorders, sensory infection, haemoglobin poisoning, asthma, bronchitis, nausea, impaired vision, dizziness, skin irritation ***Permissible limits:*** SO_x (50–80 μg/m^3), NO_x (40–80 μg/m^3), CO_2 and CO (2–4 μg/m3), hydrocarbons (500–1000 ppm), ash and soot (70 μg/m^3)	Nur et al. (2019), Primo & García (2013)
Transesterification	Organic substances, oils	***Membrane filters and activated charcoal*** for oil and degradation of organics	Ill functioning of body systems, digestive disorders ***Permissible limits:*** organics (10–1000 ppm), oils and grease (10 mg/l) in effluent water	Vedaraman et al. (2011)

advantages over conventional fuels. As the same biomass feedstock can be utilised via several energy conversion methods (thermochemical and biochemical), it is critical to figure out which route is the most cost-effective. Cherubini and coworkers (Cherubini et al., 2009) found that producing heat and electricity from biomass reduces GHG emissions and saves energy compared to the biofuel production route. The waste/residue biomass has the best environmental efficiency because it avoids both the impacts of energy-intensive crop processing and the emissions from waste management (Taheri et al., 2014). In a nutshell, it can be concluded that bioenergy has lower environmental impacts than traditional energy production schemes be it fossil fuel based or nuclear fuel based as renewable biomass is CO_2-neutral when burnt. The resource base can be preserved if harvested biomass is produced again, and residues can be value added in shorter time.

13 Conclusion

The main objective of this chapter is to provide detailed information about the recent trends in biomass to energy conversion technologies. The two major technologies currently used for biomass to energy conversion, namely, thermochemical and biochemical processes and their merits and demerits, have been discussed in detail. The selection of a conversion technology depends upon the form of output energy required. The estimated potential and current installed capacities of plants in India were discussed. A significant effort has been put to compare global biomass energy generation plants with the Indian scenario in general. Challenges to implementing biomass to energy projects due to lack of regulations, proper financial investment and socio-political and technological issues in implementing the project have been widely discussed. Authors have attempted to identify the major air pollutants and associated health effects from such pollutants. Life cycle analysis (LCA) of biomass energy conversion technologies has been discussed to assess the environmental impact of such a project compared to its benefits.

Hence to summarize, bioenergy has lower environmental impacts than traditional (fossil and nuclear) energy sources as the biomass energy conversion process is carbon-neutral. On the whole, the environmental impact caused by bioenergy plants is much lesser than that of fossil fuel-based power plants. If biomass is properly utilized and financial benefits can be improved, biomass energy conversion plants can become a major asset to reduce India's dependency on imported fossil fuels.

References

Abbaszaadeh, A., et al. (2012). Current biodiesel production technologies: A comparative review. *Energy Conversion and Management, 63*, 138–148.

Abowei, M. F. N., Ayotamuno, M. J., and Eze, C. L. (2009). Comparative evaluation of batch and continuous anaerobic digesters in biogas production from municipal solid waste using mathematical models. *Agricultural Engineering International: CIGR Journal.*

Agbor, E., Zhang, X., & Kumar, A. (2014). A review of biomass co-firing in North America. *Renewable and Sustainable Energy Reviews, 40*, 930–943.

Aghbashlo, M., et al. (2016). Exergy analysis for decision making on operational condition of a continuous photobioreactor for hydrogen production via WGS reaction. *International Journal of Hydrogen Energy, 41*(4), 2354–2366.

Akkerman, I., et al. (2002). Photobiological hydrogen production: photochemical efficiency and bioreactor design. *International Journal of Hydrogen Energy, 27*(11-12), 1195–1208.

Almutairi, K., Thoma, G., & Durand-Morat, A. (2018). Ex-ante analysis of economic, social and environmental impacts of large-scale renewable and nuclear energy targets for global electricity generation by 2030. *Sustainability, 10*(8), 2884.

Anastasakis, K., et al. (2018). Continuous hydrothermal liquefaction of biomass in a novel pilot plant with heat recovery and hydraulic oscillation. *Energies, 11*(10), 2695.

Anderson, E., et al. (2016). Economic screening of renewable energy technologies: Incineration, anaerobic digestion, and biodiesel as applied to waste water scum. *Bioresource Technology, 222*, 202–209.

Archer, E., et al. (2005). *MBT: A Guide for Decision Makers-Processes, Policies and Markets. Juniper Consulting Ltd.*

Bacenetti, J. (2020). Economic and environmental impact assessment of renewable energy from biomass. 5619.

Balagurumurthy, B., Singh, R., & Bhaskar, T. (2015). Catalysts for thermochemical conversion of biomass. In *Recent Advances in Thermo-Chemical Conversion of Biomass* (pp. 109–132). Elsevier. Retrieved from https://doi.org/10.1016/B978-0-444-63289-0.00004-1

Bansal, M., Saini, R. P., & Khatod, D. K. (2013). Development of cooking sector in rural areas in India—A review. *Renewable and Sustainable Energy Reviews, 17*, 44–53.

Barik, D. (Ed.). (2018). *Energy from toxic organic waste for heat and power generation.* Woodhead Publishing.

Basu, P. (2010). *Biomass gasification and pyrolysis: practical design and theory.* Academic press.

Basu, P. (2018). *Biomass gasification, pyrolysis and torrefaction: practical design and theory.* Academic press.

Bello, B. Z., Nwokoagbara, E., & Wang, M. (2012). Comparative techno-economic analysis of biodiesel production from microalgae via transesterification methods. *Computer Aided Chemical Engineering, 30.* Elsevier, 132–136.

Berstad, D. O., Stang, J. H., & Nekså, P. (2010). Large-scale hydrogen liquefier utilising mixed-refrigerant pre-cooling. *International Journal of Hydrogen Energy, 35*(10), 4512–4523. https://doi.org/10.1016/j.ijhydene.2010.02.001

Bhatia, R. (1990). Diffusion of renewable energy technologies in developing countries: A case study of biogas engines in India. *World Development, 18*(4), 575–590.

Bhatia, S. C. (Ed.). (2014). *Advanced renewable energy systems,(Part 1 and 2).* CRC Press.

Blenkinsopp, T., Coles, S. R., & Kirwan, K. (2013). Renewable energy for rural communities in Maharashtra, India. *Energy Policy, 60*, 192–199.

Boardman, R. D., et al. (2014) *Logistics, costs, and GHG impacts of utility scale cofiring with 20% biomass.* No. PNNL-23492. Pacific Northwest National Lab.(PNNL), Richland, WA (United States).

C.P.C. BOARD (Ed.) (2013). Ministry of Environment Forests & Climate Change, New Delhi.

Canter, L. W. (1996). *Environmental impact assessment* (2nd ed.). McGraw-Hill. (McGraw-Hill series in water resources and environmental engineering).

Capellán-Pérez, I., et al. (2014). Fossil fuel depletion and socio-economic scenarios: An integrated approach. *Energy, 77*, 641–666.

Carneiro, M. L. N. M., et al. (2017). Potential of biofuels from algae: Comparison with fossil fuels, ethanol and biodiesel in Europe and Brazil through life cycle assessment (LCA). *Renewable and Sustainable Energy Reviews, 73*, 632–653.

Castello, D., Pedersen, T. H., & Rosendahl, L. A. (2018). Continuous hydrothermal liquefaction of biomass: A critical review. *Energies, 11*(11), 3165.

Chen, D., et al. (2014). Pyrolysis technologies for municipal solid waste: a review. *Waste Management, 34*(12), 2466–2486.

Chen, H., & Wang, L. (2016). *Technologies for biochemical conversion of biomass*. Academic Press.

Cheng, H., & Yuanan, H. (2010). Curbing dioxin emissions from municipal solid waste incineration in China: Re-thinking about management policies and practices. *Environmental Pollution, 158*(9), 2809–2814.

Cherubini, F., et al. (2009). Energy-and greenhouse gas-based LCA of biofuel and bioenergy systems: Key issues, ranges and recommendations. *Resources, Conservation and Recycling, 53*(8), 434–447.

Chundawat, S. P. S., Beckham, G. T., Himmel, M. E., & Dale, B. E. (2011). Deconstruction of lignocellulosic biomass to fuels and chemicals. *Annual Review of Chemical and Biomolecular Engineering, 2*(1), 121–145. https://doi.org/10.1146/annurev-chembioeng-061010-114205

Czajczyńska, D., et al. (2017). Potential of pyrolysis processes in the waste management sector. *Thermal Science and Engineering Progress, 3*, 171–197.

da Silva, C., & Vladimir Oliveira, J. (2014). Biodiesel production through non-catalytic supercritical transesterification: current state and perspectives. *Brazilian Journal of Chemical Engineering, 31*(2), 271–285.

Dasgupta, C. N., et al. (2010). Recent trends on the development of photobiological processes and photobioreactors for the improvement of hydrogen production. *International Journal of Hydrogen Energy, 35*(19), 10218–10238.

De Boer, K., & Bahri, P. A. (2011). Supercritical methanol for fatty acid methyl ester production: A review. *Biomass and Bioenergy, 35*(3), 983–991.

de Luna, Mark Daniel, G., et al. (2017). Transesterification of soybean oil using a novel heterogeneous base catalyst: synthesis and characterization of Na-pumice catalyst, optimization of transesterification conditions, studies on reaction kinetics and catalyst reusability. *Fuel, 209*, 246–253.

De Mes, T. Z. D., et al. (2003). Methane production by anaerobic digestion of wastewater and solid wastes. *Bio-Methane & Bio-Hydrogen*, 58–102.

de Titto, E., & Savino, A. (2019). Environmental and health risks related to waste incineration. *Waste Management & Research, 37*(10), 976–986.

Delivand, M. K., Barz, M., & Gheewala, S. H. (2011). Logistics cost analysis of rice straw for biomass power generation in Thailand. *Energy, 36*(3), 1435–1441.

Demirbas, A. (2008). Liquefaction of biomass using glycerol. *Energy Sources, Part A, 30*(12), 1120–1126.

Deshpande, A., et al. (2010). Supercritical biodiesel production and power cogeneration: technical and economic feasibilities. *Bioresource Technology, 101*(6), 1834–1843.

Donoso-Bravo, A., et al. (2011). Model selection, identification and validation in anaerobic digestion: a review. *Water Research, 45*(17), 5347–5364.

Duan, N., et al. (2012). High-solid anaerobic digestion of sewage sludge under mesophilic conditions: feasibility study. *Bioresource Technology, 104*, 150–156.

Edjabou, M. E., et al. (2015). Municipal solid waste composition: Sampling methodology, statistical analyses, and case study evaluation. *Waste Management, 36*, 12–23.

Energy Statistics, C.S. Office (Ed.) (2014), Ministry of Statistics and Programme Implementation,

Erdinc, O., & Uzunoglu, M. (2012). Optimum design of hybrid renewable energy systems: Overview of different approaches. *Renewable and Sustainable Energy Reviews, 16*(3), 1412–1425.

Fillion, E., et al. (2005). Meldrum's acids as acylating agents in the catalytic intramolecular Friedel−Crafts reaction. *The Journal of Organic Chemistry, 70*(4), 1316–1327.

Foong, S. Y., et al. (2020). Valorization of biomass waste to engineered activated biochar by microwave pyrolysis: progress, challenges, and future directions. *Chemical Engineering Journal, 389*, 124401.

Foston, M., & Ragauskas, A. J. (2012). Biomass characterization: recent progress in understanding biomass recalcitrance. *Industrial Biotechnology, 8*(4), 191–208.

Freedman, B. E. H. P., Pryde, E. H., & Mounts, T. L. (1984). Variables affecting the yields of fatty esters from transesterified vegetable oils. *Journal of the American Oil Chemists Society, 61*(10), 1638–1643.

Garhyan, P., & Elnashaie, S. S. E. H. (2004). Utilization of mathematical models to investigate the bifurcation and chaotic behavior of ethanol fermentors. *Mathematical and Computer Modelling, 39*(4-5), 381–427.

Gautam, V., Thapar, R., & Sharma, M. (2010). Biomedical waste management: Incineration vs. environmental safety. *Indian Journal of Medical Microbiology, 28*(3), 191.

Ghimire, P. C. (2013). SNV supported domestic biogas programmes in Asia and Africa. *Renewable Energy, 49*, 90–94.

Ghirardi, M. L., et al. (2009). Photobiological hydrogen-producing systems. *Chemical Society Reviews, 38*(1), 52–61.

Gollakota, A. R. K., Kishore, N., & Sai, G. (2018). A review on hydrothermal liquefaction of biomass. *Renewable and Sustainable Energy Reviews, 81*, 1378–1392.

Goodrich, P. R., and Schmidt, D. (2002). Anaerobic digestion for energy and pollution control. *2002 ASAE Annual Meeting*. American Society of Agricultural and Biological Engineers.

Hermansen, J. E., et al. (2017). *Green biomass-protein production through bio-refining*. DCA-Nationalt Center for Fødevarer og Jordbrug.

Hiloidhari, M., & Baruah, D. C. (2011). Crop residue biomass for decentralized electrical power generation in rural areas (part 1): Investigation of spatial availability. *Renewable and Sustainable Energy Reviews, 15*(4), 1885–1892.

Hiloidhari, M., Baruah, D. C., Kumari, M., Kumari, S., & Thakur, I. S. (2019). Prospect and potential of biomass power to mitigate climate change: A case study in India. *Journal of Cleaner Production, 220*, 931–944.

Hjelmar, O., Johnson, A., & Comans, R. (2010). Incineration: solid residues. *Solid Waste Technology & Management, 1*, 430–462.

Hofbauer, H., & Materazzi, M. (2019). Waste gasification processes for SNG production. In *Substitute natural gas from waste* (pp. 105–160). Academic Press.

Iso, M., et al. (2001). Production of biodiesel fuel from triglycerides and alcohol using immobilized lipase. *Journal of Molecular Catalysis B: Enzymatic, 16*(1), 53–58.

Jadhao, S. B., et al. (2017). Bury, burn, or gasify: assessing municipal solid waste management options in Indian megacities by exergy analysis. *Clean Technologies and Environmental Policy, 19*(5), 1403–1412.

Jahirul, M. I., et al. (2012). Biofuels production through biomass pyrolysis—a technological review. *Energies, 5*(12), 4952–5001.

Jana, K., Mahanta, P., & De, S. (2018). Role of biomass for sustainable energy solution in India. In *Sustainable energy and transportation* (pp. 209–233). Springer.

Jena, U., & Das, K. C. (2011). Comparative evaluation of thermochemical liquefaction and pyrolysis for bio-oil production from microalgae. *Energy & Fuels, 25*(11), 5472–5482.

Ji, L., et al. (2016). Levels and profiles of dioxins from circulating fluidised bed incineration. *International Journal of Environment and Pollution, 60*(1-4), 136–155.

Jiao, F., et al. (2016). Study on the species of heavy metals in MSW incineration fly ash and their leaching behavior. *Fuel Processing Technology, 152*, 108–115.

Kachrimanidou, V., et al. (2020). Increasing the volumetric productivity of fermentative ethanol production using a fed-batch vacuferm process. *Biomass Conversion and Biorefinery*, 1–8.

Kandasamy, R., et al. (2020). Anaerobic biovalorization of leather industry solid waste and production of high value-added biomolecules and biofuels. *Biovalorisation of Wastes to Renewable Chemicals and Biofuels*. Elsevier, 3–25.

Kaniyamparambil, J. S., & Krantzberg, D. G. (2011). *A look at India's biogas energy development programafter three decades, is it useful (doing what it should) and should it be continued. McMaster University*.

Karavalakis, G., et al. (2012). Impacts of ethanol fuel level on emissions of regulated and unregulated pollutants from a fleet of gasoline light-duty vehicles. *Fuel, 93*, 549–558.

Kardos, L., et al. (2011). Comparing of mesophilic and thermophilic anaerobic fermented sewage sludge based on chemical and biochemical tests. *Applied Ecology and Environmental Research, 9*(3), 293–302.

Katakojwala, R., et al. (2020). Hydrothermal liquefaction of biogenic municipal solid waste under reduced H2 atmosphere in biorefinery format. *Bioresource Technology, 310*, 123369.

Köse, Ö., Tüter, M., & Ayşe Aksoy, H. (2002). Immobilized Candida antarctica lipase-catalyzed alcoholysis of cotton seed oil in a solvent-free medium. *Bioresource Technology, 83*(2), 125–129.

Kruse, O., et al. (2005). Improved photobiological H2 production in engineered green algal cells. *Journal of Biological Chemistry, 280*(40), 34170–34177.

Kurella, S., Bhukya, P. K., & Meikap, B. C. (2017). Removal of H2S pollutant from gasifier syngas by a multistage dual-flow sieve plate column wet scrubber. *Journal of Environmental Science and Health, Part A, 52*(6), 515–523.

Laciak, M., et al. (2016). The analysis of the underground coal gasification in experimental equipment. *Energy, 114*, 332–343.

Lam, C. H. K., et al. (2010). Use of incineration MSW ash: a review. *Sustainability, 2*(7), 1943–1968.

Libra, J. A., et al. (2011). Hydrothermal carbonization of biomass residuals: a comparative review of the chemistry, processes and applications of wet and dry pyrolysis. *Biofuels, 2*(1), 71–106.

Lin, Y., et al. (2012). Factors affecting ethanol fermentation using Saccharomyces cerevisiae BY4742. *Biomass and Bioenergy, 47*, 395–401.

Lin, Y., et al. (2017). Hydrothermal carbonization of typical components of municipal solid waste for deriving hydrochars and their combustion behavior. *Bioresource Technology, 243*, 539–547.

Liu, L., Li, J., & Xie, J. (2017). The role of biomass in deeply decarbonizing China's power generation: implications for policy design and implementation. *Carbon Management, 8*(2), 191–205.

Liu, Y., et al. (2020). Electrochemically mediated carbon dioxide separation with quinone chemistry in salt-concentrated aqueous media. *Nature Communications, 11*(1), 1–11.

Lohri, C. R., Sweeney, D., & Rajabu, H. M. (2015). *Carbonizing urban biowaste for low-cost char production in developing countries. A review of knowledge, practices and technologies*. Retrieved from http://hdl.handle.net/20.500.11810/3161

Lu, X., Jordan, B., & Berge, N. D. (2012). Thermal conversion of municipal solid waste via hydrothermal carbonization: Comparison of carbonization products to products from current waste management techniques. *Waste Management, 32*(7), 1353–1365.

Lüschen, A., and Madlener, R. (2010). Economics of biomass co-firing in new hard coal power plants in Germany.

Madanayake, B. N., et al. (2017). Biomass as an energy source in coal co-firing and its feasibility enhancement via pre-treatment techniques. *Fuel Processing Technology, 159*, 287–305.

Mahmoudi, A. H. (2015). *Prediction of heat-up, drying and gasification of fixed and moving beds by the Discrete Particle Method (DPM)*. Diss. University of Luxembourg, Luxembourg.

Mamun, A., Rashed, M., & Torii, S. (2017). Enhancement of methane concentration by removing contaminants from biogas mixtures using combined method of absorption and adsorption. *International Journal of Chemical Engineering, 2017*, 7906859.

Manisalidis, I., et al. (2020). Environmental and health impacts of air pollution: a review. *Frontiers in Public Health, 8*, 14.

Mannucci, P. M., & Franchini, M. (2017). Health effects of ambient air pollution in developing countries. *International Journal of Environmental Research and Public Health, 14*(9), 1048.

Marchuk, A., Likhanov, V. A., & Lopatin, O. P. (2019). Alternative energy: methanol, ethanol and alcohol esters of rapeseed oil as eco-friendly biofuel. *Теоретическая и прикладная экология, 3*, 80–86.
McKinlay, J. B., & Harwood, C. S. (2010). Photobiological production of hydrogen gas as a biofuel. *Current Opinion in Biotechnology, 21*(3), 244–251.
Menon, V., & Rao, M. (2012). Trends in bioconversion of lignocellulose: biofuels, platform chemicals & biorefinery concept. *Progress in Energy and Combustion Science, 38*(4), 522–550.
Milieu en Natuur Planbureau, & Shukla, P. R. (2007). *Biomass energy strategies for aligning development and climate goals in India*. MNP.
Mittal, S., Ahlgren, E. O., & Shukla, P. R. (2018). Barriers to biogas dissemination in India: A review. *Energy Policy, 112*, 361–370.
MNRE (2015). Annual Report, 2015–16. In: Energy, M.O.N.A.R. (Ed.), New Delhi.
Moh, Y. C., & Manaf, L. A. (2014). Overview of household solid waste recycling policy status and challenges in Malaysia. *Resources, Conservation and Recycling, 82*, 50–61.
Muhtadin, M., et al. (2020). Investigation of biomass combustion and conceptual design of a fluidized-bed. *Journal of Advanced Research in Fluid Mechanics and Thermal Sciences, 74*(1), 81–97.
Mwirigi, J., et al. (2014). Socio-economic hurdles to widespread adoption of small-scale biogas digesters in Sub-Saharan Africa: A review. *Biomass and Bioenergy, 70*, 17–25.
National Research Council (2000). Incineration processes and environmental releases. *Waste Incineration & Public Health*. National Academies Press (US).
Nur, M., Sumariyah, S., and Suseno, A. (2019). Removal of emission gas COx, NOx and SOx from automobile using non-thermal plasma. *IOP Conference Series: Materials Science and Engineering*. 509. 1. IOP Publishing.
Ortuño, N., et al. (2014). Pollutant emissions during pyrolysis and combustion of waste printed circuit boards, before and after metal removal. *Science of the Total Environment, 499*, 27–35.
Pandiyan, K., et al. (2019). Technological interventions for utilization of crop residues and weedy biomass for second generation bio-ethanol production. *Renewable Energy, 132*, 723–741.
Patel, N. K., et al. (2017). Engine performance, emission and fluorescence study of fuel from plant oils. *Journal of Mechanical Engineering Research and Developments, 40*(4), 673–691.
Pena, M. S. B., & Rollins, A. (2017). Environmental exposures and cardiovascular disease: a challenge for health and development in low-and middle-income countries. *Cardiology Clinics, 35*(1), 71–86.
Phillips, J. (2006). Different types of gasifiers and their integration with gas turbines. *The Gas Turbine Handbook* 1
Planning Commission. (2012). *Twelfth Five Year Plan (2012–2017). Economic Sectors* (Vol. II). Retrieved from http://14.139.60.153/bitstream/123456789/1456/1/12th%205%20year%20 plan_vol_2.pdf
Poe, N. E., et al. (2020). Compositional variability of food wastes and its effects on acetone-butanol-ethanol fermentation. *Waste Management, 107*, 150–158.
Primo, A., and García, H. (2013). New and future developments in catalysis: 425–449.
Prince, R. C., & Kheshgi, H. S. (2005). The photobiological production of hydrogen: potential efficiency and effectiveness as a renewable fuel. *Critical Reviews in Microbiology, 31*(1), 19–31.
Puig-Arnavat, M., Bruno, J. C., & Coronas, A. (2010). Review and analysis of biomass gasification models. *Renewable and Sustainable Energy Reviews, 14*(9), 2841–2851.
Raha, D., Mahanta, P., & Clarke, M. L. (2014). The implementation of decentralised biogas plants in Assam, NE India: The impact and effectiveness of the National Biogas and Manure Management Programme. *Energy Policy, 68*, 80–91.
Rahimi, M., et al. (2020). Carbon dioxide capture using an electrochemically driven proton concentration process. *Cell Reports Physical Science, 1*(4), 100033. https://doi.org/10.1016/j.xcrp.2020.100033

Rao, M., Narayana, et al. (2016). *Solid and hazardous waste management: science and engineering*. Butterworth-Heinemann.
Reay, D. S., et al. (2018). Methane and global environmental change. *Annual Review of Environment and Resources, 43*, 165–192.
Rivera, E. C., et al. (2017). Effect of temperature on sugarcane ethanol fermentation: Kinetic modeling and validation under very-high-gravity fermentation conditions. *Biochemical Engineering Journal, 119*, 42–51.
Roni, M. S., et al. (2017). Biomass co-firing technology with policies, challenges, and opportunities: A global review. *Renewable and Sustainable Energy Reviews, 78*, 1089–1101.
Roopnarain, A., & Adeleke, R. (2017). Current status, hurdles and future prospects of biogas digestion technology in Africa. *Renewable and Sustainable Energy Reviews, 67*, 1162–1179.
Rupf, G. V., et al. (2015). Barriers and opportunities of biogas dissemination in Sub-Saharan Africa and lessons learned from Rwanda, Tanzania, China, India, and Nepal. *Renewable and Sustainable Energy Reviews, 52*, 468–476.
Sakurai, H., et al. (2013). Photobiological hydrogen production: bioenergetics and challenges for its practical application. *Journal of Photochemistry and Photobiology C: Photochemistry Reviews, 17*, 1–25.
Samar, E. K. K., & Meena, E. G. L. (2016). The solid state biogas plant: A boon for water scarce areas. *Akshay Urja, 9*, 4.
Seepana, S., et al. (2018). Evaluation of feasibility of pelletized wood co-firing with high ash Indian coals. *Journal of the Energy Institute, 91*(6), 1126–1135.
Sharifian, R., et al. (2021). Electrochemical carbon dioxide capture to close the carbon cycle. *Energy & Environmental Science, 14*(2), 781–814. https://doi.org/10.1039/D0EE03382K
Sharma, A., and Arya, S. K. (2019). Photobiological production of biohydrogen: Recent advances and strategy. *Prospects of Renewable Bioprocessing in Future Energy Systems*. 89–116.
Shelef, M. (1995). Selective catalytic reduction of NOx with N-free reductants. *Chemical Reviews, 95*(1), 209–225.
Sivan, A. (2011). New perspectives in plastic biodegradation. *Current Opinion in Biotechnology, 22*(3), 422–426.
Speight, J. G. (2015). Waste gasification for synthetic liquid fuel production. In *Gasification for Synthetic Fuel Production* (pp. 277–301). Woodhead Publishing.
Speight, J. G. (2019). *Handbook of industrial hydrocarbon processes*. Gulf Professional Publishing.
Szwaja, S., et al. (2019). Theoretical and experimental analysis on co-gasification of sewage sludge with energetic crops. *Energies, 12*(9), 1750.
Tabatabaei, M., et al. (2010). Importance of the methanogenic archaea populations in anaerobic wastewater treatments. *Process Biochemistry, 45*(8), 1214–1225.
Taheri, M., et al. (2014). Environmental impact assessment of municipal solid waste disposal site in Tabriz, Iran using rapid impact assessment matrix. *Impact Assessment and Project Appraisal, 32*(2), 162–169.
Teter, S. A., Brandon Sutton, K., & Emme, B. (2014). *Enzymatic processes and enzyme development in biorefining. Advances in Biorefineries* (pp. 199–233). Woodhead Publishing.
Theegala, C. S., & Midgett, J. S. (2012). Hydrothermal liquefaction of separated dairy manure for production of bio-oils with simultaneous waste treatment. *Bioresource Technology, 107*, 456–463.
Tian, H., et al. (2012). Temporal trends and spatial variation characteristics of hazardous air pollutant emission inventory from municipal solid waste incineration in China. *Environmental Science & Technology, 46*(18), 10364–10371.
Tradler, S. B., et al. (2018). Hydrothermal carbonization as an all-inclusive process for food-waste conversion. *Bioresource Technology Reports, 2*, 77–83.
Tumuluru, J. S., et al. (2012). Formulation, pretreatment, and densification options to improve biomass specifications for co-firing high percentages with coal. *Industrial Biotechnology, 8*(3), 113–132.

Vedaraman, N., et al. (2011). Preparation of palm oil biodiesel and effect of various additives on NOx emission reduction in B20: An experimental study. *International Journal of Green Energy, 8*(3), 383–397.
Verma, M., et al. (2012). Biofuels production from biomass by thermochemical conversion technologies. *International Journal of Chemical Engineering, 2012*, 542426.
Verma, M., et al. (2017). Drying of biomass for utilising in co-firing with coal and its impact on environment–A review. *Renewable and Sustainable Energy Reviews, 71*, 732–741.
Virmond, E., et al. (2013). Valorization of agroindustrial solid residues and residues from biofuel production chains by thermochemical conversion: a review, citing Brazil as a case study. *Brazilian Journal of Chemical Engineering, 30*(2), 197–230.
Voskian, S., and Alan Hatton, T. (2018). "Faradaic electro-swing reactive adsorption for carbon capture at low concentrations." *14th Greenhouse Gas Control Technologies Conference Melbourne.*
Wang, W.-j., & You, X.-y. (2021). Benefits analysis of classification of municipal solid waste based on system dynamics. *Journal of Cleaner Production, 279*, 123686.
Wang, Z., et al. (2018). Study on the thermal liquefaction of Xilinguole lignite in solvent at high temperature. *Fuel Processing Technology, 176*, 167–173.
Williams, C. L., et al. (2016). Sources of biomass feedstock variability and the potential impact on biofuels production. *BioEnergy Research, 9*(1), 1–14.
Williams, C. L., Emerson, R. M., and Tumuluru, J. S. (2017). Biomass compositional analysis for conversion to renewable fuels and chemicals. *Biomass Volume Estimation and Valorization for Energy*: 251–270.
Wu, Y., et al. (2018). Bioenergy production and environmental impacts. *Geoscience Letters, 5*(1), 1–9.
Yang, J., & Yang, L. (2019). A review on hydrothermal co-liquefaction of biomass. *Applied Energy, 250*, 926–945.
Yousuf, A., et al. (2016). Financial sustainability of biogas technology: Barriers, opportunities, and solutions. *Energy Sources, Part B: Economics, Planning, and Policy, 11*(9), 841–848.
Yuan, X., et al. (2020). Environmental impact of the natural gas liquefaction process: An example from China. *Applied Sciences, 10*(5), 1701. https://doi.org/10.3390/app10051701
Zaman, C. Z., et al. (2017). *Pyrolysis: a sustainable way to generate energy from waste* (p. 1). *Pyrolysis; IntechOpen.*
Zhang, L., et al. (2013). Upgrading of bio-oil from biomass fast pyrolysis in China: A review. *Renewable and Sustainable Energy Reviews, 24*, 66–72.
Zhang, Y., et al. (2019). Gasification technologies and their energy potentials. In *Sustainable resource recovery and zero waste approaches* (pp. 193–206). Elsevier.
Zhou, H., et al. (2014). Classification and comparison of municipal solid waste based on thermochemical characteristics. *Journal of the Air & Waste Management Association, 64*(5), 597–616.
Ziemiński, K., & Frąc, M. (2012). Methane fermentation process as anaerobic digestion of biomass: Transformations, stages and microorganisms. *African Journal of Biotechnology, 11*(18), 4127–4139.

Pyrolysis of Waste Biomass Using Solar Energy for Clean Energy Production

Leena Kapoor, Jay Mant Jha, Dan Bahadur Pal, Dinesh Kumar Pandey, and Amit Kumar

Abstract Biomass is a sustainable source of energy that is widely available and has the potential to meet the daily energy demands. Solar thermal biomass pyrolysis is one of the emerging technologies for converting waste biomass into bioenergy. Production of bio-oil via pyrolysis has attracted many researchers due to its high energy density and direct utilization of oil in combustion engines. The composition of bio-oil depends mainly on the type of feedstock, reactor, and operating conditions used for carrying out pyrolysis. Conversion of biomass via pyrolysis to biofuels is an endothermic process. In conventional pyrolysis method, the heat energy input is in the form of electrical energy that is uneconomical and has a harmful effect on the environment. In this paper the production of pyrolysis products, namely, bio-oil, char, and gases, in a continuous solar thermal pyrolyzer is highlighted. Scant knowledge and scattered information for the utilization of solar energy for continuous pyrolysis process of waste biomass are reviewed and amalgamated to make proper utilization of continuous solar biomass pyrolysis for various applications. The main operating parameters for the continuous production of bio-oil using solar energy are identified. This paper addresses the current state of the continuous solar energy-based pyrolysis reactor.

L. Kapoor (✉)
Department of Chemical Engineering, UPES, Dehradun, India
e-mail: lkapoor@ddn.upes.ac.in

J. M. Jha
Department of Chemical Engineering, Maulana Azad National Institute of Technology, Bhopal, Madhya Pradesh, India

D. B. Pal
Department of Chemical Engineering, Birla Institute of Technology, Mesra, Ranchi, Jharkhand, India

D. K. Pandey
Chemical Engineering School of Technology, GSFC University, Vadodara, India

A. Kumar
Department of Chemical Engineering, Institute of Technology, Nirma University, Ahmedabad, India

D. B. Pal, J. M. Jha (eds.), *Sustainable and Clean Energy Production Technologies*, Clean Energy Production Technologies,
https://doi.org/10.1007/978-981-16-9135-5_6

Keywords Solar pyrolysis · Auger reactor · Operating variables · Feedstock · Challenges

1 Introduction

Presently nearly, 80% of world energy needs is met by fossil fuels. With increasing population, the global energy needs will be increased to 40% by year 2050 (Mohanty et al., 2011). To cater the increasing energy needs despite of the fact of lowering CO_2 emissions, alternate energy sources like solar and biomass energy need more attention. Presently, the population of India is around 1.4 million, and based on the recent policies, the demand for energy in India would increase to double in 2040 with increasing electricity demands nearly to triples due to an increase in appliance ownership and cooling needs (Kumar et al., 2015). India would have to add a huge amount of power generation plants in order to meet the demands of nearly 1 billion air-conditioning units that the country will need by year 2050 (Yadav & Banerjee, 2016). Solar thermal biomass pyrolysis is an emerging technology for converting waste into energy. Pyrolysis is a conversion of biomass thermally in the complete absence of oxygen to produce liquid (bio-oil), solid (char), and uncondensed gases. Production of bio-oil via pyrolysis has attracted many researchers as the bio-oil produced can be directly used in combustion engine. The composition of bio-oil depends mainly on feedstock type, reactor, and operating conditions used for carrying out pyrolysis. Various biomass feedstocks such as date seed, waste wood pellets, beech wood, jatropha seeds, and *Pinus pinaster* (maritime pine) have been used by various researchers. From the literature it is found that through pyrolysis process around 50% of the biomass can be converted to oil (Zeng et al., 2017). For example, Solar et al. (2016) performed pyrolysis experiments with *Pinus pinaster* biomass at 900 °C temperature using solar energy source that resulted in 20–30 wt.% biochar, 6–22wt.% liquid, and greater than 47wt.% gases. Similarly Zeng et al. (2015) used beech wood pellets in a vertical axis solar furnace at a temperature of 900 °C that resulted in 13.3 wt.% biochar and 57.7wt.% bio-oil. Joardder et al. (2014) used date seed biomass in a stainless steel externally heated fixed-bed reactor system with parabolic solar reflector at a temperature of 500 °C that resulted in 50 wt.% of bio-oil. Researchers also worked with feedstocks such as orange peels (Morales et al., 2014), mahua seeds, rice husk (Maximino, 2013), and karanja seeds (Rony et al., 2019).

According to literature the yield of pyrolysis products produced is dependent on temperature, residence time, catalyst used, heating rate, and nitrogen flowrate as the operating parameters (Kapoor et al., 2020). The literature available on pyrolysis is concentrated much on fast pyrolysis of biomass (Rahman et al., 2021). Bhattacharya et al. (2009) reported high carbon and hydrogen-based bio-oil produced at a temperature of 773 K with wood/plastic blend biomass feedstock. Further many researchers have attained good yield of bio-oil in a screw reactor operated in continuous mode (Brown & Brown, 2012). Thangalazhy-Gopakumar et al. (2012) performed pyrolysis at low temperatures of 400–500 °C of pinewood chip biomass.

Veses et al. (2015) carried out catalytic pyrolysis using Mg- and Ca-based catalyst at 500 °C temperature to produce improved bio-oil quality in a continuous pyrolyzer. Puy et al. (2011) performed pyrolysis in screw-based pyrolyzer at a temperature of 800 °C using pinewood chips in order to find the influence of various operating parameters on product yield. Ferreira et al. (2015) have found from their studied temperature to be the main influencing parameter for driving the pyrolysis reactions. Several researchers (Kan et al., 2016) have conducted experiments in the operating range of 600–2000 °C and concluded that high-temperature pyrolysis favors formation of gaseous products, whereas low-temperature pyrolysis favors liquid products (Brassard & Raghavan, 2017). At high temperature, secondary reaction occurs resulting in the formation of hydrogen and carbon monoxide, whereas at low temperature, the hydrogen and carbon monoxide content is lower. This may be due to the lower tar decomposition reaction (Brassard & Raghavan, 2017).

Conversion of biomass via pyrolysis to biofuels is an endothermic process. In traditional method of pyrolysis, the heat input is supplied via electrical source which is uneconomical and has a harmful effect on the environment (Bridgwater, 1999). Solar biomass pyrolysis is a sustainable process, which does not affect the environment. This process utilizes the low energy density feedstock, i.e., biomass and solar energy, into a high energy density fuel. The solar biomass pyrolysis process eradicates the need of storage of solar energy, which in turn can now be stored as biofuel or tar. Heating via using solar energy provides an environmentally friendly heating process as compared to traditional heating methods (Ndukwu et al., 2020). For concentrating solar energy and for proper utilization, additional systems like concentrators, etc. are required. The type of concentrators to be used depends on the temperature of pyrolysis (Morales et al., 2014). The reactor orientation for the solar biomass pyrolysis can be vertical or horizontal depending on the type of pyrolysis. Very little research work has been done for continuous solar thermal pyrolysis. Solar et al. (2016) did pioneering work in solar thermal auger screw reactor. They have conducted experiments over three different temperature profiles referred to as 573–1023 K, 573–1173 K, and 1173–1173 K to know the effect of temperature. Experiments were also performed at two different residence times of 32 min and 64 min in order to determine the effect of residence time on pyrolysis process. They concluded that with high temperature and residence time the solid product yield is more with low volatile content (6–10 wt.% dry basis) and high fixed carbon (87 wt. % dry basis). Also high residence time and temperature result in the formation of high hydrogen and carbon monoxide gas and lower CO_2 content. Similarly, Zeng et al. (2015) used beech wood biomass for pyrolysis in a vertical axis solar furnace with parabolic mirror and heliostat. Chintala et al. (2017) used jatropha seed biomass in a batch-scale solar thermal pyrolyzer. Similarly, many other lab-scale equipment have been used for carrying out solar thermal pyrolyzer using parabolic concentrator for producing gaseous products from waste biomass (Morales et al., 2014; Zeng et al., 2015; Li et al., 2016; Perkins, 2018; Rony et al., 2019).

Limited literature studies on continuous solar pyrolysis of waste biomass are available. Recently, Ndukwu et al. (2020) reviewed the solar biomass pyrolysis system configuration and reactor orientation highlighting the available optical concentrating device and reactor orientation comparing existing configurations. For efficiently utilization of solar energy, the factors like feedstock type, type of concentrators, type of reactor and its orientation, and type of concentrators must be considered. The reactor configuration and orientation significantly influence the production of pyrolysis products.

Zeng et al. (2017) studied the recent developments in solar pyrolysis highlighting the mechanisms and experimental demonstrations. Chintala et al. (2017) reviewed the manufacture, upgradation, and application of solar thermal biomass pyrolysis. Xie et al. (2015) reviewed the biochar production and utilization methods. These review papers presented the production and characterization of pyrolysis products. However, none of the review study is available on the process of continuous solar thermal screw reactor pyrolysis. Moreover, the information on the operational parameters of continuous reactor, product yield, and utilization of products obtained is scanty. Hence, in this review the continuous solar-assisted biomass pyrolysis in screw reactor considering various biomass feedstocks is done. The scattered information on this area has been brought together in a single platform to have proper insight in the field of continuous solar pyrolysis technology. In addition, the feasibility issues are also presented in the current study.

2 Continuous Pyrolysis Reactors

2.1 Recent Developments in Solar Energy-Assisted Auger Reactor

The dependence of India on foreign oil to meet its energy demands has increased to nearly about 84% from 82.9% in 2017–2018. With a growing concern on the use of fossil fuel all across the world, it is high time especially for a growing economy nation like India to start contemplating and making efforts in using renewable energy sources.

Relatively few solar pyrolysis systems have been tested experimentally. Miranda et al. (2009) proposed three parts of solar pyrolysis system, namely, the reactor, the concentrators, and the condensers. Morales et al. (2014) proposed a thermo-solar system for performing solar biomass pyrolysis. The solar concentrators focuses the solar radiations on to the solar collector, which is nothing but the reactor (Morales et al., 2014). For mounting solar concentrators and solar collectors, supporting structures are used. Six types of concentrators are commonly being used by various researchers for solar pyrolysis, which are tabulated in Table 1. The advantages, characteristics, and disadvantages of these concentrators are tabularized in Table 1. Out of the listed solar concentrators, researchers mostly used the parabolic trough concentrator, which is an ideal concentrator for attaining the required temperature for biomass pyrolysis.

Table 1 Solar concentrators' advantages/characteristics and disadvantages

Concentrator	Advantages/characteristics	Disadvantages	References
Parabolic dish	Two-axis solar tracking system, parabolic dish-shaped mirrors, operating temperature is 1800K, concentration ratio in the range of 1000 and 5000K, used in photovoltaic application	Difficult to build, bulky, and high-cost device	Joardder et al. (2017)
Heliostat	Uses direct solar radiation. Uses two-axis tracking system, operating temperature is 1600K, concentration ratios of 300–1500. Easy to build	Very expensive system	Joardder et al. (2017)
Parabolic trough	Direct solar radiation is used. Uses one-axis tracking system, operating temperature in the range of 500–700K, concentration ratio of 30–100. Ideal for cylindrical surface heating	All radiation falls parallel to the axis, ideal for low-temperature processes	Fernández-García et al. (2010)
Compound parabolic	Used for focusing diffuse and direct solar radiation. Uses one-axis tracking system, simple device, ideal concentrator that can collect and focuses on larger sunlight area onto small area with least losses	Used only for low-temperature processes	Patel et al. (2018)
Linear Fresnel	Uses direct solar radiation. Uses one-axis tracking, low cost, operating temperature 150–550 °C, concentration ratio 30–80. Easy to build	Used only for low-temperature process	Sun (2021)
Fresnel lens	Uses direct solar radiation. Uses two-axis tracking system	Fragile and expensive	Joardder et al. (2017)

Concentrated solar energy has many advantages in biomass pyrolysis such as the following:

(1) Pyrolysis reactors can be rapidly heated to high temperatures due high heat flux.
(2) Lesser secondary reactions will occur due to small focal area.
(3) Solar energy is a renewable energy source which lowers the cost of heating the reactor.
(4) No harmful emissions of gases to the environment due to no fossil fuel burning.

Several researchers, i.e., Joardder et al. (2014), Zeng et al. (2015), and Chintala et al. (2017), used parabolic concentrators to achieve the desired temperature for biomass pyrolysis. Joardder et al. (2017) proposed a parabolic dish reflector and presented a unique solar pyrolysis reactor with sliding concentrators and rotating reactor. They worked on the solar and biomass resources available in Bangladesh region. Morales et al. (2014) used orange peel biomass in a parabolic trough solar concentrator for performing pyrolysis experiments. They tested two different particle sizes of biomass in continuous solar pyrolyzer. There findings show that when particle size is decreased the final conversion rate increases and vice versa. This is due to the fact that as the particle size decreases the surface area increases. Zeng et al. (2015) used shutter system in a two-stage heliostat concentrator for regulating temperature and heating rate of a pyrolysis reactor. Zeaiter et al. (2015) used two-axis Fresnel lens tracking system for the solar pyrolysis of waste rubber at temperature of 550 °C. Chintala et al. (2017) used Scheffler dish concentrator for focusing the solar radiation on pyrolysis reactor.

2.1.1 Potential Feedstock for Solar Pyrolysis

Only, few authors have studied solar pyrolysis of carbonaceous materials. However many researchers have used coal or cellulose as the raw materials (Zeng et al., 2017). The various experiments using different feedstocks in solar reactor along with temperature and product yield are listed in Table 2. Boutin et al. (2002) carried out experiments using xenon lamp with cellulose biomass. Morales et al. used an agro-industrial waste biomass for solar pyrolysis. Chintala et al. (2017) performed solar pyrolysis in a fixed-bed circular reactor using jatropha seeds as feedstock. Zeaiter et al. (2015) and Zeng et al. (2015) used beech wood biomass to perform solar pyrolysis in tubular fixed-bed type of reactor and Pyrex reactor. Li et al. (2016) used agricultural and forestry biomass residue in a transparent Pyrex balloon reactor. Joardder et al. (2014) performed solar pyrolysis in stainless steel reactor which is

Table 2 Experiments carried out with continuous solar pyrolysis reactor: temperature and product yield

Biomass	Solar pyrolysis reactor	Temperature (°C)	Liquid (wt.%)	Solid (wt. %)	Gas (wt. %)	References
Rice husk	Horizontal cylindrical silica glass tube with dish concentrator	800	38.7	35.8	25.48	Weldekidan et al. (2018)
Beech wood	Vertical furnace, parabolic mirror concentrator	1200	8.92	37.5	53.60	Zeng et al. (2015)
Orange peel	Horizontal borosilicate glass tube, parabolic trough concentrator	900	77.6	21	1.4	Morales et al. (2014)
Date palm	Vertical steel reactor, double parabolic dish	500	50	–	–	Joardder et al. (2014)
Forestry and agricultural biomass residues	Pyrex reactor	1200	–	–	63.5	Li et al. (2016)

heated externally by using date seed as feedstock. Weldekidan et al. (2018) pyrolyzed rice straw biomass using dish concentrator in a cylindrical silica glass tube reactor. Grassmann et al. (2015) used wheat straw biomass in a rotary stainless cylindrical kiln horizontal retort-type reactor. Ayala-Cortés et al. (2019) used tomato waste and agave leaves as feedstock in a spherical-shaped reactor with vertical orientation using parabolic trough-type concentrator. Thus from the above literature studies, it can be concluded that the quality of bio-oil produced via solar pyrolysis is greatly dependent on the feedstock type and composition. The bio-oil produced via pyrolysis process consists of nearly about 300 compounds that are mixtures of phenols, esters, and aromatic derivatives (Chintala et al., 2017). Bio-oil consists of mainly carbon, nitrogen, oxygen, sulfur, and hydrogen (Nanda et al., 2014). The biomass feedstock composition plays an important role in the quality and composition of bio-oil produced. It is found that the percentage of carbon in bio-oil is usually less as compared to diesel fuels. Depending upon the pyrolysis technology, pyrolysis temperature, and biomass composition, the percentage of carbon varies from 40% to 80% in bio-oil. Hydrogen percentage in bio-oil generally varies from 4% to 12% based on the pyrolysis conditions. Bio-oil generally has high oxygen content (50%) as compared to biodiesel which results in lowering of high heating value (HHV) of bio-oil (Jahirul et al., 2012). Many studies on upgradation of bio-oil were conducted, but there is scant knowledge on improvement of the entire pyrolysis products in the reactor itself.

Brassard et al. (2017) reported wood to be the mostly considered biomass for continuous pyrolysis. Nearly 19 studies among the 29 reported in the literature review for pyrolysis were done on forest biomass. About nine studies were done on agricultural biomasses such as crop residues. Two of the studies were done using manure as biomass (Brassard & Raghavan, 2017). The type of biomass used as a feedstock greatly affects the yield and properties of pyrolysis products (McKendry, 2002). It is observed that biomass having high lignin content offers higher char formation (Tsai et al., 2006). Demirbaş and Arin (2002) reported higher biochar yield with olive husk having high lignin (50.6%) in comparison with oak wood and wheat straw. High ash content in biomass feedstock also favors biochar yield and less bio-oil yield (Demirbaş & Arin, 2002).

2.1.2 Feasibility Assessment

Reactor is the heart of the pyrolysis process. Several efforts were made for using solar energy for pyrolysis, but no industrial implementation of the process is yet successful. The main consideration is in the design of auger solar pyrolysis reactor. Appropriate reactor design is essential for proper heat transfer during the pyrolysis process. Sole solar heating can be highly effective for pyrolysis (Perkins, 2018). This method of heating via solar energy for pyrolysis is cost-effective as it minimizes heating cost by using renewable energy source for heating reactor. In this method, the reactor along with the feed material is heated to an elevated temperature by

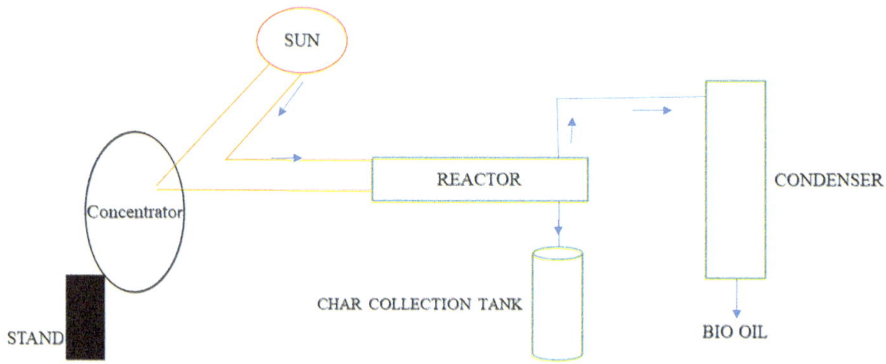

Fig. 1 Theoretical design of solar pyrolysis in a continuous based reactor

concentrating solar energy to the reactor. A theoretical design of solar pyrolysis is represented in Fig. 1.

Solar pyrolysis of waste biomass has several advantages over the traditional pyrolysis methods, but attaining uniform temperature and heating across the reactor is still a challenge (Fassinou et al., 2009). The stationary fixed-bed solar pyrolysis reactors result in single heating point on reactor surface. The auger reactor design proposed in Fig. 1 will resolve this problem. With rotating screw inside the reactor, the uniform heat transfer will be there with even decomposition of feed materials. The concept of continuous solar pyrolysis is in developing stage so there is no sufficient literature available for comparing continuous solar pyrolysis with fixed-bed pyrolysis. The octagonal mirror type reactor using an electric motor in order to keep the focal point stationary throughout the day is being used by many researchers. Caputo and Mašek (2021) proposed a novel approach based on a Cassegrain optics configuration. They used Cassegrain optics parabolic dish system by integrating it into the solar pyrolysis reactor with a manual tracking system and optically optimized with a Monte Carlo ray tracing methodology. The proposed system proves to be cost-effective with the potential to obtain optical efficiencies of up to 72% under 1.5 degree angle tracking error. The products obtained are biochar and electricity. Chintala et al. (2017) demonstrated a novel approach of PLC (programmable logic controller) solar pyrolysis system for concentrating solar radiation on fixed-bed reactor installed with automatic positioning system. Maintaining a constant temperature in solar pyrolysis reactor is a thought-provoking task. On the other hand, not much literature studies are available in the field of solar pyrolysis. Most of the studies related to solar pyrolysis are restricted to lab scale only where the temperature of the operating system can be maintained easily with minimal heat losses. In case of continuous screw reactor, the heat loss from surface, choking of reactors, and non-uniform het flux affects the reactor performance and thus results in low yield of pyrolysis products (Kapoor et al., 2020) (Kapoor, 2019). The reactor should be designed in such a way to overcome these problems.

2.1.3 Operating Parameters

Continuous screw pyrolysis reactors are of two types, namely, twin- and single-screw reactors. Based on the desired products needed, these reactors are selected. Brassad et al. (2017) in the review reported that twin-screw pyrolyzer can produce more bio-oil in comparison to a single-screw reactor. This may be due to the better mixing and more efficient heat transfer resulting into the complete carbonization of biomass with in the reactor itself. Brassad et al. (2017) concluded that about 54.5 wt.% of bio-oil yield was obtained with twin-screw reactor on pyrolysis of forest residue biomass, while only 37 wt.% of bio-oil yield was reported using single-screw reactor. Similar results were obtained with other biomasses also. On contrary, biochar yield is generally lower in twin-screw reactor. From the literature, it is reported that with an increased capacity of twin-screw reactor the bio-oil yield decreases due to lower heat transfer in larger tubes (Kapoor et al., 2020).

Temperature is one of the major influencing parameters for the product yield (Solar et al., 2016). With increase in the reactor temperature, the biomass conversion increases. As reported in the literature for continuous auger reactor, the bio-oil yield increases up to a certain temperature and then decreases, whereas biochar yield decreases with an increase in temperature. Lédé (2010) highlighted a difference in temperature between the reactor wall and biomass bed. It is found that the temperature and heating rate of the particle differ from that of the reactor. Bosong et al. (2014) also found temperature difference between the reactor wall and biomass particle. They studied the time vs temperature profile for pinewood biomass and found that when the temperature of the reactor is 600 °C the temperature of biomass inside the reactor was approximately 100 °C low. Thus, it is difficult to know the exact temperature of biomass particles inside the screw reactor. Brown (2009) illustrated that the vapor existing the auger reactor can be useful for analyzing the actual reaction temperature. Zeng et al. (2017) from the experimental studies concluded that with an increasing temperature the gas yield increases (at 1600 °C gas yield was 63 wt.%), while bio-oil and char yield decreases (28.8 wt.% and 8.4 wt.%). Chintala et al. (2017) reviewed that heating rates and heat and energy transfer mechanism significantly affect the biomass temperature inside the reactor. Chintala et al. (2017) reported that with increasing heating rate the biochar yield decreases, whereas the liquid and gaseous yield increases. Zeng et al. (2017) from the experimental studies concluded that product yield varies considerably until 50 °C/min, beyond which the product yield varies at a moderate rate. In addition to the heating rate, the size of biomass particle affects the product yield and characteristics (Wu, 2015). Brassad et al. (2017) reviewed that in an auger reactor the smaller particle size of feedstock produces more bio-oil due to the shorter residence time of vapors inside the solid and hence minimizes the possibility of secondary reactions. Various other researchers (Demirbaş & Arin, 2002; Sirijanusorn et al., 2013; Pattiya & Suttibak, 2012) also reported that with the increase in particle size the liquid yield decreases, while char formation increases. In an auger reactor, carrier gas flowrate significantly affects the residence time of vapor inside the reactor. Brown (2009) reported that with increasing carrier gas flowrate the biochar yield decreases due to decreasing vapor residence time in the reactor, which in turn favors bio-oil formation. Antal et al. (1983) reported that with low carrier gas flowrate in an auger reactor the reaction of volatile matter with solid carbonaceous residue increases and thus produces more char. Several other researchers (Sirijanusorn et al., 2013; Fassinou et al., 2009; Liaw et al., 2013) also concluded that with increased carrier gas flowrate from 4 to 10 L/min the bio-oil yield steadily decreased. Solid residence time inside the reactor is also an important operating parameter for pyrolysis reactions (Babu & Chaurasia, 2004). Various experiments on auger reactor have been performed with a solid residence time of 11 s to 120 min, but most of the studies used a solid residence time of 8 min (Brassard & Raghavan, 2017). Funke et al. (2017) reported that the solid residence time for an auger reactor could be determined by the rotational speed of the screw, length of the tube, and the screw pitch. Di Blasi (2002) reported increased residence time of biomass in an auger reactor increases the biochar yield keeping constant temperature. This may be due to the formation of secondary reactions. Researchers also found that residence time of biomass in the reactor affects the physical and chemical characteristics of the pyrolysis product formed (Solar et al., 2016). Brown and Brown (2012) used a heat carrier material (sand) in the auger reactor for carrying out fast pyrolysis reactions. Introduction of a heat carrier material helps in improving the heat transfer rate with increased liquid yield but only if the contact time of heat carrier material and biomass is short. Solar et al. (2016) demonstrated via experiments by comparing product yield at two different residence times of 64 min vs 32 min in an auger reactor and concluded that with increased residence time the char yield decreases, while bio-oil yield increases.

3 Upgradation of Bio-oil

A considerable amount of work has been done on biomass pyrolysis. Many different biomass feedstocks were chosen for pyrolysis, e.g., bark, wood, nuts, forestry and agricultural residues, seeds and nuts, algae, grass, lignin, and cellulose. All these feedstocks were tested under various pyrolysis conditions and in different pyrolysis reactors. A major portion of research has been done using olive bagasse, wheat straw (agro-biomass), and pinewood (woody biomass) biomasses as feedstocks (Dickerson & Soria, 2013). The intermediate pyrolysis of rapeseed as reported in the literature (Zabeti et al., 2012) was studied in fixed-bed reactor at a temperature of 550 °C with a heating rate 300 °C/min using 0.6–0.85 mm particle size under nitrogen flowrate of 100 cm^3/min. It resulted in about 65% of bio-oil yield. Neem seed pyrolysis was studied in semi-batch reactors. It resulted in 38% yield of bio-oil operated at a temperature range of 400–500 °C with a heating rate of 20 °C/min (Onay & Kockar, 2003). The intermediate pyrolysis of groundnut as a feedstock was studied in a fluidized-bed reactor with a temperature range of 400–475 °C. It resulted in 63.48 wt.% of bio-oil (Ateş et al., 2019). From all of these studies, the four basic parameters of the resulting oil were identified, i.e., viscosity, pH, density, and low heating value (LHV) (24.56 cSt, 4.20, 1.2 kg/m^3, and 31.07 MJ/kg (Demirbaş & Arin, 2002). The bio-oil thus produced from waste biomass consists of various impurities such as high water and ash content, high viscosity, high solid content, high corrosiveness, chemical instability, and low heating value. The presence of these impurities restricts the use of bio-oil as a fuel in combustion engines. The unprocessed bio-oil also consists of oxygenates in the form of acids, esters, alcohols, ketones, and aldehydes (Gollakota et al., 2016). The presence of these compounds results in acidic character in bio-oil with pH of 2–3. Unprocessed bio-oil also consists of various organic compounds such as phenols, sugars, lignin, cellulose, furans, hemicellulose, hydrocarbons, and water (Kan et al., 2016). The presence of high moisture and oxygen in unprocessed bio-oil lowers its heating value. Hence, bio-oil needs to be upgraded.

The common mechanism used by various researchers for bio-oil upgrading is zeolite upgradation, steam reforming, catalytic cracking, hydrodeoxygenation (HDO), supercritical extraction, esterification, and emulsification (Kan et al., 2016). Among various techniques, the catalytic hydrodeoxygenation is a convincing approach in order to get petroleum-like fuels. In this technique, the oxygen present in bio-oil is removed through water formation (Cheng et al., 2017). Catalytic hydrodeoxygenation process for upgrading bio-oil has proved to be a promising method for improving the properties of bio-oil (Sharma et al., 2015). Through this process, a high-quality bio-oil with increased carbon content can be obtained.

The catalysts used for catalytic hydrodeoxygenation process are transition metal sulfides, carbides, nitrides, oxynitrides, metal oxides, phosphides, precious metals, and noble metals. Ni-Mo and Co-Mo supported on Al_2O_3 are the conventional metal sulfides used for upgrading of bio-oil(Gollakota et al., 2016). Bharath et al. (2020) developed new bimetallic Ni_3Fe catalysts for hydrodeoxygenation of bio-oil using date seed as biomass (Bharath et al., 2020). This iron-supported nicked-based bimetallic catalyst has shown improved catalytic activity and high magnetic properties. Due to high magnetic property, the Ni_3Fe catalyst can be regenerated effectively. Ni_3Fe catalyst has an octagon morphology with high magnetization and diameter of 12 nm (Bharath et al., 2020). The bimetallic Ni_3Fe was found to be an effective catalyst for the upgradation of bio-oil using HDO treatment. Using this catalyst the HHV of bio-oil was increased from 29.06 to 36.78 MJkg^{-1}, while viscosity, oxygen content, and density abridged from 18.5 to 4.7 cSt , 28.5 to 7.3%, and 1015 to 681 kgm-3, respectively. The hydrocarbon content in bio-oil increased from 12.021 to 23.787 wt.% (Guo et al., 2020). In addition, the HDO of major components of bio-oil such as acetic acid, 5-hydroxoymethylfurfural, D-glucose, m-cresol, and 2-furfuraldehyde resulted into the formation of ethane; ethyl acetate; 2,5-dimethylfuran; D-sorbitol; D-mannitol; methylcyclohexane, and furfural alcohol (Attia et al., 2020). The Ni_3 Fe catalyst showed better stability and reusability along structural integrity; it could be used for five cycles of bio-oil upgradation (Bharath et al., 2020).

Many researchers have used a wide range of acidic and basic catalysts for improving the yield of pyrolysis products (Kan et al., 2016). Some common catalysts used for the study were zeolites, silica, NaOH, Zno, and MgO (Din et al., 2019).

Some researchers have also used magnetic nanoparticles for pyrolysis. The magnetic nanoparticles have the advantage that they can be fabricated rapidly using single-step biosynthetic route. Cheng et al. (2017) reported a maximum yield of 55.9% bio-oil was achieved at 500 °C in the presence of 2g of nanocatalyst. It is reported that maximum conversion of pyrolytic product was achieved at 0.46–0.56 mm particle size (Saravana Sathiya Prabhahar et al., 2020). Without nanocatalyst, the yield of bio-oil was 45.4%, and with nanocatalyst, the bio-oil yield was 54.3% (Bhol et al., 2020). Biomass is a complex mixture of hundreds of compounds and hence differs in their characteristics. Therefore, it is not easy to say which particular catalyst will behave for each biomass type.

4 Challenges for Continuous Auger Reactor Solar Pyrolysis

Solar pyrolysis using auger reactor is a promising technology for continuous mode of pyrolysis. Continuous pyrolysis process suffers from both technical and economic challenges. Technical challenges for continuous solar pyrolysis include design of right solar collector, weather conditions, uniform heat flux in the reactor, and heat loss consideration. The economic challenges include the cost of solar concentrators, heliostats, and solar dishes. The main issues of continuous auger reactor solar pyrolysis are described below:

4.1 Choice of Solar Collector

The choice of solar concentrator is one of the challenging processes. The solar energy is cheap and renewable on the earth, but the application of high-temperature heat application is a great challenge. In a continuous solar pyrolysis, the biomass feedstock is heated generally by convection and radiation from the wall of the reactor by an intermediate heat carrier (Brown & Brown, 2012). Solar concentrator performance depends on the solar radiation intensity, position of the sun, the incident angle of solar radiation, and the reactor to be heated. Rahman et al. in their recent review have highlighted the application of various solar concentrators in various research/projects. They concluded that parabolic dish or trough concentrator system is mostly used for solar pyrolysis. According to the review, mostly heliostat field collector is used for large projects (Rahman et al., 2021). For short-term development technology, the use of parabolic trough concentrators would be the best choice (Morales et al., 2014). It is also stated that heliostat field concentrators with a tower has low operating cost and offers high capacity (Kalogirou, 2014). Rahman et al. (2021) reviewed that dishes offer high conversion efficiencies in comparison to trough and tower concentrators. Therefore, the choice of solar concentrators is dependent on the design of the reactor, operating parameters, and the feedstock type.

4.2 Weather Conditions

Uneven heating or interrupted heating during the daytime due to weather conditions is a challenge for continuous pyrolysis process. The uneven heating results in incomplete pyrolysis resulting in lowering yield of pyrolysis products. In addition, the high wind speed results in more heat losses from the reactor surface. Rahman et al. (2021) stated that for this intermittence nature of solar energy there is a need to store energy in the absence of radiation. Therefore, for such cases heat transfer by fluid by using indirect heating will be helpful.

4.3 Uniform Heat Flux in the Reactor

Uniform heat distribution inside the reactor for biomass feedstock is a major challenge. The high-temperature solar beam concentrated into the reactor is generally focused on the feedstock, which is placed in the middle of the reactor creating a non-uniform temperature gradient in the reactor (Yadav & Banerjee, 2016). In order to overcome this, the reactor material selected should have high thermal conductivity, and the feed particle size should be small enough for uniform heat transfer (Guedes et al., 2018).

4.4 Heat Losses

Heat losses by convention or radiation were observed during direct heating of reactor (Brassard & Raghavan, 2017). With the increase in temperature, the heat losses from the reactor increase. A well-insulating techniques need to be developed to reduce heat losses from the reactor.

5 Discussions

In this paper, continuous solar thermal pyrolysis with the combination of solar concentrators is reviewed. The previous studies on solar pyrolysis were done using fixed-bed solar reactor. According to the literature studies, the following conclusions were made:

- Continuous solar pyrolysis could be successfully used for the continuous production of biofuels. For continuous removal of char from a high-temperature reactor, the concept of twin-screw reactor is crucial. As hot char inside the reactor is highly catalytically active and may result in secondary cracking and contaminate the pyrolysis vapors, its constant removal from the reactor is a must.

- In the auger reactor, mostly studies on forest and agricultural biomasses were done with particle size of 5 mm for uniform heat transfer inside the reactor.
- The optimum temperature for pyrolysis ranged from 330 to 900 °C in an auger reactor for total conversion of biomass. The residence time of solids in the reactor in most of the research studies has been reported to be 8 min with maximum of 120 min and minimum of 11 s, respectively. In most studies, nitrogen has been used as an inert gas/carrier gas at a flowrate between 0.19 and 20 L/min.
- In continuous solar pyrolysis, the operating parameters, biomass feedstock type, solid residence time, temperature, carrier gas flowrate, and particle size, are the most influencing parameters on the product yield and properties.

6 Conclusion

The continuous feeding, removal, and collecting of products are always the critical tasks in solar fixed-bed pyrolysis, which is easier in continuous solar pyrolysis. Although the continuous solar pyrolysis in an auger reactor is superior to solar thermal fixed bed pyrolysis but still due to the lack of experimental studies at lab scale constrains its implementation to an industrial level. Variation in solar flux with respect to time throughout the season, reactor design for effective thermochemical conversion of biomass feedstock, heat losses from the reactor surface, and high cost of solar concentrators are the major challenges which need to be overcome for commercial-scale applications of continuous solar pyrolysis in an auger reactor. Bio-oil produced from waste biomass consists of various impurities. Catalytic hydrodeoxygenation process for upgrading bio-oil has proved to be a promising method for improving the properties of bio-oil. Through this process, a high-quality bio-oil with increased carbon content can be obtained.

Declaration of Competing Interest The authors declare that it has no known competing financial interests or personal relationships that could have appeared to influence the work reported in this paper.

References

Antal, M. J., et al. (1983). Design and operation of a solar fired biomass flash pyrolysis reactor. *Solar Energy, 30*(4), 299–312. https://doi.org/10.1016/0038-092X(83)90185-8

Ateş, F., et al. (2019). Comparison of non-catalytic and catalytic fast pyrolysis of pomegranate and grape marcs under vacuum and inert atmospheres. *Fuel, 255*, 115788. https://doi.org/10.1016/j.fuel.2019.115788

Attia, M., Farag, S., & Chaouki, J. (2020). Upgrading of oils from biomass and waste: Catalytic hydrodeoxygenation. *Catalysts, 10*(12), 1–30. https://doi.org/10.3390/catal10121381

Ayala-Cortés, A., et al. (2019). Solar pyrolysis of agave and tomato pruning wastes: Insights of the effect of pyrolysis operation parameters on the physicochemical properties of biochar. *AIP Conference Proceedings, 2126*. https://doi.org/10.1063/1.5117681

Babu, B. V., & Chaurasia, A. S. (2004). Pyrolysis of biomass: Improved models for simultaneous kinetics and transport of heat, mass and momentum. *Energy Conversion and Management, 45*(9–10), 1297–1327. https://doi.org/10.1016/j.enconman.2003.09.013

Bharath, G., et al. (2020). Catalytic hydrodeoxygenation of biomass-derived pyrolysis oil over alloyed bimetallic Ni3Fe nanocatalyst for high-grade biofuel production. *Energy Conversion and Management, 213*, 112859. https://doi.org/10.1016/j.enconman.2020.112859

Bhattacharya, P., et al. (2009). Wood/plastic copyrolysis in an auger reactor: Chemical and physical analysis of the products. *Fuel, 88*(7), 1251–1260. https://doi.org/10.1016/j.fuel.2009.01.009

Bhol, P., et al. (2020). Modern chemical routes for the controlled synthesis of anisotropic bimetallic nanostructures and their application in catalysis. *Frontiers in Chemistry, 8*. https://doi.org/10.3389/fchem.2020.00357

Bosong, L., et al. (2014). Pyrolysis and catalytic upgrading of pine wood in a combination of auger reactor and fixed bed. *Fuel, 129*, 61–67.

Boutin, O., Ferrer, M., & Lédé, J. (2002). Flash pyrolysis of cellulose pellets submitted to a concentrated radiation: Experiments and modelling. *Chemical Engineering Science, 57*(1), 15–25. https://doi.org/10.1016/S0009-2509(01)00360-8

Brassard, P., & Raghavan, V. (2017). ScienceDirect Pyrolysis in auger reactors for biochar and bio-oil production : A review, *1*. https://doi.org/10.1016/j.biosystemseng.2017.06.020

Brassard, P., Tymko, M. M., & Ainslie, P. N. (2017). Sympathetic control of the brain circulation: Appreciating the complexities to better understand the controversy. *Autonomic Neuroscience, 207*, 37–47.

Bridgwater, A. V. (1999). Principles and practice of biomass fast pyrolysis processes for liquids. *Journal of Analytical and Applied Pyrolysis, 51*(1), 3–22. https://doi.org/10.1016/S0165-2370(99)00005-4

Brown, J. N. (2009). Development of a lab-scale auger reactor for biomass fast pyrolysis and process optimization using response surface methodology. Dissertation, Iowa State University.

Brown, J. N., & Brown, R. C. (2012). Process optimization of an auger pyrolyzer with heat carrier using response surface methodology. *Bioresource Technology, 103*(1), 405–414. https://doi.org/10.1016/j.biortech.2011.09.117

Caputo, C., & Mašek, O. (2021). Spear (Solar pyrolysis energy access reactor): Theoretical design and evaluation of a small-scale low-cost pyrolysis unit for implementation in rural communities. *Energies, 14*, 8. https://doi.org/10.3390/en14082189

Cheng, S., et al. (2017). Upgrading pyrolysis bio-oil through hydrodeoxygenation (HDO) using non-sulfided Fe-Co/SiO2 catalyst. *Energy Conversion and Management, 150*(August), 331–342. https://doi.org/10.1016/j.enconman.2017.08.024

Chintala, V., et al. (2017). Solar thermal pyrolysis of non-edible seeds to biofuels and their feasibility assessment. *Energy Conversion and Management, 153*(October), 482–492. https://doi.org/10.1016/j.enconman.2017.10.029

Demirbaş, A., & Arin, G. (2002). An overview of biomass pyrolysis. *Energy Sources, 24*(5), 471–482. https://doi.org/10.1080/00908310252889979

Di Blasi, C. (2002). Modeling intra- and extra-particle processes of wood fast pyrolysis. *AIChE Journal, 48*(10), 2386–2397. https://doi.org/10.1002/aic.690481028

Dickerson, T. and Soria, J. (2013). 'Catalytic fast pyrolysis: A review', pp. 514–538. doi: https://doi.org/10.3390/en6010514.

Din, M. I., et al. (2019). Fabrication of magnetite nanoparticles (Fe 3 O 4 -NPs) for catalytic pyrolysis of nutshells biomass. *Soft Materials, 17*(1), 24–31. https://doi.org/10.1080/1539445X.2018.1542315

Fassinou, W. F., et al. (2009). Pyrolysis of Pinus pinaster in a two-stage gasifier: Influence of processing parameters and thermal cracking of tar. *Fuel Processing Technology, 90*(1), 75–90. https://doi.org/10.1016/j.fuproc.2008.07.016

Fernández-García, A., et al. (2010). Parabolic-trough solar collectors and their applications. *Renewable and Sustainable Energy Reviews, 14*(7), 1695–1721. https://doi.org/10.1016/j.rser.2010.03.012

Ferreira, S. D., et al. (2015). Pyrolysis of Medium Density Fiberboard (MDF) wastes in a screw reactor. *Energy Conversion and Management, 92*, 223–233. https://doi.org/10.1016/j.enconman.2014.12.032

Funke, A., et al. (2017). Dimensional Analysis of Auger-Type Fast Pyrolysis Reactors. *Energy Technology, 5*(1), 119–129. https://doi.org/10.1002/ente.201600095

Gollakota, A. R. K., et al. (2016). A review on the upgradation techniques of pyrolysis oil. *Renewable and Sustainable Energy Reviews, 58*, 1543–1568. https://doi.org/10.1016/j.rser.2015.12.180

Grassmann, H., et al. (2015). Solar biomass pyrolysis with the linear mirror II. *Smart Grid and Renewable Energy, 06*(07), 179–186. https://doi.org/10.4236/sgre.2015.67016

Guedes, R. E., Luna, A. S., & Torres, A. R. (2018). Operating parameters for bio-oil production in biomass pyrolysis: A review. *Journal of Analytical and Applied Pyrolysis, 129*, 134–149. https://doi.org/10.1016/j.jaap.2017.11.019

Guo, D., et al. (2020). Catalytic pyrolysis of rain tree biomass with nano nickel oxide synthetized from nickel plating slag: A green path for treating waste by waste. *Bioresource Technology, 315*(May), 123831. https://doi.org/10.1016/j.biortech.2020.123831

Jahirul, M. I., et al. (2012). Biofuels production through biomass pyrolysis- A technological review. *Energies, 5*(12), 4952–5001. https://doi.org/10.3390/en5124952

Joardder, M. U. H., et al. (2014). Solar assisted fast pyrolysis: A novel approach of renewable energy production. *Journal of Engineering (United Kingdom), 2014*, 252848. https://doi.org/10.1155/2014/252848

Joardder, M. U. H. *et al.* (2017). *Solar pyrolysis: Converting waste into asset using solar energy, Clean Energy for Sustainable Development: Comparisons and Contrasts of New Approaches*. Elsevier Inc. doi: https://doi.org/10.1016/B978-0-12-805423-9.00008-9.

Kalogirou, S. A. (2014). 'Chapter 3 – Solar Energy Collectors', *Solar Energy Engineering (Second Edition)*, pp. 125–220. doi: https://doi.org/10.1016/B978-0-12-374501-9.00003-0.

Kan, T., Strezov, V., & Evans, T. J. (2016). Lignocellulosic biomass pyrolysis: A review of product properties and effects of pyrolysis parameters. *Renewable and Sustainable Energy Reviews, 57*, 1126–1140. https://doi.org/10.1016/j.rser.2015.12.185

Kapoor, L. (2019). Experimental investigation of the pyrolysis of biomass to produce green fuels. *International Journal of Sustainable Energy, 38*(7), 692–700. https://doi.org/10.1080/14786451.2019.1566235

Kapoor, L., Bose, D., & Mekala, A. (2020). Biomass pyrolysis in a twin-screw reactor to produce green fuels. *Biofuels, 11*(1), 101–107. https://doi.org/10.1080/17597269.2017.1345360

Kumar, A., et al. (2015). A review on biomass energy resources , potential , conversion and policy in India. *Renewable and Sustainable Energy Reviews, 45*, 530–539. https://doi.org/10.1016/j.rser.2015.02.007

Lédé, J. (2010). Biomass pyrolysis: Comments on some sources of confusions in the definitions of temperatures and heating rates. *Energies, 3*(4), 886–898. https://doi.org/10.3390/en3040886

Li, R., et al. (2016). Product distribution from solar pyrolysis of agricultural and forestry biomass residues. *Renewable Energy, 89*, 27–35. https://doi.org/10.1016/j.renene.2015.11.071

Liaw, S. S., et al. (2013). Effect of pretreatment temperature on the yield and properties of bio-oils obtained from the auger pyrolysis of Douglas fir wood. *Fuel, 103*, 672–682. https://doi.org/10.1016/j.fuel.2012.08.016

Maximino, R. (2013). 'Pyrolysis of biomass residues in a screw reactor', (October), pp. 1–10.

McKendry, P. (2002). Energy production from biomass (part 1): Overview of biomass. *Bioresource Technology, 83*(1), 37–46. https://doi.org/10.1016/S0960-8524(01)00118-3

Miranda, R., et al. (2009). Pyrolysis of sweet orange (Citrus sinensis) dry peel. *Journal of Analytical and Applied Pyrolysis, 86*(2), 245–251. https://doi.org/10.1016/j.jaap.2009.06.001

Mohanty, P., et al. (2011). Fuel production from biomass: Indian perspective for pyrolysis oil. *Journal of Scientific and Industrial Research, 70*(8), 668–674.

Morales, S., et al. (2014). Solar biomass pyrolysis for the production of bio-fuels and chemical commodities. *Journal of Analytical and Applied Pyrolysis, 109*, 65–78. https://doi.org/10.1016/j.jaap.2014.07.012

Nanda, S., et al. (2014). Physico-chemical properties of bio-oils from pyrolysis of lignocellulosic biomass with high and slow heating rate, *4*(3), 21–32. https://doi.org/10.5539/eer.v4n3p21

Ndukwu, M. C., et al. (2020). Review of solar-biomass pyrolysis systems: Focus on the configuration of thermal-solar systems and reactor orientation. *Journal of King Saud University - Engineering Sciences*. https://doi.org/10.1016/j.jksues.2020.05.004

Onay, O., & Kockar, O. M. (2003). Slow, fast and flash pyrolysis of rapeseed. *Renewable Energy, 28*(15), 2417–2433. https://doi.org/10.1016/S0960-1481(03)00137-X

Patel, D. K., Brahmbhatt, P. K., & Panchal, H. (2018). A review on compound parabolic solar concentrator for sustainable development. *International Journal of Ambient Energy, 39*(5), 533–546. https://doi.org/10.1080/01430750.2017.1318786

Pattiya, A., & Suttibak, S. (2012). Production of bio-oil via fast pyrolysis of agricultural residues from cassava plantations in a fluidised-bed reactor with a hot vapour filtration unit. *Journal of Analytical and Applied Pyrolysis, 95*, 227–235. https://doi.org/10.1016/j.jaap.2012.02.010

Perkins, G. (2018). Techno-economic comparison of the levelised cost of electricity generation from solar PV and battery storage with solar PV and combustion of bio-crude using fast pyrolysis of biomass. *Energy Conversion and Management, 171*(May), 1573–1588. https://doi.org/10.1016/j.enconman.2018.06.090

Puy, N., et al. (2011). Valorisation of forestry waste by pyrolysis in an auger reactor. *Waste Management, 31*(6), 1339–1349. https://doi.org/10.1016/j.wasman.2011.01.020

Rahman, M. A., Parvej, A. M., & Aziz, M. A. (2021). Concentrating technologies with reactor integration and effect of process variables on solar assisted pyrolysis: A critical review. *Thermal Science and Engineering Progress, 25*(May), 100957. https://doi.org/10.1016/j.tsep.2021.100957

Rony, A. H., et al. (2019). Kinetics, thermodynamics, and physical characterization of corn stover (Zea mays) for solar biomass pyrolysis potential analysis. *Bioresource Technology, 284*(March), 466–473. https://doi.org/10.1016/j.biortech.2019.03.049

Saravana Sathiya Prabhahar, R., Nagaraj, P., & Jeyasubramanian, K. (2020). Promotion of bio oil, H2 gas from the pyrolysis of rice husk assisted with nano silver catalyst and utilization of bio oil blend in CI engine. *International Journal of Hydrogen Energy, 45*(33), 16355–16371. https://doi.org/10.1016/j.ijhydene.2020.04.123

Sharma, A., Pareek, V., & Zhang, D. (2015). Biomass pyrolysis—A review of modelling, process parameters and catalytic studies. *Renewable and Sustainable Energy Reviews, 50*, 1081–1096. https://doi.org/10.1016/j.rser.2015.04.193

Sirijanusorn, S., Sriprateep, K., & Pattiya, A. (2013). Pyrolysis of cassava rhizome in a counter-rotating twin screw reactor unit. *Bioresource Technology, 139*, 343–348. https://doi.org/10.1016/j.biortech.2013.04.024

Solar, J., et al. (2016). Influence of temperature and residence time in the pyrolysis of woody biomass waste in a continuous screw reactor. *Biomass and Bioenergy, 95*, 416–423. https://doi.org/10.1016/j.biombioe.2016.07.004

Sun, J. (2021). *Hybrid solar power system, Advances in Clean Energy Technologies*. Elsevier Inc. https://doi.org/10.1016/b978-0-12-821221-9.00011-6.

Thangalazhy-Gopakumar, S., Adhikari, S., & Gupta, R. B. (2012). Catalytic pyrolysis of biomass over H +ZSM-5 under hydrogen pressure. *Energy and Fuels, 26*(8), 5300–5306. https://doi.org/10.1021/ef3008213

Tsai, W. T., Lee, M. K., & Chang, Y. M. (2006). Fast pyrolysis of rice straw, sugarcane bagasse and coconut shell in an induction-heating reactor. *Journal of Analytical and Applied Pyrolysis, 76*(1–2), 230–237. https://doi.org/10.1016/j.jaap.2005.11.007

Veses, A., et al. (2015). Production of upgraded bio-oils by biomass catalytic pyrolysis in an auger reactor using low cost materials. *Fuel, 141*, 17–22. https://doi.org/10.1016/j.fuel.2014.10.044

Weldekidan, H., et al. (2018). Production and analysis of fuels and chemicals obtained from rice husk pyrolysis with concentrated solar radiation. *Fuel, 233*(June), 396–403. https://doi.org/10.1016/j.fuel.2018.06.061

Wu, S. (2015). Estimation and comparison of bio-oil components from different pyrolysis conditions, *3*(June), 1–11. https://doi.org/10.3389/fenrg.2015.00028

Xie, T., et al. (2015). Characteristics and applications of biochar for environmental remediation: A review. *Critical Reviews in Environmental Science and Technology, 45*(9), 939–969. https://doi.org/10.1080/10643389.2014.924180

Yadav, D., & Banerjee, R. (2016). A review of solar thermochemical processes. *Renewable and Sustainable Energy Reviews, 54*, 497–532. https://doi.org/10.1016/j.rser.2015.10.026

Zabeti, M., et al. (2012). In situ catalytic pyrolysis of lignocellulose using alkali-modified amorphous silica alumina. *Bioresource Technology, 118*, 374–381. https://doi.org/10.1016/j.biortech.2012.05.034

Zeaiter, J., et al. (2015). Design of an automated solar concentrator for the pyrolysis of scrap rubber. *Energy Conversion and Management, 101*, 118–125. https://doi.org/10.1016/j.enconman.2015.05.019

Zeng, K., et al. (2015). Solar pyrolysis of wood in a lab-scale solar reactor: Influence of temperature and sweep gas flow rate on products distribution. *Energy Procedia, 69*, 1849–1858. https://doi.org/10.1016/j.egypro.2015.03.163

Zeng, K., et al. (2017). Solar pyrolysis of carbonaceous feedstocks: A review. *Solar Energy, 156*, 73–92. https://doi.org/10.1016/j.solener.2017.05.033

Biomass Digestion for Production of Electrical Energy

Vikrant Sharma and Ravikant Gupta

Abstract Currently, energy security and environmental protection are two of the major concerns the world is facing. Electricity generation, being one of the fastest growing and most polluting sectors, needs to be improved with incorporation of efficient, sustainable, cost-effective and eco-friendly fuels and methods. Biomass can be such a fuel as it is abundantly available and carbon neutral. For generation of electricity from biomass, combustion, gasification and digestion can be utilized. This chapter focusses on the biomass digestion (BD) for generation of electricity. It consists of four steps, namely, pretreatment, anaerobic digestion, gas cleaning and electricity generation. Pretreatment consists of physical, chemical and/or biological operations to expose sites for microbial attack. In anaerobic digestion, biomass is converted into biogas, mainly consisting methane (CH_4), in the absence of air by the action of microorganisms. Anaerobic digestion mainly follows four major steps, i.e. hydrolysis, acidogenesis, acetogenesis and methanogenesis. The undesirable components in resultant biogas, such as hydrogen sulphide (H_2S), oxygen (O_2), nitrogen (N_2), water (H_2O), carbon dioxide (CO_2) and particle traces, can cause corrosion in the downstream equipment and can reduce thermal efficiency of the process. Therefore, gas cleaning and upgradation is employed to overcome such challenges. The chemical energy of the biogas can be converted into electricity using it in external and internal combustion engines and/or fuel cells. Alternatively, pressurized steam can be produced by biogas combustion to drive steam turbines. In addition, methane-rich biogas can be utilized in a reformer to produce syngas, a mixture of carbon monoxide (CO) and hydrogen (H_2), to drive combined cycle and/or to produce variety of chemicals including gaseous and liquid fuels.

V. Sharma
Department of Chemical Engineering, Indian Institute of Technology Roorkee, Haridwar, Uttarakhand, India
e-mail: vsharma@ch.iitr.ac.in

R. Gupta (✉)
Department of Chemical Engineering, Banasthali Vidyapith, Tonk, Jaipur, Rajasthan, India
e-mail: ravikantg@banasthali.in

D. B. Pal, J. M. Jha (eds.), *Sustainable and Clean Energy Production Technologies*, Clean Energy Production Technologies,
https://doi.org/10.1007/978-981-16-9135-5_7

Keywords Biomass · Digestion · Biogas · Renewable energy · Mathematical model · Life cycle assessment · Hydrolysis · Acidogenesis · Acetogenesis · Methanogenesis

1 Introduction

Current trends in energy supply and use are economically, environmentally and socially unsustainable (IEA, 2012). In 2019, the total global energy consumption was 583.9 exajoules with the three most consumed fuels as oil, coal and natural gas with their respective shares of 33.1%, 27% and 24.2% (BP, 2020). The share of fossil fuels in total energy consumption (80%) is nearly constant for the past 30 years. The reserves of oil, natural gas and coal in the world will last for 49.9, 49.8 and 132 years, respectively, with the current (2019) production rate (BP, 2020). Although the proved reserves of fossil fuels are increasing due to technological development, it would not be able to fulfill the exploding demand of energy due to increase in population and industrialization. Further, utilization of these fossil fuels leads to irreversible environmental damage, in terms of global warming and climate change as well as health issues due to emissions of greenhouse gases (GHG) (CO_2, CH_4), acid gases (NO_x, SO_x), toxic effluents (CO, heavy metals) and particulate matter. Global CO_2 emissions in 2019 were 34,169 million tonnes, increasing at a rate of approximately 1% per year (BP, 2020). Energy-related CO_2 emissions are projected to more than double by 2050 (IEA, 2012). Furthermore, these resources are not equitably distributed around the world, thereby causing price fluctuations, dependence on foreign imports and unavailability. Therefore, research and development of renewable, efficient, environment-friendly and sustainable fuels and methodologies is of utmost importance.

In 2019, renewable energy contributed 28.98 exajoules (5% of the total energy consumption), increasing at a growth rate of 13.7% per year in the past decade (BP, 2020). Renewables contributed the largest (41%) to the increase in energy demand in 2019 (BP, 2020). Biomass is the oldest and largest (69.5%) contributor to the renewable energy which is mostly utilized in traditional and unsustainable manner in rural areas, especially in developing countries. The efficiency of such use is very low (10–20%), and it causes several health and environmental issues. Biomass can be defined as any organic matter derived from plants or living organisms, excluding fossil fuels, available on a renewable basis such as agricultural and forest residue, municipal and industrial organic wastes, etc. (IEA, 2012; Maneesha & Bhaskarwar, 2012). As the carbon emitted from utilization of biomass can be utilized by growing plants (considering sustainable utilization), biomass is considered carbon neutral. Further, biomass can be available on demand, in contrast to solar or wind energy (Moustakas et al., 2020). As it is a carbon-neutral, abundant and widely distributed energy source, it can play an important role in energy security and environmental protection. To limit the global rise in temperature to 2 °C by the end of this century, it has been estimated that biomass-derived energy, i.e., bioenergy, should provide 24% (160 exajoules) of total primary energy supply and 7.5% (3100 TWh) of total electricity generation in 2050 (IEA, 2012). Further, bioenergy should bring about 2 Gt CO_2-equivalent emission savings per year in 2050. Moreover, bioenergy can also be an important step towards waste mitigation. However, bioenergy is costlier than conventional fossil fuel energy. Therefore, for large-scale deployment of bioenergy, a strong policy framework, for emission regulation, carbon taxes, subsidies, sustainable production and utilization of biomass, competition with existing uses of biomass, pressure on irrigation, changes in land use, soil degradation, loss of biodiversity, etc., is required to maximise social, environmental and economic benefits.

Biomass can be utilized to produce energy in terms of heat, electricity and various liquid and gaseous fuels through combustion, gasification, fermentation and/or digestion. Currently, electricity from biomass is produced through combustion followed by power generation in steam turbines, especially using co-firing of biomass with coal (IEA, 2012). However, as the bulk density and calorific value of biomass are lower than fossil fuels, transportation, storage and combustion of the former are more expensive. Biomass digestion (BD) is one of the most promising and cost-competitive technologies for biomass energy recovery, greenhouse gas (GHG) and pollution reduction and waste management (Aberilla et al., 2019; Adams & McManus, 2019; Kheybari et al., 2019; Hasan & Ammenberg, 2019; Freitas et al., 2019; Shirzad et al., 2019; Mohammad & Ardebili, 2020; Singh et al., 2020).

In 2018, 59.3 billion m^3 of biogas was produced globally with an annual growth rate of 9% (2000–2018) (WBA, 2020). Further, 637 TWh of biopower was generated in 2018 with an annual growth rate of 8% from 2000 to 2018, and the share of biogas was merely 14% (WBA, 2020). Currently, Germany (2700 MW) followed by Italy (1400 MW) has the most biogas power plants in the world (IEA, 2012; Battista et al., 2019). Recently, Moustakas et al. (2020) estimated that digestion can utilize 707,164 tonnes per year of agricultural biomass produced in Greece to generate 619 GWh of electricity and 895 GWh of thermal energy (maximum per year).

BD occurs naturally in the environment, but its design, development and control are challenging due to varying physical and chemical properties of biomass and complex metabolic pathways (Rasapoor et al., 2020). BD comprises decay of biomass in the absence of oxygen to produce biogas, a mixture of CH_4 (50–75%), CO_2 (25–50%), H_2, H_2O, NH_3, H_2S, etc., and digestate, solid residue with inorganic matter (Zheng et al., 2014). The biogas, after required cleaning and upgrading according to intended use, can have several uses such as direct combustion for cooking and heating, production of electricity, as a fuel in gas-operated vehicles and/or production of variety of chemicals. The digestate can be used as a high-quality organic fertilizer which is more advantageous than undigested fertilizer. However, the digestate can lead to acidification and eutrophication (Moustakas et al., 2020). Other benefits of BD include reduced odours and improved agricultural practices. Various biomasses have been utilized for BD such as cattle waste (Chakravarthi, 1997), maize (Adams & McManus, 2019), algae (Vassalle et al., 2020), grass (Oleszek et al., 2014), food waste (Rosas-Mendoza et al., 2020), municipal organic waste (Filho et al., 2020), chicken manure (Jurgutis et al., 2020), etc.

2 Biomass

Biomass can be defined as any organic matter derived from plants or living organisms, excluding fossil fuels, available on a renewable basis (Maneesha & Bhaskarwar, 2012). The producer species, i.e. plants, take solar energy and convert it to chemical energy through photosynthesis process, whose representative equation can be given as follows:

$$6CO_2 + 12H_2O \rightarrow C_6H_{12}O_6 + 6H_2O + 6O_2 \quad (1)$$

The primary consumers, i.e. herbivores, utilize energy stored in plants to generate energy and biomass. Similarly, secondary consumers, i.e. carnivores, use energy stored in herbivores for their growth. Thus, biomass stores and transfers energy from the sun in the food chain. The chemical energy in the biomass can be converted into mechanical, thermal and/or electrical energy, and it is referred to as bioenergy. Biomass resources have traditionally been used by direct burning for cooking and heating.

Biomass is a carbon-neutral fuel as the CO_2 released by their combustion can be utilized by plants to grow, provided it is utilized sustainably (Sherwood, 2020). Further, if carbon capture, utilization and storage are incorporated, biomass can be a carbon-negative source of energy. Therefore, bioenergy can be an important step towards solving environmental issues related to waste management and disposal, emissions of GHG and other pollutants and establishing circular economy (Sherwood, 2020; Pampillón-González & Canepa, 2017). However, bioenergy competes against other demands of biomass such as the production of food, industrial resources and/or cultural goods and services, among others. McKendry (2002a, 2002b) and Perea-Moreno et al. (2019) provided excellent reviews on current research and development regarding biomass resources and its energy conversion technologies.

3 Biomass Conversion Technologies

Biomass can be converted to electricity through combustion, gasification and digestion.

3.1 Biomass Combustion

Biomass combustion is the oldest and most utilized conversion technology. It involves oxidation of biomass in air (more than stoichiometric requirement) (Pampillón-González & Canepa, 2017). Traditional burning of biomass, especially in rural regions of the developing countries, is very inefficient (10–20%) and leads to several health and environmental problems due to emissions of smoke and particulate matter (IEA, 2012). In contrast, biomass combustion in power plants is more efficient and controlled. In biomass combustion power plants, the energy released from biomass combustion is utilized to produce steam which is utilized in steam turbine to generate electricity. The overall reaction of biomass combustion can be given as follows (assuming that biomass contains C, H, O, N and S-containing species and all other species are inert):

$$C_aH_bO_cN_dS_e + f\,O_2 \rightarrow g\,CO_2 + h\,H_2O + i\,NO_x + j\,SO_y \tag{2}$$

3.2 *Biomass Gasification*

Biomass gasification is a thermochemical process in which biomass is converted into syngas, a mixture of CO, H_2, CH_4, CO_2 etc., in oxygen-lean environment of air and/or H_2O. The syngas can be directly burned, used to produce electricity in combined cycle and/or combined heat and power (CHP) plants and utilized as raw material for production of variety of chemicals including liquid and gaseous fuels. Biomass integrated gasification combined cycle (BIGCC) power plants utilize the syngas in gas turbine and residual heat to produce steam to be used in steam turbine to generate electricity. Thus, the efficiency of BIGCC power plants is higher than conventional combustion-based power plants. Further, the pollutant removal from syngas is more effective and economical than flue gas from traditional combustion-based power plants. Several BIGCC plants are in various developmental stages in Europe, the United States, Japan and India. The gasification reaction of biomass can be represented as follows:

$$C_aH_bO_cN_dS_e + f\,H_2O + g\,O_2 \rightarrow h\,CO + i\,H_2 + j\,CO_2 + k\,CH_4 + l\,NH_3 + m\,H_2S \tag{3}$$

3.3 *Biomass Digestion*

Biomass digestion (BD) is a natural biological process in which biomass is broken down, by microorganisms and without oxygen, into biogas, consisting of CH_4 (50–70%), CO_2 (25–45%) and minor constituents (H_2, H_2O, NH_3, H_2S, volatile organic compounds (VOC), SO_2, N_2, etc.), and a solid residue (digestate), a mixture of inorganic compounds (of N, P, K, Ca, etc.) (IRENA, 2012; Reith et al., 2003). As CH_4 has the global warming potential (GWP) 25 times higher than CO_2, capturing and utilizing such methane emissions would be beneficial to the environment. Considering biomass as an organic compound ($C_aH_bO_cN_dS_e$), BD can be represented by Eq. (4). The stoichiometric coefficients, i.e. the theoretical amount of the gases, can be calculated by Buswell equation.

$$C_aH_bO_cN_dS_e + f\,H_2O \rightarrow g\,CH_4 + h\,CO_2 + i\,NH_3 + j\,H_2S \quad (4)$$

Biogas has a calorific value of 20–30 MJ/Nm^3 according to the concentration of CH_4 (Reith et al., 2003; Banu & Kannah, 2019). The biogas can be easily stored and transported and can be directly combusted, similar to CH_4, for cooking and heating, for electricity and as fuel in gas-operated vehicles. The solid residue (i.e. digestate) can be used as a high-quality organic fertilizer (Pampillón-González & Canepa, 2017; IRENA, 2012).

3.4 *Selection of a Biomass Conversion Technology*

A suitable biomass conversion technology should be selected based on quality and quantity of biomass, economics and desired product. Further, the quality and quantity of biomass is dependent on weather, soil, geography, population and their economic activities, making the technology dependent on the region (Pampillón-González & Canepa, 2017). Li et al. (2017) compared methane production by gasification-methanation and BD and concluded that energy efficiency of both the processes is similar. However, biomass digestion has the following advantages in comparison with other biomass technologies (Reith et al., 2003; IRENA, 2012):

- BD is a natural process occurring in landfills, marshes, etc. and generates CH_4 which is a potent greenhouse gas. Waste utilization and removing this CH4 from environment would be useful.
- BD systems are technologically simpler and inexpensive than combustion or gasification plants as well as consume little energy and require small space.
- Digestate can be utilized as a high-quality fertilizer with almost complete retention of nutrients (N, P and K). Therefore, solid waste is smaller compared to aerobic processes.
- For biomass with high water content, BD does not require as extensive dewatering as other processes.
- Relatively odour-free end product.

Due to the environmental advantages of BD, such as reduction in GHG emissions, waste management and fertilizer by-product, BD comes out to be the most suitable technology for biomass valorization to electricity (Rasapoor et al., 2020).

4 Biomass Digestion

Biomass digestion (BD) is a complex process which involves many microorganisms interacting with each other. In general, it can be divided into four major steps, i.e. hydrolysis, acidogenesis, acetogenesis and methanogenesis, each requiring its own characteristic group of microorganisms, as shown in Fig. 1 (Reith et al., 2003; Banu & Kannah, 2019). However, all these steps usually occur simultaneously and in a coupled manner without any accumulation of intermediate compounds for stable operation.

4.1 Hydrolysis

Hydrolysis is the first step of BD process as well as the slowest or rate-limiting step, i.e., base for design of digesters. During hydrolysis, complex organic compounds such as proteins, carbohydrates, lipids, polysaccharides and fat are converted into simple digestible organic compounds (e.g. amino acids, monosaccharides, glycerol and fatty acids) in the presence of hydrolytic bacteria and anaerobic fungi (Reith et al., 2003; Banu & Kannah, 2019). It can be represented as follows:

$$C_6H_{10}O_4 + 2H_2O \rightarrow C_6H_{12}O_6 + H_2 \tag{5}$$

4.2 Acidogenesis

In acidogenesis, acidogenic microbes convert simple organic compounds produced by hydrolysis into C_2H_5OH, CH_3COOH, H_2, CO_2 and NH_3 as well as small organic acids (propionic, formic, lactic, butyric, succinic acids) and alcohols

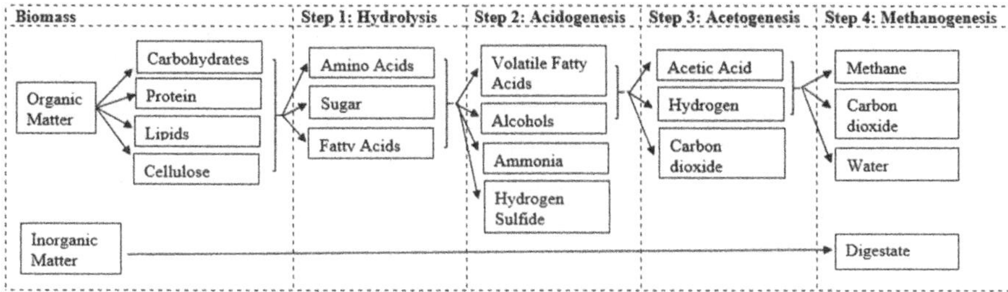

Fig. 1 Scheme of biomass digestion

(Reith et al., 2003; Banu & Kannah, 2019). Eqs. (6) and (7) show some of the reactions occurring during acidogenesis:

$$C_6H_{12}O_6 \leftrightarrow 2CH_3CH_2OH + 2CO_2 \tag{6}$$

$$C_6H_{12}O_6 + 2H_2 \leftrightarrow 2CH_3CH_2COOH + 2H_2O \tag{7}$$

4.3 Acetogenesis

In acetogenesis, acetogenic microbes degrade long-chain (LCFA) and volatile fatty acids (VFA) and alcohols generated during acidogenesis into CH_3COOH, H_2 and CO_2 (Reith et al., 2003; Banu & Kannah, 2019). The steps followed during acetogenesis can be represented as follows:

$$CH_3CH_2COOH + 3H_2O \leftrightarrow CH_3COOH + H_2CO_3 + 3H_2 \tag{8}$$

$$C_6H_{12}O_6 + 2H_2O \leftrightarrow 2CH_3COOH + 2CO_2 + 4H_2 \tag{9}$$

$$CH_3CH_2OH + 2H_2O \rightarrow CH_3COOH + 3H_2 \tag{10}$$

4.4 Methanogenesis

Methanogenesis is the final step of the BD and involves conversion of CH_3COOH and H_2 into CH_4 and CO2 by methanogenic microbes, as given in Eqs. (11)–(13) (Reith et al., 2003; Banu & Kannah, 2019):

$$CH_3COOH \rightarrow CH_4 + CO_2 \tag{11}$$

$$CO_2 + 4H_2 \rightarrow CH_4 + 2H_2O \tag{12}$$

$$2CH_3CH_2OH + CO_2 \rightarrow CH_4 + 2CH_3COOH \tag{13}$$

Biomass digestion is performed in digesters on a faster time scale than natural process. Digesters can be categorized into batch or continuous digesters according to operation; into psychrophilic (5–20 °C), mesophilic (30–40 °C) or thermophilic (50–60 °C) digesters according to temperature; into high (>10%) or low (<10%) solid digesters based on solid content; into passive, low rate or high rate digesters based on organic rate load (OLR); and into single- or multistage digesters according to number of stages.

Batch systems are the simplest system and used for high solid content. Generally, 10–15% of the product is left in the batch reactor for faster growth of microorganisms in the next batch. The CSTR digesters are the most common low-solid reactor where typical retention time is 2–4 weeks (Banu & Kannah, 2019). Plug flow digesters are suitable for semi-solid waste (solid concentration $> 10\%$), and steps of BD are separated along the length of these digesters (IRENA, 2012; IEA, 2012). For dealing with high solid content ($>30\%$), high solid systems are developed which are advantageous in terms of lower reactor size, water requirement and heating costs. Generally, high-rate systems are used to treat wastewater. In contrast, low-rate systems are usually used to digest slurries and solid wastes. In multistage systems, different steps of BD or different biochemical reactions are affected in different stages so that optimal operating conditions for a particular set of biochemical reactions can be maintained in a stage (Reith et al., 2003). Thus, faster reaction rate and higher biogas yield can be obtained in multistage digesters in comparison with single-stage digesters.

Obileke et al. (2020) provided a review of biomass digestion technology. Ziemiński and Frąc (2012) presented an excellent review of stages of biomass digestion. Sikora et al. (2018) listed enzymes responsible for the key pathways of biomass digestion based on the Kyoto Encyclopedia of Genes and Genomes database. Li et al. (2019b) reviewed the methods to enhance each step of BD.

5 Biogas Cleaning and Upgrading

Biogas contains several undesirable components (CO_2, H_2S, O_2, N_2, H_2O, siloxanes and particulate matter) which decrease its calorific value and/or are corrosive to downstream equipment. Different biogas treatments (cleaning and/or upgrading) are necessary according to the use of the biogas. It has been suggested that the biogas should contain at least 95% CH_4 in order to be economically viable (Song et al., 2014). If biogas can be cleaned and upgraded to high enough CH_4 concentration (98%), it will have the same properties as natural gas. It is referred to as biomethane and can be directly utilized in natural gas grid, gas-operated vehicles and other applications of natural gas. However, it must be delivered to the grid as compressed natural gas (CNG) (250 bar) with little or no particulate matter and fluid contaminants (Wellinger & Lindberg, 1999; Grontmij, 1997). Heat production in gas heater/boiler systems does not require a high gas quality (Wellinger & Lindberg, 1999). Kuo and Dow (2017) analysed biogas and its contaminants produced by co-digestion of food waste and municipal wastewater sludge along with air pollutants from a biogas-fueled IC engine.

CO_2 (25–45%), being the largest dilutant in biogas, lowers its energy content, and it can corrode downstream equipment. It can be removed by pressure swing adsorption (PSA), water scrubbing, organic physical scrubbing, chemical absorption and polymeric membrane separation (Pampillón-González & Canepa, 2017). Water in the biogas can corrode pipelines by forming carbonic acid (Petersson, 2013). It is generally removed by cooling the biogas followed by adsorption on activated carbon or SiO_2 (Pampillón-González & Canepa, 2017). Hydrogen sulphide (H_2S) gets converted to corrosive acids by condensation and combustion. Therefore, it should be removed by precipitation and/or adsorption with activated carbon or iron oxide (Pampillón-González & Canepa, 2017). Particularly, for use of biogas in engines, sulphur content should be minimized (below 1000 ppm). Injection of little air in the headspace of the digester is frequently used to oxidize H_2S by microorganisms and, in turn, removal of approximately 95% of sulphur from biogas. Siloxanes negatively affect downstream equipment due to formation of silicon oxide (Pampillón-González & Canepa, 2017). Therefore, it is removed from biogas by adsorption on activated carbon, activated aluminium or silica gel (Petersson, 2013).

6 Power Generation

The biogas is mostly used for direct combustion for cooking and heating and its use for electricity production is not very common. Currently, Germany (2700 MW) followed by Italy (1400 MW) has the most biogas power plants in the world (IEA, 2012; Battista et al., 2019). The mean calorific value of biogas is about 20–30 MJ/m^3, i.e. around 5.5–8.5 kWh/m^3. However, 1 m^3 of biogas can only be converted into approximately 1.7 kWh electricity due to losses during conversion. Murphy et al. (2010) showed that from 9495 tonnes of silage, 1.88 million m^3 biogas, 3992 MWh electricity and 1697 MWh thermal energy can be produced.

To produce electricity from biogas, it is usually used in combustion engines which convert it into mechanical energy, powering an electric generator to produce electricity. This is a well-known technology, and electric generators of appropriate sizes can be found easily. Combustion engines are of two types, i.e. external and internal combustion engines. In external combustion engines, heat generated by combustion of biogas is transferred to the working fluid through engine wall or heat exchanger. Therefore, the working fluid expands and, through the engine mechanism, produces mechanical work. Stirling motor is an external combustion engine which is flexible in terms of fuel quality but is expensive and has low efficiency. In contrast, in internal combustion engine, biogas is combusted with an oxidizer in the combustion chamber. The flue gas from combustion works as working fluid and produces mechanical work. It has smaller size and weight than external combustion engines. Therefore, internal combustion engines are mostly used in biogas power plants. Examples of internal combustion engines include diesel engines, gas engines, gas turbines, etc. (Moustakas et al., 2020).

Diesel engines operating on biogas require small (2%) amount of ignition oil or gas, therefore, work in dual-fuel mode only. Gas motors can work with biogas only except requiring a small amount of gasoline to start the engine (spark ignition). Gas motors can work entirely on renewable energy, but up to 200 kW, diesel engines are preferred due to higher efficiency and lower cost. Gas turbines work on Brayton cycle and are expensive due to high speeds and temperature. Further, gas turbines suffer a large decrease in efficiency due to reduction in gas inflow (Grontmij, 1997).

In contrast to natural gas, biogas is characterized by high knock resistance and, in turn, can work with high compression rates. However, internal combustion engines require gas with high purity, especially regarding corrosive compounds such as H_2S. Bigger biogas plants are generally more cost efficient than smaller ones. The payback period of around 6–9 years is estimated for medium or large plants (>50 kW). It should be noted that a large-scale plant would need approximately 8000–9000 tonnes of municipal solid waste per MW per year (IRENA, 2012).

Cleaned biogas can be efficiently utilized in CHP plants, mainly based on steam cycles, to produce heat (50%) and electricity (35%) (Malico et al., 2019; Moustakas et al., 2020). Energy losses are about 15% (Reith et al., 2003). Electrical efficiency of 29% (spark ignition) and 31% (dual-fuel engine) has been reported for small-scale CHP systems (< 45 kWe) (Reith et al., 2003). In contrast, larger engines can achieve an electrical efficiency of around 38% (Wellinger & Lindberg, 1999). The heat can be recovered from the exhaust gas (high temperature heat) and the cooling water and oil (low temperature heat) (Grontmij, 1997). CHP systems based on gas engines are more efficient and require lower specific capital cost than that based on gas turbine (Reith et al., 2003). However, maintenance costs for gas engines are higher than for turbines (Reith et al., 2003). CHP systems based on gas turbines are more economical in large-scale applications.

Biogas can also be converted directly into electricity and heat using fuel cells where electrochemical conversion of biogas with oxygen takes place. Water and CO_2 are the main by-products, and electrical and thermal efficiency of about 40–50 and 30–40% can be obtained. Two fuel cells, i.e. phosphoric acid fuel cells (PAFC) and solid oxide fuel cells (SOFC), are useful for electricity generation from biogas. PAFC operates at approx. 200 °C (Wellinger & Lindberg, 1999). SOFC operate at temperatures greater than 900 °C. While both PAFC and SOFC require near-complete removal of sulphides and halogens, SOFC has slightly higher tolerance for impurities (Reith et al., 2003). Therefore, expensive biogas cleaning and upgrading is required for its use in fuel cells. Further, the cost of fuel cell is high; therefore, substantial cost reduction of fuel cells is required for large-scale application.

7 Effect of Operating Conditions

Biomass digestion (BD) is a complex biological process with many microorganisms which have different metabolic and growth rates (Banu & Kannah, 2019). Consequently, it depends on the properties of feedstock and operating conditions such as pretreatment, temperature, pH, hydraulic retention time (HRT) and organic load rate (OLR), nutrients and salinity (Karthikeyan and Visvanathan, 2013).

7.1 Feedstock

The properties of different biomasses can be very different from each other, requiring different pretreatment strategies and equipment size and configuration, and, thereby, provide different biogas composition. Traditionally, cattle manure and sewage sludge have been used for biogas production, and biomasses with high moisture content can be processed efficiently and cost-effectively with BD (Franco et al., 2018; Raheem et al., 2018). Liotta (2013) studied the effects of water in biomass digestion of food waste, rice straw and carrot waste and concluded that final methane yield in dry condition is approximately 50% lower than wet conditions. However, dry BD can be advantageous in terms of smaller digester size, lesser water requirement, digestate production and pretreatment needed. Protein- and fat-rich biomasses have high energy content and, in turn, high biogas potential but can lead to process inhibition due to formation of NH_3, long-chain fatty acids (LCFA), etc. or foaming (He et al., 2017). The acetoclastic methanogens are considered to be most sensitive to ammonia (Rajagopal et al., 2013). Ammonia-tolerant *M. bourgensis* is shown to improve adaptation to higher ammonia concentrations (Fotidis et al., 2014). Incorporation of *Coprothermobacter proteolyticus* improves hydrolysis and fermentation in waste-activated sludge which is protein enriched (Sasaki et al., 2011). The degradation of fats can be improved by addition of lipases or *Syntrophomonas zehnderi* and *Clostridium lundense* (Nzila, 2017; Cirne et al., 2007). Lignocellulosic biomass does not pose any risk of process disturbance but can require long time for degradation (Banu & Kannah, 2019). For the degradation of lignocellulosic biomass, addition of cellulose-degrading bacteria, hydrolytic enzymes and anaerobic fungi can improve biogas production (Speda et al., 2017; Öner et al., 2018). Carbohydrate-rich biomasses are difficult to be utilized for biogas production due to high carbon-to-nitrogen (C/N) ratio (Banu & Kannah, 2019). Further, cattle manure is undesirable due to low C/N ratio (Neshat et al., 2017). The optimum C/N ratio for BD suggested to be is 15–30 (Jankowska et al., 2017; Wirth et al., 2018). In these cases, co-digestion of different biomasses can improve process performance and biogas yield (Xu et al., 2018b; Ziels et al., 2018). Animal manure is usually co-digested with straw (Banu & Kannah, 2019).

Co-digestion of multiple feedstocks can be used to achieve optimal biogas yield and process stability. Adjustment of both macro- and micro-elements can decrease digestion time (Rasapoor et al., 2020). Solé-Bundó et al. (2019) analysed co-digestion of microalgae and primary sludge and concluded that CH_4 yield increased by 65% in comparison with microalgae mono-digestion. Further, Piñas et al. (2019) suggested that the optimum sizes of mono-digestion and co-digestion biogas plant are 740 and 1000 kW, respectively, based on economic models of centralized and decentralized technological schemes for mono-digestion and co-digestion systems of cattle manure, maize silage and grass silage. Different feedstocks can be evaluated by a parameter called anaerobic biodegradability which is measured by the total amount of CH_4 produced during retention time of at least 50 days (Sanders, 2001).

7.2 *Pretreatment*

The feedstock usually needs some form of pretreatment to maximise methane production (IRENA, 2012). Particularly, lignocellulosic biomass, due to its complex structure, requires pretreatment step to improve its biodegradability and, consequently, achieves higher biogas rate and yield (Koupaie et al., 2019; Abraham et al., 2020). Pretreatment enhances the digestibility of biomass by disturbing the structure, increasing the porosity, increasing the surface area, decreasing the particle size and improving access to the attack sites (Banu & Kannah, 2019). It can be broadly classified into physical (thermal and microwave), chemical (alkaline and acidic), biological (enzyme and fungal) and mechanical (high-pressure homogenizer and ultrasonic) nanoparticles and combinative pretreatment methods (Koupaie et al., 2019; Abraham et al., 2020). Paudel et al. (2017), Zheng et al. (2014) and Abraham et al. (2020) reviewed the pretreatment methods for various lignocellulosic feedstocks such as cereal residue, fruit and vegetable wastes and grasses. A set of optimum pretreatment methodologies will depend on the characteristics of feedstock and economics of the process (Paudel et al., 2017; Abraham et al., 2020).

Schwede et al. (2013) showed that thermal pretreatment significantly improved biodegradability and methane yield of algal biomass under both batch and semi-continuous digestion. Alkali and steam explosion leads to an enhanced anaerobic digestion performance (Paudel et al., 2017). Koupaie et al. (2019) presented an excellent review of the enzymatic pretreatment methods for augmentation of CH_4 production from lignocellulosic biomass. They observed that cellulose, β-glucosidase and xylanase were usually utilized, and there are large variations in methane improvement among various studies. Janke et al. (2019) observed that ensiling of sugar cane with lactic acid bacteria improved the digestibility of the cellulosic biomass for methane production. Mechanical pretreatments break the structure of lignocellulosic biomass and, consequently, reduce the lag time and improve methane yield (Dahunsi, 2019). Zhu et al. (2020) studied the effect of pretreatment of highly lignified wheat straw with NiO_x nanoparticles and visible light (350–780 nm). They observed that pretreatment with 3 wt% NiO_x at 4 h light exposure increased methane yield by 40% during BMP assays and by 30% in CSTRs. Passos and Ferrer (2015) concluded that hydrothermal pretreatment enhanced microalgae biodegradability. Further, increase in organic matter solubilisation (8–13%), digestion rate (30–90%) and final methane yield (17–39%) is observed due to pretreatment. Bohutskyi et al. (2019) showed that hydrothermal pretreatment of lipid-extracted algal residue increased the methane production rate by 15–30%. Further, methane yield increased through thermal pretreatment by 15%, dilute acid by 5%, dilute alkali by 17% and acid- and alkali-assisted thermochemical pretreatments by 23 and 27%, respectively. Xiao et al. (2020a) conducted life cycle assessment (LCA) for biogas production from microalgae biomass with hydrothermal pretreatment and concluded that hydrothermal pretreatment improved the biogas yield and energy gain as well as reduced the levelised cost of energy.

Rasapoor et al. (2020) showed that activated carbon, biochar and phenazine augment direct inter-species electron transfer and, in turn, biogas production due to improvement in digestibility. Johnravindar et al. (2020) presented the effects of granular activated carbon on co-digestion of food waste and brewery sludge and found 45% increase in methane yield with 1.5% granular activated carbon due to enhanced electron transfer efficiency for methanogenesis. Pan et al. (2019) reviewed the application of biochar to increase biomass digestion and concluded that biochar enhanced direct electron transfer and, in turn, methanogenesis. Zhang et al. (2020) showed that addition of algal biochar to co-digestion of algal biomass and food waste (75:25) increased average methane yield by 12–54%. Xiao et al. (2020b) studied the effect of addition of bottom ash (0.37–2.96%) from biomass power plant, containing about 35 minerals and alkalinity, to rice straw for 7 days on BD. Abdelsalam and Samer (2019) reviewed various trace elements and nanoparticles for enhancement of biogas and methane production. Further, the recommended concentrations of trace elements and nanoparticles in anaerobic digesters were suggested.

7.3 *Starting Inoculum and Bioaugmentation*

Bioaugmentation is a process in which some microorganisms are added to speed up the degradation or to make it efficient. A starting inoculum, i.e. set of microorganisms added at the start of BD, improves the degradation rate and CH_4 yield (Liu et al., 2017; Wojcieszak et al., 2017). The inocula from wastewater treatment, agricultural biogas and/or municipal organic waste plants are commonly used due to its highly varied and dynamic community (Raposo et al., 2011). Further, inoculum from a similar BD plant is found to be the most suitable in terms of small time for start-up and stable operation (Liu et al., 2017; Koch et al., 2017). A mixture of different hemicellulose and cellulolytic bacteria has been found to increase biogas yield (Poszytek et al., 2016; Nzila, 2017). At start-up, incorporation of *Methanosarcina* sp. can increase activity of the methanogenic organisms and their stability (Lins et al., 2014). Rumen microorganisms are found to be effective for digesting lignocellulosic biomass due to their intrinsic ability to degrade substrate rich in cellulosic fibre (Sawatdeenarunat et al., 2015).

7.4 *Temperature*

Temperature strongly affects the microbial communities and time required for its growth and, in turn, performance and stability of the process (Westerholm et al., 2018). BD can be divided into three categories based on temperature regimes, namely, psychrophilic (5–20 °C), mesophilic (25–45 °C) and thermophilic (45–65 °C) (Banu & Kannah, 2019; Reith et al., 2003). In general, increase in temperature improves enzymatic activity of microorganisms and metabolic rates, and the rate of CH_4 production is maximum at around 35 °C (Chen & Chang, 2017; Jankowska et al., 2017). As the rate of growth of microorganisms and biochemical reactions decrease with decrease in temperature, BD at lower temperature requires longer retention time and larger digesters. As energy required for biomass heating increases with increase in temperature, thermophilic digestion is usually applied when the biomass is already at high temperature or when pathogen removal is required (Reith et al., 2003). Maximum methanogenic activity occurs at about 55 °C or higher (Reith et al., 2003). However, BD becomes sensitive to disturbances at high temperature, and it can cause increase in inhibitors (Labatut et al., 2014). At high temperature, the equilibrium between NH_4^+ and NH_3 shifts towards the latter, and increased hydrolysis of nitrogenous compounds can increase NH_3 concentration (Ward et al., 2014). Zhang et al. (2020) observed that there is a narrow distribution of VFAs due to reduction in the production of branched VFAs at high temperature. The most suitable temperature regime will depend on feedstock, relationship between energy requirement and biogas yield and economics.

7.5 Organic Loading Rate

The organic load rate (OLR) refers to the amount of *organic* material per unit volume of the digester retained for a given unit time. A high OLR is desirable as it will lead to small equipment and high biogas production and, consequently, low cost. However, with increase in OLR, hydrolytic and acidogenetic microorganisms increase along with accumulation of inhibitors such as NH_3 and VFAs (Xu et al., 2018a; Tonanzi et al., 2018). Consequently, biogas production efficiency decreases. The OLR is generally 1–6 g VS/L day (Wirth et al., 2018). In general, with increase in OLR, the genus *Methanosarcina*, important for continued and efficient CH_4 production, increases due to efficient degradation of acetate and robust stress management (De Vrieze et al., 2012). For protein-rich waste, the *Thermoanaerobacteriales* are found to increase with increase in OLR, whereas the *Bacteroidetes* decreases (Kovács et al., 2013). Magdalena et al. (2019) studied the effect of organic loading rate (OLR) on BD of microalgae biomass (*Chlorella vulgaris*) and stepwise increase in OLR shown to increase volatile fatty acids. However, acidogenic inhibition is observed at high OLR.

7.6 Hydraulic Retention Time

The hydraulic retention time (HRT) can be defined as the average time for which biomass is retained in the digester. A short HRT is desirable as it will lead to fast waste treatment and low capital cost. However, HRT lower than time required for doubling microorganisms can result in washout of their community and acidification of the digester due to low occurrence of methanogenesis (Miron et al., 2000; Ward et al., 2014). In contrast, a high HRT could cause ammonia inhibition (Klassen et al., 2016). The typical values of HRT are 10–30 days (Wirth et al., 2018). A HRT of 15 days is suggested for sufficient hydrolysis, acidogenesis and methanogenesis at 25 °C (Miron et al., 2000). Further, HRT should be higher at lower temperatures as the growth rate of microorganisms decrease with temperature. Liang et al. (2018) showed that decreasing HRT, from 20 to 3 days, increases *Firmicutes* and decreases *Thermotogae* and *Chloroflexi* during thermophilic BD of lignocellulose. Further, increasing OLR (3–7 g VS/L.day) and HRT (15–20 days) increases *Methanosarcinaceae* and *Methanosaetaceae* in mesophilic BD of food waste (Xu et al., 2018a).

7.7 Nutrients

Nutrients in biomass can also affect microbial activity and, consequently, stability and methane yield of the BD. The nutrients such as Co, Ni, Fe, Mo and W are essential, especially for acetogenesis and methanogenesis (Azim et al., 2017). They improve methanogenic community such as the genus *Methanoculleus*, *Methanosarcinales* and genus *Methanobrevibacter* (Molaey et al., 2018). Several micronutrients, such as, Na^+, K^+ and Ca^{++}, heavy metals and sulphides, can have adverse effect on methanogenesis (Reith et al., 2003). Sulphide is an essential nutrient and supplied by degradation of proteins, but at high concentration (200 mg/L), it can form metal complexes and decrease the nutrients essential for growth of microorganisms (Ward et al., 2014; Minh et al., 2016).

7.8 pH

The pH in the digester affects growth of microorganisms and activity of enzyme, and consequently, appropriate pH is essential for effective biogas production. Generally, the pH is maintained from 7 to 8 in biomass digesters (Jankowska et al., 2017). However, each step of BD has different optimum pH level. The steps of BD, except methanogenesis, can withstand a wide range of pH. Methanogenesis can only proceed with a neutral pH, i.e. 6.5–7.5 (Lettinga & Haandel, 1993). Hydrogen carbonate (or bicarbonate alkalinity) helps to maintain the optimal pH range for methanogenesis (Reith et al., 2003). High pH values can cause inhibitors, such as NH_3 and VFA, to increase. The equilibrium of NH_3 and NH_4^+ shifts towards ammonia at higher pH (Banu & Kannah, 2019). Accumulation of volatile fatty acids (VFA) decrease the pH, and in turn, methanogenesis decreases which reduces pH further (Reith et al., 2003).

7.9 *Salinity*

High salinity can obstruct the BD process. The effects of salinity on BD have been studied by Ward et al. (2014) and Wirth et al. (2018). Mottet et al. (2014) investigated the adaptations of microorganisms with change in salinity.

8 Mathematical Modelling of Biomass Digestion

The mathematical models can aid in the design, development and optimization of digesters. Mathematical modelling of biomass digestion seeks to establish a relationship among characteristics of biomass, process conditions and concentrations of various microbial groups and desired products to predict the behaviour and performance of the process (Banu & Kannah, 2019).

As discussed earlier, biomass digestion follows four steps, i.e. hydrolysis, acidogenesis, acetogenesis and methanogenesis, occurring through different microbial groups. Each microbial group takes the products of the previous step as substrate. The growth of a microorganism in a batch digester can be given by Fig. 2 as a curve of cell concentration versus time (Banu & Kannah, 2019).

Initial phase, where the concentration of microorganism is small and increases very slowly as it adapts to the environment, is referred to as lag phase or lethargy. Thereafter, a sharp increase in cell concentration is observed in the growth phase. As soon as equilibrium is established (i.e. growth rate is equal to death rate), the stationary phase is achieved. The stationary phase is followed by death phase

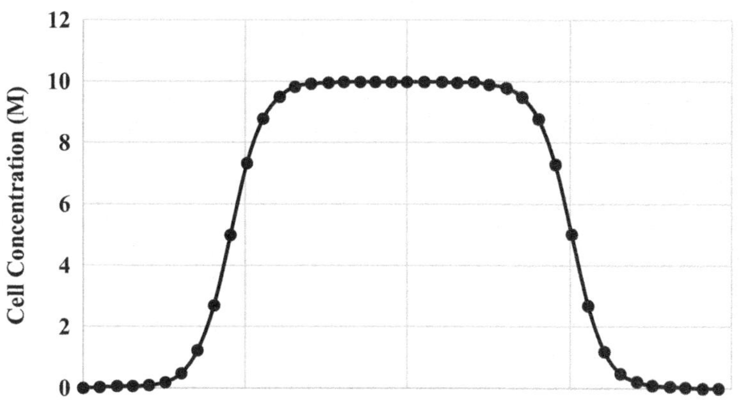

Fig. 2 Growth of a microorganism in a batch digester (Banu & Kannah, 2019)

where death rate is higher than growth rate due to small concentration of the biomass, and consequently, a sharp drop in the cell concentration is observed in the death phase. As biomass digestion involves several such microbial groups, the final concentration of desired products will be based on various parallel chemical and biological reactions occurring simultaneously.

The simplest model of growth phase in BD assumes that the rate of cell reproduction is proportional to the cell concentration with constant cell growth rate (μ), leading to an exponential growth in cell concentration. However, it leads to large errors as cell growth rate is variable due to various inhibitions such as deficiency of nutrients and competition among microorganisms, etc. (Banu & Kannah, 2019). Therefore, Monod (1942) suggested to calculate the cell growth rate as follows:

$$\mu = \frac{\mu_{max} S}{K_s + S} \tag{14}$$

where μ_{max} is the maximum rate of cell growth, K_s is a constant referred to as saturation and S is the concentration of the substrate. They are dependent on microbial species and operating conditions (Banu & Kannah, 2019). These models were easy to apply but did not predict the concentrations of products satisfactorily, especially in the region close to stationary phase. Therefore, Winsor (1932) proposed Gompertz (1825) model for human demography to be used for BD given as follows:

$$\mu = c \ln\left(\frac{a}{X}\right) \tag{15}$$

where X is the cell concentration and a and c are constants. Recently, this equation has been used by Bayrakdar et al. (2018), Juárez et al. (2018) and Li et al. (2018).

Further, some kinetic models were proposed which can be classified into four groups: single-step first-order reaction (Zhang et al., 2018), two-step first-order reaction (Shin & Song, 1995), single-step first-order reaction with two rates (Dennehy et al., 2016) and two-step first-order reaction with two rates (Brulé et al., 2014). Zahan et al. (2018) compared the aforementioned models and found only small differences among them with large root mean square errors.

Currently, the most comprehensive and utilized model for biomass digestion is Anaerobic Digestion Model No. 1 (ADM1) developed by the International Water Association (IWA) Task Group for mathematical modelling of anaerobic digestion processes in 2002 (Batstone et al., 2002). ADM1 is highly complex and includes 19 chemical and biological conversion processes with 24 dynamic state variables for description of the reactions occurring in a perfectly mixed BD. Recently, ADM1 was applied by Zhao et al. (2019), Li et al. (2019a), Bedoić et al. (2019), etc.

9 Life Cycle Assessment

Life cycle assessment (LCA) is a useful tool for analysing the environmental performance of a biogas production and associated power plant. There have been several LCA studies on biogas production and electricity generation from biogas. Kimming et al. (2011) compared CHP plants based on biomass combustion, gasification and digestion with that based on fossil fuel using LCA. They concluded that biomass digestion has the best performance with respect to global warming potential and use of the by-product as a fertilizer. Čuček et al. (2011) devised a mixed integer programming mathematical model for LCA and concluded that biogas production with auxiliary facilities is profitable for large-scale applications. Meester et al. (2012) studied the environmental sustainability of BD for electricity production and showed that energy efficiency of 22.6–33.3% and resource saving of 90% can be achieved by BD. However, operation without heat valorization results in loss in performance and use of digestate as fertilizer can cause increased nitrogen and methane emissions. Bacenetti and Fiala (2014) evaluated the environmental performance of electricity production (< 300 kW) from four BD plants using animal slurry using LCA method. They concluded that BD plants have better environmental performances than fossil fuel–based power plants in terms of global warming and resource depletion. Moreover, the recovery and the valorization of heat are essential for an eco-friendly operation. Ertem et al. (2017) performed comparative LCA for energy crops and macroalgae at a real biogas plant and suggested that microalgae can lead to sustainable energy production and environmental benefits if microalgae are regionally available. Bedoić et al. (2019) studied the environmental effects of the biogas production for combined heat and power plant from co-digestion of riverbank grass with cattle slurry and maize silage using LCA. They evaluated four categories, i.e. human health, ecosystem quality, climate change and resources and global warming potential (GWP). Slorach et al. (2019) performed LCA for 1 MWh electricity production from household food waste and compared it with other electricity generation options based on 19 environmental impacts. BD was found to have lower environmental impacts, but use of digestate as fertilizer increases eutrophication, acidification and particulate matter. Aberilla et al. (2019) compared combustion, gasification and digestion for small-scale power generation based on LCA using 18 environmental impacts and concluded that global warming potential of digestion is lower than other technologies. Lijó et al. (2019) reviewed the LCA methodology and related issues for bioenergy production. Aui et al. (2019) studied farm scale CHP plants (950 kWe) with respect to internal rate of return (IRR) and GHG emissions. The IRR were found to vary between 3.51% and 5.57% and GHG emissions changed from −82.6 to 498.52 g CO2/kWh. Further, while power efficiency, operating capacity and waste generation per cattle affected the IRR the most, GHG emissions from manure was the most important factor for GHG emissions. Hijazi et al. (2020) studied the environmental impact (global warming, GHG emissions, acidification, eutrophication, resource depletion, human toxicity potential and ozone layer depletion potential) of utilizing different nanoparticles (Ni, Co, Fe and Fe_3O_4)

for biogas production. They showed that Co nanoparticles are the most suitable in terms of the highest reduction in GHG emissions from electricity production.

10 Barriers to Growth

Although electricity generation from BD will be an important technology for environment, there are some barriers to its large-scale implementation. The most important factors that hinder its growth are as follows:

- Economical: Electricity from BD, which costs relatively higher, must compete with conventional technologies. Further, economic data is not available to make informed decisions. It has been suggested that all the products, i.e. heat, power and fertilizer, has to be marketed for the plant to be economical. Therefore, market access and fair pricing of all these products need to be available.
- Technical: Low energy yield, reduction of prices, integration to the grid and optimum and mature technology of electricity from BD will require technical capabilities to reduce dependence on foreign import of technology and for effective utilization of local biomass.
- Social: Electricity from BD should improve its social acceptance which is lacking due to unsuccessful plants, unhygienic work environment and awareness about its environmental and agricultural advantages.
- Administrative: Policy makers have to improve cooperation among parties working on various aspects of electricity generation from BD and remove any legal and administrative obstacles for commercialization and access to energy grid. If need be, subsidies should be implemented in view of the environmental benefits of these projects.

However, in future, non-availability and, in turn, rising prices of fossil fuels and implementation of carbon taxes will aid in its growth.

11 Conclusion

Biomass is renewable, widely available and equitably distributed energy source which is easy to store and transport. Biomass digestion (BD) is a promising technology for ensuring energy security and environmental protection as well as waste utilization. Biomass digestion (BD) is a natural biological process in which biomass is broken down by several microorganisms into biogas, mainly consisting of CH_4 and CO_2, and digestate, solid residue with mineral substances. Biogas produced by BD can be utilized, after thorough cleaning and upgradation, as natural gas as fuel for cooking, heating and in gas operated vehicles as well as to generate heat and electricity and produce variety of chemicals. Further, digestate can be used as a high-quality fertilizer. Reduced odours and improved agriculture practices are also important benefits of the BD. To produce electricity from BD, four steps, i.e. pretreatment, BD, biogas cleaning and upgrading and power generation, are followed. In general, BD comprises of four steps, namely, hydrolysis, acidogenesis, acetogenesis and methanogenesis. The composition of biogas depends on feedstock, starting inoculum, temperature, organic loading rate, hydraulic retention time, trace elements, pH and salinity. Mathematical models could be an important tool to predict the behaviour and performance of BD and advancement in research and development of BD. Currently, Anaerobic Digestion Model No. 1 (ADM1) is the most comprehensive and utilized model. The ADM1 describes BD as a five-stage anaerobic digestion process with 19 subprocesses. Life cycle assessment can be used to analyse the environmental performance of the BD and associated energy production processes. The growth of electricity generation from BD is hindered by technical, social, economical and administrative issues.

References

Abdelsalam, E. M., & Samer, M. (2019). Biostimulation of anaerobic digestion using nanomaterials for increasing biogas production. *Reviews in Environmental Science and Biotechnology, 18*, 525–541.

Aberilla, J. M., Gallego-Schmid, A., & Azapagic, A. (2019). Environmental sustainability of small-scale biomass power technologies for agricultural communities in developing countries. *Renewable Energy, 141*, 493–506.

Abraham, A., Mathewa, A. K., Park, H., Choi, O., Sindhu, R., Parameswaran, B., Pandey, A., Park, J. H., & Sang, B.-I. (2020). Pretreatment strategies for enhanced biogas production from lignocellulosic biomass. *Bioresource Technology, 301*, 122725.

Adams, P. W. R., & McManus, M. C. (2019). Characterisation and variability of greenhouse gas emissions from biomethane production via anaerobic digestion of maize. *Journal of Cleaner Production, 218*, 529–542.

Aui, A., Li, W., & Wright, M. M. (2019). Techno-economic and life cycle analysis of a farm-scale anaerobic digestion plant in Iowa. *Waste Management, 89*, 154–164.

Azim, A. A., Pruckner, C., Kolar, P., Taubner, R.-S., Fino, D., Saracco, G., Sousa, F. L., & Rittmann, S. K.-M. R. (2017). The physiology of trace elements in biological methane production. *Bioresource Technology, 241*, 775–786.

Bacenetti, J. and Fiala, M. (2014). 'Electricity production from anaerobic digestion of animal slurries in farm scale plants', The 4th World Sustainability Forum, 1–30 November, 2014.

Banu, J. R., & Kannah, R. Y. (2019). *Anaerobic digestion*. IntechOpen.

Batstone, D. J., Keller, J., Angelidaki, I., Kalyuzhnyi, S. V., Pavlostathis, S. G., Rozzi, A., Sanders, W. T. M., Siegrist, H. and Vavilin, V. A. (2002) Anaerobic digestion model no. 1 Technical Report Report No. 13. London: IWA Publishing.

Battista, F., Frison, N., & Bolzonella, D. (2019). Energy and nutrients' recovery in anaerobic digestion of agricultural biomass: An Italian perspective for future applications. *Energies, 12*, 3287.

Bayrakdar, A., Sürmeli, R. Ö., & Çalli, B. (2018). Anaerobic digestion of chicken manure by a leach-bed process coupled with side-stream membrane ammonia separation. *Bioresource Technology, 258*, 41–47.

Bedoić, R., Čuček, L., Ćosić, B., Krajnc, D., Smoljanić, G., Kravanja, Z., Ljubas, D., Pukšec, T., & Duić, N. (2019). Green biomass to biogas – A study on anaerobic digestion of residue grass. *Journal of Cleaner Production, 213*, 700–709.

Bohutskyi, P., Phan, D., Spierling, R. E., Kopachevsky, A. M., Bouwer, E. J., Lundquist, T. J., & Betenbaugh, M. J. (2019). Production of lipid-containing algal-bacterial polyculture in wastewater and biomethanation of lipid extracted residues: Enhancing methane yield through hydrothermal pretreatment and relieving solvent toxicity through co-digestion. *Science of the Total Environment, 653*, 1377–1394.

BP (2020) Statistical review of world energy.

Brulé, M., Oechsner, H., & Jungbluth, T. (2014). Exponential model describing methane production kinetics in batch anaerobic digestion: A tool for evaluation of biochemical methane potential assays. *Bioprocess and Biosystems Engineering, 37*(9), 1759–1770.

Chakravarthi, J. (1997, July 27–August 1). *Biogas and energy production from cattle waste*. 32nd Intersociety Energy Conversion Engineering Conference (IECEC-97), Honolulu, HI, USA.

Chen, H., & Chang, S. (2017). Impact of temperatures on microbial community structures of sewage sludge biological hydrolysis. *Bioresource Technology, 245*, 502–510.

Cirne, D. G., Paloumet, X., Bjornsson, L., Alves, M. M., & Mattiasson, B. (2007). Anaerobic digestion of lipid-rich waste—Effects of lipid concentration. *Renewable Energy, 32*, 965–975.

Čuček, L., Drobež, R., Pahor, B., & Kravanja, Z. (2011). Sustainable LCA-based MIP synthesis of biogas processes. *Computer Aided Chemical Engineering, 29*, 1999–2003.

Dahunsi, S. O. (2019). Mechanical pretreatment of lignocelluloses for enhanced biogas production: Methane yield prediction from biomass structural components. *Bioresource Technology, 280*, 18–26.

Dennehy, C., Lawlor, P. G., Croize, T., Jiang, Y., Morrison, L., Gardiner, G. E., & Zhan, X. (2016). Synergism and effect of high initial volatile fatty acid concentrations during food waste and pig manure anaerobic co-digestion. *Waste Management, 56*, 173–180.

De Vrieze, J., Hennebel, T., Boon, N., & Verstraete, W. (2012). Methanosarcina: The rediscovered methanogen for heavy duty biomethanation. *Bioresource Technology, 112*, 1–9.

Ertem, F. C., Neubauer, P., & Junne, S. (2017). Environmental life cycle assessment of biogas production from marine macroalgal feedstock for the substitution of energy crops. *Journal of Cleaner Production, 140*, 977–985.

Filho, D. A. S., Oliveira, L. R. G., Fraga, T. J. M., Sobrinho, M. A. M., & Jucá, J. F. T. (2020). Development of a horizontal reactor with radial agitation to synthesize bio-methane from biomass waste and domestic sewage sludge. *Journal of Cleaner Production, 257*, 120616.

Franco, R. T., Buffiere, P., & Bayard, R. (2018). Co-ensiling of cattle manure before biogas production: Effects of fermentation stimulants and inhibitors on biomass and methane preservation. *Renewable Energy, 121*, 315–323.

Fotidis, I. A., Wang, H., Fiedel, N.-R., Luo, G., & Karakashev, D. B. (2014). Bioaugmentation as a solution to increase methane production from an ammonia-rich substrate. *Environmental Science and Technology, 48*, 7669–7676.

Freitas, F. F., DeSouza, S. S., Ferreira, L. R. A., Otto, R. B., Alessio, F. J., DeSouza, S. N. M., Venturini, O. J., & Ando-Junior, O. H. (2019). The Brazilian market of distributed biogas generation: Overview, technological development and case study. *Renewable and Sustainable Energy Reviews, 101*, 146–157.

Gompertz, B. (1825). On the nature of the function expressive of the law of human mortality, and on a new mode on determining the value of live contingencies. *Philosophical Transactions of the Royal Society of London, 115*, 513–585.

Grontmij (1997) Guidebook on landfill gas extraction and utilisation, Novem contract number 258320/0020. G.C. engineers.

Hasan, A. S. M. M., & Ammenberg, J. (2019). Biogas potential from municipal and agricultural residual biomass for power generation in Hazaribagh, Bangladesh—A strategy to improve the energy system. *Renewable Energy Focus, 29*, 14–23.

He, Q., Li, L., Zhao, X., Qu, L., Wu, D., & Peng, X. (2017). Investigation of foaming causes in three mesophilic food waste digesters: Reactor performance and microbial analysis. *Science Reporter, 7*, 13701.

Hijazi, O., Abdelsalam, E., Samer, M., Attia, Y. A., Amer, B. M. A., Amer, M. A., Badr, M., & Bernhardt, H. (2020). Life cycle assessment of the use of nanomaterials in biogas production from anaerobic digestion of manure. *Renewable Energy, 148*, 417–424.

International Energy Agency (IEA). (2012). *Technology roadmap: Bioenergy for heat and power*. IEA.

International Renewable Energy Agency (IRENA) (2012) Biomass for power generation, renewable energy technologies: Cost analysis series, Volume 1: Power Sector, Issue 1/5.

Janke, L., McCabe, B. K., Harris, P., Hill, A., Lee, S., Weinrich, S., Marchuk, S., & Baillie, C. (2019). Ensiling fermentation reveals pre-treatment effects for anaerobic digestion of sugarcane biomass: An assessment of ensiling additives on methane potential. *Bioresource Technology, 279*, 398–403.

Jankowska, E., Sahu, A. K., & OleskowiczPopiel, P. (2017). Biogas from microalgae: Review on microalgae's cultivation, harvesting and pretreatment for anaerobic digestion. *Renewable and Sustainable Energy Reviews, 75*, 692–709.

Johnravindar, D., Liang, B., Fu, R., Luo, G., Meruvu, H., Yang, S., Yuan, B., & Fei, Q. (2020). Supplementing granular activated carbon for enhanced methane production in anaerobic co-digestion of post-consumer substrates. *Biomass and Bioenergy, 136*, 105543.

Juárez, J. M., Pastor, E. R., Sevilla, J. M. F., Torre, R. M., García-Encina, P. A., & Rodríguez, S. B. (2018). Effect of pretreatments on biogas production from microalgae biomass grown in pig manure treatment plants. *Bioresource Technology, 257*, 30–38.

Jurgutis, L., Slepetiene, A., Volungevicius, J., & Amaleviciute-Volunge, K. (2020). Biogas production from chicken manure at different organic loading rates in a mesophilic full scale anaerobic digestion plant. *Biomass and Bioenergy, 141*, 105693.

Karthikeyan, O. P., & Visvanathan, C. (2013). Bio-energy recovery from high-solid organic substrates by dry anaerobic bioconversion processes: A review. *Reviews in Environmental Science and Bio/Technology, 12*, 257–284.

Kheybari, S., Rezaie, F. M., Naji, S. A., & Najafi, F. (2019). Evaluation of energy production technologies from biomass using analytical hierarchy process: The case of Iran. *Journal of Cleaner Production, 232*, 257–265.

Kimming, M., Sundberg, C., Nordberg, Å., Baky, A., Bernesson, S., Norén, O., & Hansson, P.-A. (2011). Biomass from agriculture in small-scale combined heat and power plants – A comparative life cycle assessment. *Biomass and Bioenergy, 35*(4), 1572–1581.

Klassen, V., Blifernez-Klassen, O., Wobbe, L., Schlueter, A., Kruse, O., & Mussgnug, J. H. (2016). Efficiency and biotechnological aspects of biogas production from microalgal substrates. *Journal of Biotechnology, 234*, 7–26.

Koch, K., Lippert, T., & Drewes, J. E. (2017). The role of inoculum's origin on the methane yield of different substrates in biochemical methane potential (BMP) tests. *Bioresource Technology, 243*, 457–463.

Koupaie, E. H., Dahadha, S., Lakeh, A. A. B., Azizi, A., & Elbeshbishy, E. (2019). Enzymatic pretreatment of lignocellulosic biomass for enhanced biomethane production—A review. *Journal of Environmental Management, 233*, 774–784.

Kovács, E., Wirth, R., Mároti, G., Bagi, Z., Rákhely, G., & Kovács, K. L. (2013). Biogas production from protein-rich biomass: Fed-batch anaerobic fermentation of casein and of pig blood and associated changes in microbial community composition. *PLoS One, 8*, 1–18.

Kuo, J., & Dow, J. (2017). Biogas production from anaerobic digestion of food waste and relevant air quality implications. *Journal of the Air & Waste Management Association, 67*(9), 1000–1011.

Labatut, R. A., Angenent, L. T., & Scott, N. R. (2014). Conventional mesophilic vs. thermophilic anaerobic digestion: A tradeoff between performance and stability? *Water Research, 53*, 249–258.

Lettinga, G., & Haandel, A. C. (1993). Anaerobic digestion for energy production and environmental protection. In T. B. Johansson, H. Kelly, A. K. N. Reddy, & R. H. Williams (Eds.), *Renewable energy; sources for fuels and electricity* (pp. 817–839). Island press.

Li, H., Mehmood, D., Thorin, E., & Yu, Z. (2017). Biomethane production via anaerobic digestion and biomass gasification. *Energy Procedia, 105*, 1172–1177.

Li, C., Wang, X., Zhang, G., Li, J., Li, Z., Yu, G., & Wang, Y. (2018). A process combining hydrothermal pretreatment, anaerobic digestion and pyrolysis for sewage sludge dewatering and co-production of biogas and biochar: Pilot-scale verification. *Bioresource Technology, 254*, 187–193.

Li, X., Yang, Z., Liu, G., Ma, Z., & Wang, W. (2019a). Modified anaerobic digestion model no.1 (ADM 1) for modelling anaerobic digestion process at different ammonium concentrations. *Water Environment Research, 91*, 700–714.

Li, Y., Chen, Y., & Wu, J. (2019b). Enhancement of methane production in anaerobic digestion process: A review. *Applied Energy, 240*, 120–137.

Liang, X., Whitham, J. M., Holwerda, E. K., Shao, X., Tian, L., Wu, Y.-W., Lombard, V., Henrissat, B., Klingeman, D. M., Yang, Z. K., Podar, M., Richard, T. L., Elkins, J. G., Brown, S. D., & Lynd, L. R. (2018). Development and characterization of stable anaerobic thermophilic methanogenic microbiomes fermenting switchgrass at decreasing residence times. *Biotechnology for Biofuels, 11*, 243.

Lijó, L., González-García, S., Lovarelli, D., Moreira, M. T., Feijoo, G., & Bacenetti, J. (2019). Life cycle assessment of renewable energy production from biomass. In R. Basosi, M. Cellura, S. Longo, & M. Parisi (Eds.), *Life cycle assessment of energy systems and sustainable energy technologies*. Springer International Publishing.

Lins, P., Reitschuler, C., & Illmer, P. (2014). Methanosarcina sp., the key to relieve the start-up of a thermophilic anaerobic digestion suffering from high acetic acid loads. *Bioresource Technology, 152*, 347–354.

Liotta, F. (2013) 'Bio-methanation tests and mathematical modelling to assess the role of moisture content on anaerobic digestion of organic waste', Ph.D. Thesis, Earth Sciences, Université Paris-Est.

Liu, T., Sun, L., Müller, B., & Schnürer, A. (2017). Importance of inoculum source and initial community structure for biogas production from agricultural substrates. *Bioresource Technology, 245*, 768–777.

Magdalena, J. A., Greses, S., & González-fernández, C. (2019). Impact of organic loading rate in volatile fatty acids production and population dynamics using microalgae biomass as substrate. *Scientific Reports, 9*, 18374.

Malico, I., Pereira, R. N., Gonçalves, A. C., & Sousa, A. M. O. (2019). Current status and future perspectives for energy production from solid biomass in the European industry. *Renewable and Sustainable Energy Reviews, 112*, 960–977.

Maneesha, P., & Bhaskarwar, A. N. (2012). Biomass conversion to energy. In C. Baskar, S. Bhaskar, & S. Ranjit (Eds.), *Biomass conversion: The Interface of the biotechnology, chemistry and material science*. Springer-Verlag Berlin Heidelberg.

McKendry, P. (2002a). Energy production from biomass (part 1): Overview of biomass. *Bioresource Technology, 83*(1), 37–46.

McKendry, P. (2002b). Energy production from biomass (part 2): Conversion technologies. *Bioresource Technology, 83*(1), 47–54.

Meester, S. D., Demeyer, J., Velghe, F., Peene, A., Langenhove, H. V., & Dewulf, J. (2012). The environmental sustainability of anaerobic digestion as a biomass valorization technology. *Bioresource Technology, 121*, 396–403.

Minh, T. P., Ketheesan, B., Yan, Z., & Stuckey, D. C. (2016). Trace metal speciation and bioavailability in anaerobic digestion: A review. *Biotechnology Advances, 34*, 122–136.

Miron, Y., Zeeman, G., van Lier, J. B., & Lettinga, G. (2000). The role of sludge retention time in the hydrolysis and acidification of lipids, carbohydrates and proteins during digestion of primary sludge in CSTR systems. *Water Resources, 34*(5), 1705–1713.

Mohammad, S., & Ardebili, S. (2020). Green electricity generation potential from biogas produced by anaerobic digestion of farm animal waste and agriculture residues in Iran. *Renewable Energy, 154*, 29–37.

Molaey, R., Bayrakdar, A., Sürmeli, R. O., & Calli, B. (2018). Influence of trace element supplementation on anaerobic digestion of chicken manure: Linking process stability to methanogenic population dynamics. *Journal of Cleaner Production, 181*, 794–800.

Monod, J. (1942). Recherches sur la croissance des cellules bactériennes, Research on the growth of bacterial cells. Ph.D. thesis, Sorbonne, Paris.

Mottet, A., Habouzit, F., & Steyer, J. P. (2014). Anaerobic digestion of marine microalgae in different salinity levels. *Bioresource Technology, 158*, 300–306.

Moustakas, K., Parmaxidou, P., & Vakalis, S. (2020). Anaerobic digestion for energy production from agricultural biomass waste in Greece: Capacity assessment for the region of Thessaly. *Energy, 191*, 116556.

Murphy, J., Braun, R., Weiland, P., & Wellinger, A. (2010). Biogas from energy crop digestion. *IEA Bioenergy Task, 37*.

Neshat, S. A., Mohammadi, M., Najafpour, G. D., & Lahijani, P. (2017). Anaerobic codigestion of animal manures and lignocellulosic residues as a potent approach for sustainable biogas production. *Renewable and Sustainable Energy Reviews, 79*, 308–322.

Nzila, A. (2017). Update on bioaugmentation in anaerobic processes for biogas production. *Anaerobe, 46*, 3–17.

Obileke, K. C., Nwokolo, N., Makaka, G., Mukumba, P., & Onyeaka, H. (2020). Anaerobic digestion: Technology for biogas production as a source of renewable energy—A review. *Energy & Environment, 32*(2), 191–225.

Oleszek, M., Król, A., Tys, J., Matyka, M., & Kulik, M. (2014). Comparison of biogas production from wild and cultivated varieties of reed canary grass. *Bioresource Technology, 156*, 303–306.

Öner, B. E., Akyol, C., Bozan, M., Ince, O., Aydin, S., & Ince, B. (2018). Bioaugmentation with clostridium thermocellum to enhance the anaerobic biodegradation of lignocellulosic agricultural residues. *Bioresource Technology, 249*, 620–625.

Pampillón-González, L., & Canepa, J. R. L. (2017). Biomass as an alternative for gas production. In H. A. Al-Megren & R. H. Altamimi (Eds.), *Advances in natural gas emerging technologies*. IntechOpen.

Pan, J., Ma, J., Zhai, L., Luo, T., Mei, Z., & Liu, H. (2019). Achievements of biochar application for enhanced anaerobic digestion: A review. *Bioresource Technology, 292*, 122058.

Passos, F., & Ferrer, I. (2015). Influence of hydrothermal pretreatment on microalgal biomass anaerobic digestion and bioenergy production. *Water Research, 68*, 364–373.

Paudel, S. R., Banjara, S. P., Choi, O. K., Park, K. Y., Kim, Y. M., & Lee, J. W. (2017). Pretreatment of agricultural biomass for anaerobic digestion: Current state and challenges. *Bioresource Technology, 245*, 1194–1205.

Perea-Moreno, M.-A., Samerón-Manzano, E., & Perea-Moreno, A.-J. (2019). Biomass as renewable energy: Worldwide research trends. *Sustainability, 11*, 863.

Petersson, A. (2013). Biogas cleaning. In A. Wellinger, J. Murphy, & D. Baxter (Eds.), *The biogas handbook: Science, production and applications* (pp. 329–341). Woodhead Publishing Series in Energy.

Piñas, J. A. V., Venturini, O. J., Lora, E. E. S., Olmo, O. A., & Roalcaba, O. D. C. (2019). An economic holistic feasibility assessment of centralized and decentralized biogas plants with mono-digestion and co-digestion systems. *Renewable Energy, 139*, 40–51.

Poszytek, K., Ciezkowska, M., Sklodowska, A., & Drewniak, L. (2016). Microbial consortium with high cellulolytic activity (MCHCA) for enhanced biogas production. *Frontiers in Microbiology, 7*, 324.

Raheem, A., Prinsen, P., Vuppaladadiyam, A. K., Zhao, M., & Luque, R. (2018). A review on sustainable microalgae based biofuel and bioenergy production: Recent developments. *Journal of Cleaner Production, 181*, 42–59.

Rajagopal, R., Massé, D. I., & Singh, G. (2013). A critical review on inhibition of anaerobic digestion process by excess ammonia. *Bioresource Technology, 143*, 632–641.

Raposo, F., Fernandez-Cegri, V., De la Rubia, M. A., Borja, R., Beline, F., Cavinato, C., Demirer, G., Fernández, B., Fernández-Polanco, M., Frigon, J. C., Ganesh, R., Kaparaju, P., Koubova, J., Méndez, R., Menin, G., Peene, A., Scherer, P., Torrijos, M., Uellendahl, H., ... de Wilde, V. (2011). Biochemical methane potential (BMP) of solid organic substrates: Evaluation of anaerobic biodegradability using data from an international interlaboratory study. *Journal of Chemical Technology and Biotechnology, 86*, 1088–1098.

Rasapoor, M., Young, B., Brar, R., Sarmah, A., Zhuang, W.-Q., & Baroutian, S. (2020). Recognizing the challenges of anaerobic digestion: Critical steps toward improving biogas generation. *Fuel, 261*, 116497.

Reith, J. H., Wijffels, R. H. and Barten, H. (2003) Bio-methane & Bio-hydrogen: Status and perspectives of biological methane and hydrogen production. The Netherlands: Dutch Biological Hydrogen Foundation, Energy Research Centre of the Netherlands (ECN).

Rosas-Mendoza, E. S., Méndez-Contreras, J. M., Aguilar-Lasserre, A. A., Vallejo-Cantú, N. A., & Alvarado-Lassman, A. (2020). Evaluation of bioenergy potential from citrus effluents through anaerobic digestion. *Journal of Cleaner Production, 254*, 120128.

Sanders, W. (2001). Anaerobic Hydrolysis during digestion of complex substrates, Ph.D. thesis, Department of Environmental Technology. The Netherlands: Wageningen University and Research Center.

Sasaki, K., Morita, M., Sasaki, D., Nagaoka, J., Matsumoto, N., Ohmura, N., & Shinozaki, H. (2011). Syntrophic degradation of proteinaceous materials by the thermophilic strains Coprothermobacter proteolyticus and Methanothermobacter thermautotrophicus. *Journal of Bioscience and Bioengineering, 112*, 469–472.

Sawatdeenarunat, C., Surendra, K. C., Takara, D., Oechsner, H., & Khanal, S. K. (2015). Anaerobic digestion of lignocellulosic biomass: Challenges and opportunities. *Bioresource Technology, 178*, 178–186.

Schwede, S., Rehman, Z.-U., Gerber, M., Theiss, C., & Span, R. (2013). Effects of thermal pretreatment on anaerobic digestion of Nannochloropsis Salina biomass. *Bioresource Technology, 143*, 505–511.

Sherwood, J. (2020). The significance of biomass in a circular economy. *Bioresource Technology, 300*, 122755.

Shin, H. S., & Song, Y. C. (1995). A model for evaluation of anaerobic degradation characteristics of organic waste: Focusing on kinetics, rate-limiting step. *Environmental Technology, 16*, 775–784.

Shirzad, M., Panahi, H. K. S., Dashti, B. B., Rajaeifar, M. A., Aghbashlo, M., & Tabatabaei, M. (2019). A comprehensive review on electricity generation and GHG emission reduction potentials through anaerobic digestion of agricultural and livestock/slaughterhouse wastes in Iran. *Renewable and Sustainable Energy Reviews, 11*, 571–594.

Sikora, A., Detman, A., Mielecki, D., Chojnacka, A., & Błaszczyk, M. (2018). Searching for metabolic pathways of anaerobic digestion: A useful list of the key enzymes. In J. R. Banu & R. Y. Kannah (Eds.), *Anaerobic digestion*. IntechOpen. (2019).

Singh, B., Szamosi, Z., Siménfalvi, Z., & Rosas-Casals, M. (2020). Decentralized biomass for biogas production. Evaluation and potential assessment in Punjab (India). *Energy Reports, 6*, 1702–1714.

Slorach, P. C., Jeswani, H. K., Cuéllar-Franca, R., & Azapagic, A. (2019). Environmental sustainability of anaerobic digestion of household food waste. *Journal of Environmental Management, 236*, 798–814.

Solé-Bundó, M., Garfí, M., Matamoros, V., & Ferrer, I. (2019). Co-digestion of microalgae and primary sludge: Effect on biogas production and microcontaminants removal. *Science of the Total Environment, 660*, 974–981.

Song, Z., Zhang, C., Yang, G., Feng, Y., Ren, G., & Han, X. (2014). Comparison of biogas development from households and medium and large-scale biogas plants in rural China. *Renewable and Sustainable Energy Reviews, 33*, 204–213.

Speda, J., Johansson, M. A., Odnell, A., & Karlsson, M. (2017). Enhanced biomethane production rate and yield from lignocellulosic ensiled forage ley by in situ anaerobic digestion treatment with endogenous cellulolytic enzymes. *Biotechnology for Biofuels, 10*, 129.

Tonanzi, B., Gallipoli, A., Gianico, A., Montecchio, D., Pagliaccia, P., Di Carlo, M., Rossetti, S., & Braguglia, C. M. (2018). Long-term anaerobic digestion of food waste at semi-pilot scale: Relationship between microbial community structure and process performances. *Biomass and Bioenergy, 118*, 55–64.

Vassalle, L., Díez-Montero, R., Machado, A. T. R., Moreira, C., Ferrer, I., Mota, C. R., & Passos, F. (2020). Upflow anaerobic sludge blanket in microalgae-based sewage treatment: Co-digestion for improving biogas production. *Bioresource Technology, 300*, 122677.

Ward, A. J., Lewis, D. M., & Green, F. B. (2014). Anaerobic digestion of algae biomass: A review. *Algal Research, 5*, 204–214.

Wellinger, A. and Lindberg, A. (1999). Biogas upgrading and utilisation, IEA Bioenergy, Task 24: Energy from biological conversion of organic waste.

Westerholm, M., Isaksson, S., Lindsjö, O. K., & Schnürer, A. (2018). Microbial community adaptability to altered temperature conditions determines the potential for process optimisation in biogas production. *Applied Energy, 226*, 838–848.

Winsor, C. P. (1932). The Gompertz curve as a growth curve. *Proceedings of the National Academy of Sciences, 18*, 1–8.

Wirth, R., Lakatos, G., Böjti, T., Maróti, G., Bagi, Z., Rákhely, G., & Kovács, K. L. (2018). Anaerobic gaseous biofuel production using microalgal biomass—A review. *Anaerobe, 52*, 1–8.

Wojcieszak, M., Pyzik, A., Poszytek, K., Krawczyk, P. S., Sobczak, A., Lipinski, L., Roubinek, O., Palige, J., Aleksandra, S., & Drewniak, L. (2017). Adaptation of methanogenic inocula to anaerobic digestion of maize silage. *Frontiers in Microbiology, 8*, 1881.

World Bioenergy Association (WBA). (2020). *Global bioenergy statistics*. WBA.

Xiao, C., Fu, Q., Liao, Q., Huang, Y., Xia, A., Chen, H., & Zhu, X. (2020a). Life cycle and economic assessments of biogas production from microalgae biomass with hydrothermal pretreatment via anaerobic digestion. *Renewable Energy, 151*, 70–78.

Xiao, Q., Chen, W., Tian, D., Shen, F., Hu, J., Long, L., Zeng, Y., Yang, G., & Deng, S. (2020b). Integrating the bottom ash residue from biomass power generation into anaerobic digestion to improve biogas production from lignocellulosic biomass. *Energy & Fuels, 34*(2), 1101–1110.

Xu, R., Yang, Z.-H., Zheng, Y., Liu, J.-B., Xiong, W.-P., Zhang, Y.-R., Lu, Y., Xue, W.-J., & Fan, C.-Z. (2018a). Organic loading rate and hydraulic retention time shape distinct ecological networks of anaerobic digestion related microbiome. *Bioresource Technology, 262*, 184–193.

Xu, R., Zhang, K., Liu, P., Khan, A., Xiong, J., Tian, F., & Li, X. (2018b). A critical review on the interaction of substrate nutrient balance and microbial community structure and function in anaerobic co-digestion. *Bioresource Technology, 247*, 1119–1127.

Zahan, Z., Othman, M. Z., & Muster, T. H. (2018). Anaerobic digestion/co-digestion kinetic potentials of different agroindustrial wastes: A comparative batch study for C/N optimisation. *Waste Management, 71*, 663–674.

Zhang, H., Luo, L., Li, W., Wang, X., Sun, Y., Sun, Y., & Gong, W. (2018). Optimization of mixing ratio of ammoniated rice straw and food waste co-digestion and impact of trace element supplementation on biogas production. *Journal of Material Cycles and Waste Management, 20*(2), 745–753.

Zhang, L., Li, F., Kuroki, A., Loha, K.-C., Wang, C.-H., Dai, Y., & Tong, Y. W. (2020). Methane yield enhancement of mesophilic and thermophilic anaerobic co-digestion of algal biomass and food waste using algal biochar: Semi-continuous operation and microbial community analysis. *Bioresource Technology, 302*, 122892.

Zhao, X., Li, L., Wu, D., Xiao, T., Ma, Y., & Peng, X. (2019). Modified anaerobic digestion model no. 1 for modelling methane production from food waste in batch and semi-continuous anaerobic digestions. *Bioresource Technology, 271*, 109–117.

Zheng, Y., Zhao, J., Xu, F., & Li, Y. (2014). Pretreatment of lignocellulosic biomass for enhanced biogas production. *Progress in Energy and Combustion Science, 42*, 35–53.

Zhu, L., Awais, M., Javed, H. M. A., Mustafa, M. S., Tlili, I., Khan, S. U., & Shadloo, M. S. (2020). Photo-catalytic pretreatment of biomass for anaerobic digestion using visible light and Nickle oxide (NiOx) nanoparticles prepared by sol gel method. *Renewable Energy, 154*, 128–135.

Ziels, R. M., Svensson, B. H., Sundberg, C., Larsson, M., Karlsson, A., & Yekta, S. S. (2018). Microbial rRNA gene expression and co-occurrence profiles associate with biokinetics and elemental composition in full-scale anaerobic digesters. *Microbial Biotechnology, 11*, 694–709.

Ziemiński, K., & Frąc, M. (2012). Methane fermentation process as anaerobic digestion of biomass: Transformations, stages and microorganisms. *African Journal of Biotechnology, 11*(18), 4127.

Solar Energy Conversion Techniques and Practical Approaches to Design Solar PV Power Station

Bobbili N. Ch. V. Chakravarthi, Lakkakula Hari Prasad, Rajya Lakshmi Chavakula, and V. V. Vijetha Inti

Abstract The sunlight is the primary energy element that controls the global environment and living system. Bridling the solar energy for high-temperature water and electricity could give a sustainable, low carbon energy source and presents an interesting way of relieving environmental change. The solar technology is to track maximum power of sun technologies which is used to produce heat, light, and power. Renewable energy sources like solar energy play a great role in providing energy solutions. As now, there are a wide variety of collectors and utilizations of sunlight-based energy. This chapter planned to overview on solar energy systems, according to types of collectors and applications used. This part of the chapter describes various solar energy systems. The different systems are classified according to the type of collector used and the type of application. Solar water heater, space heating, space cooling and refrigeration, solar cooker, dryer, concentrated solar power, and solar photovoltaic are some of the applications of solar energy.

Some of the popular technologies in the present days are electrical power generation by using renewable sources. The renewable sources like solar, wind, and tidal are contributing at higher ratio compared to the other energies throughout the world. Out of these, the solar has the leading role when compared to the wind, tidal, and other sources as they are having limitations based on the availability. But solar energy is abundantly available throughout the globe, and the installation ratio is very high than the others. At present the cost of energy production from solar is also reduced to 70% from 80% when compared to the earlier technologies, so most of the future power demand is ready to be supplied from solar power system.

The technology adopted by solar power plant is, that is, when the solar radiance strikes the semiconductor (solar cell), a flow of electrons takes place through a load

B. N. C. V. Chakravarthi (✉) · R. L. Chavakula · V. V. Vijetha Inti
Vishnu Institute of Technology, Bhimavaram, A.P., India

L. Hari Prasad
Department of Electronics and Communication Engineering, Ben-Gurion University of the Negev, Beer Sheva, Israel

D. B. Pal, J. M. Jha (eds.), *Sustainable and Clean Energy Production Technologies*, Clean Energy Production Technologies,
https://doi.org/10.1007/978-981-16-9135-5_8

(closed loop), called as transformation of energy from solar to electrical (electric power).The energy produced in this procedure is in DC nature at low voltage (LV) level so it has to increase the voltage level by using a power converter called the DC-DC converter (DC transformer). As the most of the loads and transmission system is alternating current (AC) in nature, the available DC has to be converted into AC by means of a power converter (inverter), and later it may be connected to grid (on-grid) or to the load with a battery backup (off-grid).

This chapter mainly covers with the design of a 100 kWp solar power plant, including site calculations, layout of electrical structure, estimation of cable rating (both AC and DC), choosing an inverter rating, and installation procedure for both on-grid and off-grid connections. Also the protection equipment against the unusual weather conditions and fault cases in the generation and distribution of solar electrical power is discussed.

All the erection processes have been explained by taking 100.8 kWp (Block B), solar power plant located at Vishnu Institute of Technology, Bhimavaram (16.544893, 81.521240).

Keywords Solar Thermal Technology · Solar PV Technology · DC converter · Maximum power Point Tracking · Lightning arrestors

1 Introduction

During the last decades, energy requirements are exceptionally very high because of the huge development in all kinds of technologies. This made energy in its different ways to play a major task in the worldwide economic advancement and industrialization. The incessant usage of fossil fuel leads to its depletion and made its price to increase sharply day by day. Also, the environment has been damaged through the burning of fossil fuel which emits greenhouse gases mainly carbon dioxide and chlorofluorocarbon (Ashfaq & Mi, 2011), which is the main cause of ozone layer depletion and global warming. All these factors obliged scientists and researchers to search for solutions to reduce fossil fuel usage in order to save and improve the quality of life on this planet. These solutions can be achieved by utilizing renewable energy which is environmentally friendly.

The rate of solar energy that falls on the earth is of 120 pet watt (1 pet watt = 10^{15} watt). This means that the amount of energy needed for more than 20 years can be satisfied in one day. So solar energy is witnessing scientific revolution that urges scientists to intensify their studies about it. Solar energy can be one of the effective, eco-friendly, and important approaches to assemble the limitations. Solar energy (Ramakumar et al., 1975) has probably the best potential for clean energy on the planet. It produces numerous benefits when contrasted with coal, petroleum, and nuclear power. First, the sun-based energy does not deliver the destructive gases or contaminate the climate. Second, the sun-powered energy has no geological and asset constraints and could be used securely and effectively wherever as long as there

Table 1 Various technologies of solar energy conversion

S. no.	Type	Output
1	Solar thermal energy	Heat
2	PV/T technologies	Heat
3	PV technologies	Electricity

is daylight. Subsequently innovation and utilization of sun-based energy is the most important aim of human energy improvement.

The essential objective of this section is to give a few basics of solar energy and its applications.

Solar technologies track large amounts of the sun-based energy and use this energy for the production of heat, light, and power. Solar energy can be changed over straightforwardly into power by photovoltaic cells (solar cells) and thermal power through solar collectors.

Table 1 shows the various methods of converting natural solar energy into thermal (heat) energy and electricity. From both solar thermal and photovoltaic (PV)/T technologies, heat can be generated.

1.1 Solar Thermal Technologies

The solar thermal energy (STE) is used to strap up sun energy for heat. At first this technology is used only in large-scale power plants, and thus the generated power is used for cities and communities. US Energy Information Agency described the main solar thermal collectors (STCs) as low-, medium-, or high-temperature collectors. Low-temperature collectors are used to warm pools. For residential and commercial use, medium-temperature collectors are utilized for boiling water. High-temperature collectors are used to collect sunlight with the help of lenses and mirrors which is then used for electricity generation. This is unique in relation to sunlight-based photovoltaic cell which changes over solar energy directly into electricity. Sun thermal power is the most plentiful one, and it is accessible in two structures, direct and circuitous concentrated solar power.

1.2 Non-concentrated Solar Power

A solar thermal collector (STC) collects high-temperature beams as absorbing sunlight. It is a device which consists of the solar hot water heating system. In non-concentrating gatherers, the beneficiary region is generally equivalent to the safeguard region. Non-concentrated solar force framework has no additional parts aside from the actual authority. They are regularly utilized in business structures and domestic for space warming.

A solar water heat system utilizes solar energy in order to heat water (Kannaiyan et al., 2020), for domestic and industrial purposes. In this, a working fluid is heated when it is brought in contact with a dark surface which is directed to sunlight. One of the applications of solar thermal energy is solar water pumps. The pump is driven by motor in solar water pumping system and run by solar power instead of conventional electric grid. The soar water pumping system associates with water from the drag well, open well, trench, lake, and so forth. The other alternative is directly consuming the electrical power generated by the solar plates.

1.3 Concentrated Solar Power

These technologies utilize mirrors to focus an enormous space of sunlight onto a receiver and thus generate solar power. Power is produced when the collected sunlight is changed over to heat (solar thermal energy [STE]), which drives a motor (by and large steam turbine) associated with an electrical force generator.

1.4 Two Types of Concentrated Solar Collector Technologies

Medium temperature – Line focusing (≈400 °C) (the receiver can reach 400 °C and produce electric power).

- **Parabolic trough –** Parabolic trough model uses long parabolic-shaped mirrors for collecting solar energy, where the sunlight is collected on a pipe which is loaded up with oil that runs down along with the trough. At that point, the oil gets hot which is utilized to bubble water in a steam generator for creation of electric power.

High temperature – Point focusing (>400 °C).

- **Solar power tower** (central receivers) – This system utilizes a huge field of mirrors to collect sun energy to the top of the tower, where a collector sits. Liquid salts generally moving through the collector is warmed by the concentrated sunlight. The gathered or obtained heat is transformed into power by a steam generator.
- **Parabolic dish/engine systems –** This system utilizes a large mirrored dish which is similar to that of a huge satellite dish. This surface collects sunlight directly and gathers the sun's heat on to the collector at the point of the dish. Illustrative dish frameworks can reach at a temperature of 1000 °C at the receiver and accomplish the most elevated efficiencies for creating solar energy to power.

The collector absorbs the solar heat energy and transfers it to a liquid and makes the liquid grow against a cylinder which produces mechanical force. The mechanical energy is used for running a generator which produces power.

Concentrating solar power (CSP) technologies can be utilized to create electricity for different kinds of uses, ranging from remote power systems to grid connected systems, i.e., from small few kilowatts (kW) to MW or more.

The solar thermal energy generation can take part in a major role in fulfilling the need supply for power.

Three kinds of utilizations are conceivable:

- Rural power utilizing solar dish innovation technology.
- The solar thermal power stations can be coordinated with existing enterprises.
- By the combination of solar thermal power (STP) with existing coal thermal energy stations.

2 Solar Electrical Technologies

2.1 Solar Photovoltaic (PV)

A sun-powered cell, or photovoltaic cell, is an electrical gadget that changes over sun energy into electricity, which is comprised of semiconducting materials. A basic PV cell (Luque & Hegedus, 2010; Turcek et al., 2011) can power the calculators, watches, and a number of electronic gadgets, and furthermore, it can be used to light houses. In some applications, sun-based electric frameworks are used for business structure lighting, road lighting, village and rural lighting, and so on. Individual sunlight-based cell gadgets can be consolidated to frame modules, additionally called as sun-based boards (solar panels).

A photovoltaic (PV) plant allows for the transformation of solar radiation into the electrical energy, and this conversion takes place through a so-called semiconductor devices termed as PV cells. Nowadays the most used semiconductor material is silicon, which can be monocrystalline, polycrystalline, or amorphous. In order to protect from the weather and external factors, the cells are assembled into modules. These modules are usually grouped as strings, which in turn are connected to the control or storage devices (inverter, charge controller, and like battery banks). The conversion efficiency of these cells is usually between 15% and 20%. The power of the photovoltaic cells is expressed in watts or kilowatt peak, which represents the nominal power that the unit is capable of delivering in reference standard test conditions.

2.2 Applications in Solar Energy

Solar energy offers a lot of applications in order to utilize this available renewable energy resource. Figure 1 illustrates the applications of solar energy and the types of each one.

This chapter highlights the technology involved in solar PV system and installation procedure with the proper site calculations and ratings of electrical equipment.

The mechanism behind the solar cell is converting the available or incidental sunlight into the electricity. The generated cell voltage is in DC nature and it is around 0.5–0.7 V. In real time, the required voltage level is very high compared to the volts produced by a solar cell. So, always a series of solar cells has to be connected to get considerable voltage value. But still the voltage is not adequate for power system, so the DC voltage has to be increased to higher values by using DC-DC converter (Rojas Aldana et al., 2015), and the same has to be inverted with the DC-AC inverter (Kalagotla et al., n.d.; Gupta et al., 2016) into AC as the power transmission and distribution in nature. As the energy conversion rate is very less in the solar technology compared to the many others, the extraction power at any point should be as maximum as possible to make the system more economical. In this connection, either the maximum power point tracking (MPPT) or the mechanical tracking should be the better method to extract more power from solar plates. And the preliminary requirements other than electrical technology are selection of site and power ratings of equipment.

2.3 Site Calculations

To design a solar power plant, the primary requirement is finding the location which may be ground-mounted or at the rooftop. Here all the parameters are discussed mostly suitable for all types of solar installations. In general, the maximum site

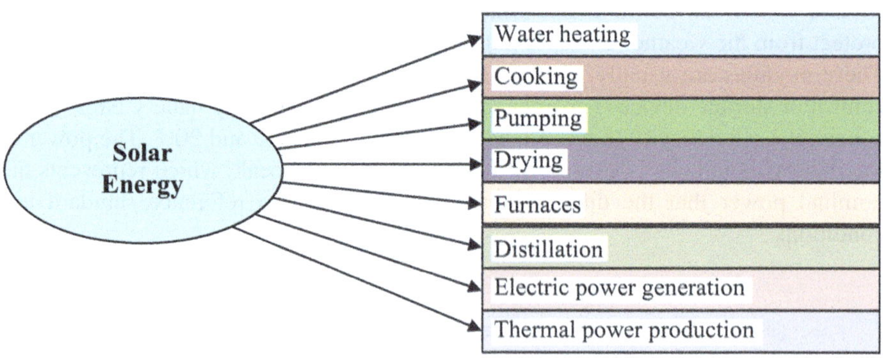

Fig. 1 Various applications of solar energy conversion

required is about 1200 sq. meters or 12,900 sq. feet for 100.8 kWp solar power plants in case of uneven surface, and around 8000–9000 sq. feet for a plain surface (Iskandar et al., 2019; Ali et al., 2018) is enough. To draw the panel layout, first, make a note on the module rating and sizes, for noncommercial use (low power applications), and these are available from 10 W at 12v, and in another case, at higher power applications, the module is designed at different ratings by manufacturers. Usually, the panels are available with 250Wp, 315Wp, 330Wp, etc., and mostly with a width of 1 m and a length of 1.8–2 m.

Firstly, you need to identify the origin of the total available site by the simple mechanism of finding the midpoint in both vertical and horizontal axis, and from that by using compass, identify the south to north position and draw a line perpendicularly through the origin. On the line at any point, the compass has to show the same angle between the line and the south-north vector. The placement panel depends on on-site location; as in northern hemisphere, the solar modules should be facing south, but the southern hemisphere modules should be facing north as the location of the plant is in India which is in the northern hemisphere, so panels are located to face south side. The position of panels can be decided by tilt angle and azimuth angle.

Tilt angle: The can be defined as between the horizontal plane and the pitch of a solar panel.

Tilt is being calculated against a horizontal plane, regardless of the slope on the roof (of the site) or the ground structure. If the panel is placed on a flat surface, the amount of solar light that strikes the panel which is the majority of the energy is reflected in the atmosphere, because the sunlight does not occur perpendicularly on the panel. So an angle should always be maintained between the panel concerning the horizontal plane, and it should be perpendicular most of the day to absorb a major portion of solar energy. So it will maximize the power generation; the angle of the panel lies around the latitude of the location. This angle describes the number of degrees that the PV modules are off of a horizontal surface. There are so many theories for positioning of a solar panel; out of those, for a flat surface area, one or two degrees less than the latitude will result in higher performance. By incorporating position tracking (Whavale & Dhavalikar, 2018; Sohag et al., 2015), the overall system performance can improve to some extent in MW and GW solar plants.

In addition to the tilt angle (Liu et al., 2018), the other important parameter is the azimuth angle; it is defined as the position of the modules in terms of how many degrees the array is from the north. To locate India in the northern hemisphere, all the modules have to face to the south side, so azimuth angle will be 180°.

In Fig. 2 (a–e), 1000 sq. mts is considered for a layout of 100.8 kWp, and the modules are rated at 315Wp and panel dimensions have a length of 1.8 m and a width of 1 meter. For calculations, the site here is considered as rectangle; if the site has uneven measurements, then divide total site into different even pieces and do the same procedure as mentioned above. One more important parameter is the shadow at the site; it drastically decreases the system performance in long-time prospect, so before drawing, consider laying out all the shadow boundaries.

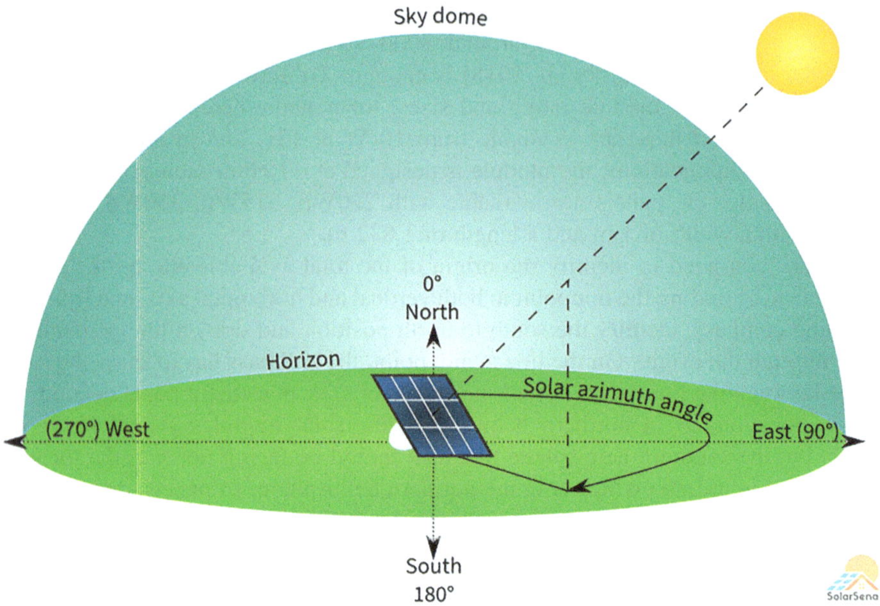

Fig. 2 (**a**) Azimuth angle (Etukudor et al., 2018) for solar panel. (**b**) Module layout for 100.8 kWp. (**c**) Block diagram of electrical circuit for grid connected solar PV system. (**d**) Flow chart P&O algorithm. (**e**) Representation of site with panel arrangement

After civil structure calculations, the second thing is about electrical supporting system which includes inverters, DC cables, AC cables, and protection equipment against both electrical short circuit and lighting during heavy rains.

2.4 Electrical Supporting System

The solar panel of the electrical circuit design is the major part in solar power generation. The basic technologies involved are DC-DC converter and DC-AC inverter and controlling circuit and battery (in the case of off-grid system). As the voltage developed in solar panels has less value, it has to be increased to a sufficient level by DC-DC converters, and later it has to be converted into AC and then integrated with the grid (on-grid) or connected to the load directly with battery support (off-grid). In both types of connections, the inverter deals with the conversion of DC into AC at a defined voltage and frequency levels by taking grid as the reference. In general, the inverter will come with multiple numbers of strings (number of solar modules are connected in series to enhance the voltage level). Some of the inverters are designed with individual MPPT circuit for each string, but

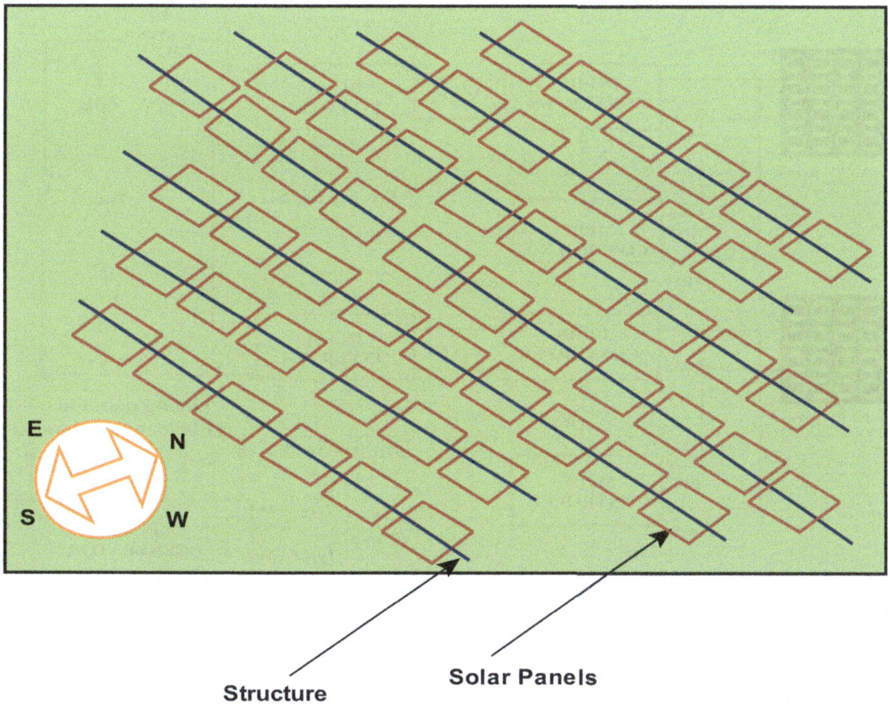

Fig. 2 (continued)

in most of the inverters, there is only one common MPPT circuit for all the strings (more economic). These strings have been connected in parallel to the inverter to increase the current value. The basic function of these strings is to take out the maximum power from the solar panel.

2.5 DC-DC Power Converter

It is a device, and it allows a DC input voltage which has less value and gives the pure DC output voltage at higher value. The output voltage can be more prominent than the information or the other way around. These are utilized to coordinate with the supply. A least complex of this circuit is comprised of a toggle used to control association and disengagement of its output with source.

This fundamental circuit (Liang & Li, 2017; Ortiz et al., 2010) comprises power moved from the output to power generated by the stockpiling appliances like passive elements using toggle similar to a semiconductor. It utilizes the direct voltage controllers or switched-type controllers. In the first type of controller, the reference

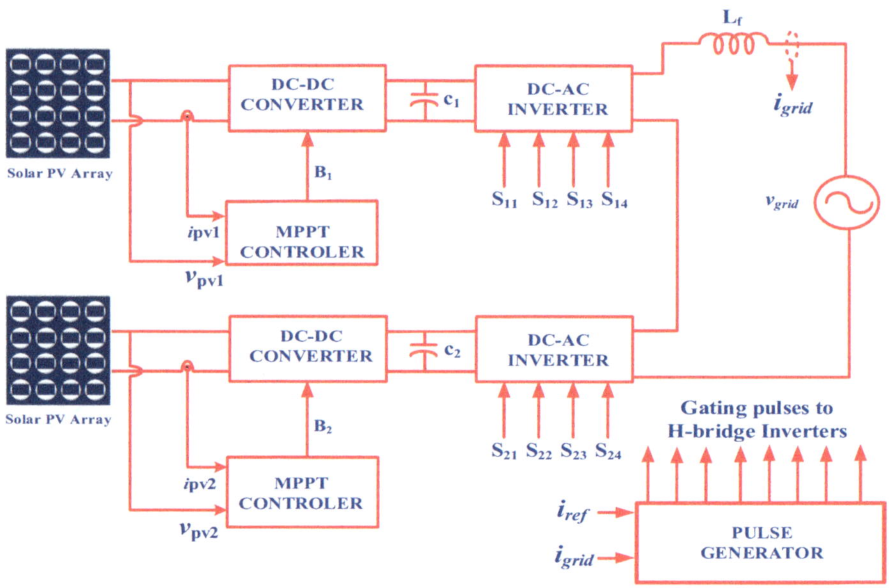

Fig. 2 (continued)

voltage of a semiconductor is determined using a manage circuit to acquire the ideal load voltage. In a switched-type controller, the semiconductor is utilized as toggle. Here, using a down converter or a buck converter, when toggle is shut, the passive device permits current, and when the toggle is opened, the inductor gives power to the load.

The DC-DC converter circuit is associated with DC source and the load. This circuit comprises switching appliances, for example, MOSFETs or thyristors, which can be used as switching devices to produce DC output voltage.

The voltage produced is constrained by changing on schedule of the switching device which helps to alter the output voltage. This exchanging technique is known as pulse width modulation (PWM) control. The DC-DC converter production may be less or more prominent than the information; furthermore, this may be fixed or variable. The DC-DC converters are mostly utilized in any type of DC drives, i.e., EV and mixture EV. These DC-DC converters are characterized into three essential sorts dependent on input and output voltage levels.

2.6 Step-Down Chopper or Buck Converter

This type of circuit creates a nominal voltage lower than the information DC voltage. Normally the step-down transformer in AC is the same as this chopper. At this point, the toggling device is a thyristor which changes the source to load while it is set off at

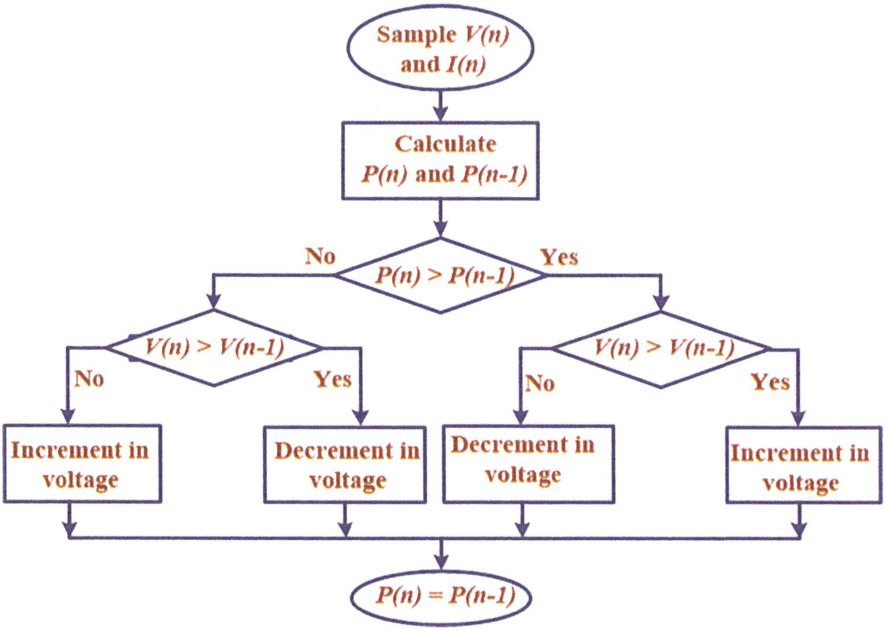

Fig. 2 (continued)

specific moments. A switching device diode goes about as unrestrictive diode which allows the output current to course from side to side when the thyristor is turned off.

The nominal input of converter is differed by managing turn ON/OFF times of thyristor. The output voltage is the same as the input supply when the thyristor is turned ON, and when the thyristor is off, the final value is zero.

The final value of voltage to (TON/T) Vin. In this way, by managing the duty ratio

$K =$ (TON/T), the output voltage will be expanded.

2.7 Step-Up Chopper or Boost Converter

In this type of circuit, the output voltage is consistently more prominent as compared to the input supply. At this point, likewise, a toggle is utilized, which corresponds to the output. This toggle is a thyristor or a SCR or a MOSFET. Like the buck converter, a diode is put in arrangement along the output which allows output current to stream when the switching device is switched OFF.

At the point when the switching device is switched ON, the diode is opposite one-sided; thus this circuit separates output from the input supply. Hence, the

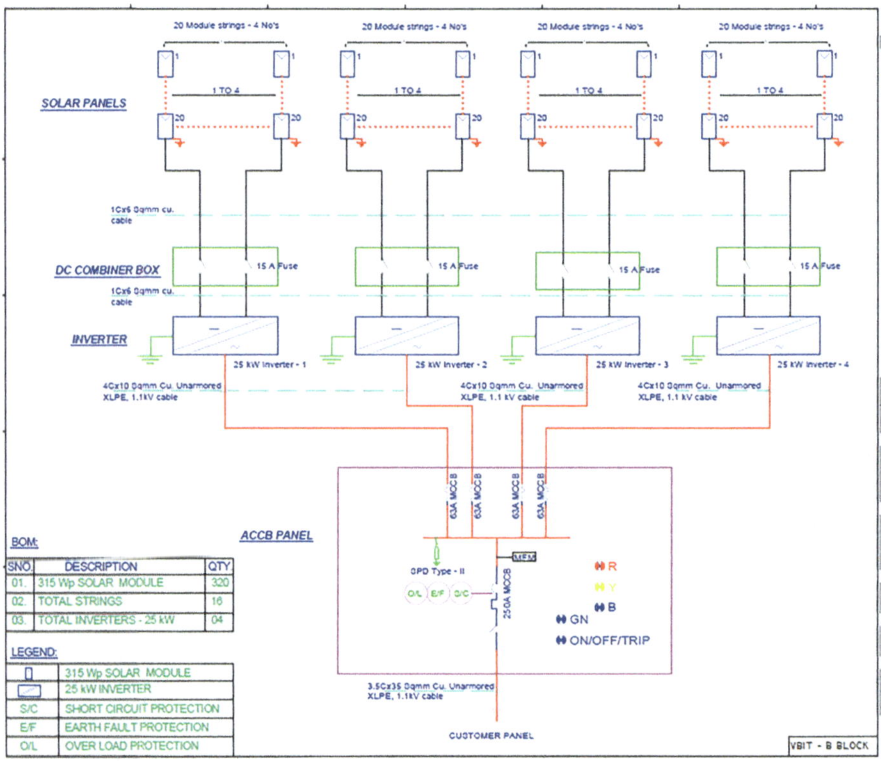

Fig. 2 (continued)

passive device which is connected charges the greatest value of the input supply. At the point when the switching device is c, the output acquires the energy from the stored voltage of the inductor. Consequently the input voltage is lesser than the output voltage.

$$\text{Inductor} : L \geq \frac{V_s * D}{f_s \Delta I_L}$$

$$\text{Capacitor} : C \geq \frac{I_{o(\max)} * D}{f_s \Delta V_C}$$

$$\text{Gain} : \frac{1}{1 - D}$$

At this point, the final value is $(1/1 - d)$ times the input voltage, where d is the duty ratio (TON/T). By shifting this obligation proportion, the output voltage will be differed until the output gets enough value of voltage. These traditional converters are having limitations like inductor value and gain parameter. At present a more new concepts are intended by instigators with enhanced technology in DC-DC converters for high-power applications.

2.8 *Buck/Boost Converter*

This circuit is able to utilize together in sync behind and venture some wake modes by constantly changing its obligation series. This arrangement of the circuit is shown in the figure below that is comprised of just one switching device, i.e., one thyristor.

Alongside a passive device along with diode, extra capacitor is associated in corresponding by means of this circuit.

At the point while the switching device is switched ON, the flow of current is through the inductor, thyristor, and initiates the output in the inductor. At the point when the switching device is OFF, the flow of current through the inductor will in general diminish with the actuated EMF turning around extremity. The output value of the circuit stays consistent as the capacitor is associated with the load.

The value of output voltage is lower than the information voltage which regularly reaches $0 \geq k > 0.5$, which is obtained by changing the evaluation of duty ratio to a specific worth. Also, the input voltage is less than the resultant voltage if the duty ratio is in the scope of $0.5 > K \geq 1$, which along these lines goes about as a boost converter. A wide variety of high gain boost (Chakravarthi & Siva Krishna Rao, 2020a; Khanfh, 2007) converters are realized in the recent papers, and the performance of these converters is more reliable than the traditional converters.

2.9 *Maximum Power Point Tracking (MPPT)*

A lot of various methods are available to extract maximum possible power (Esram & Chapman, 2007; Elgendy et al., 2012) from the solar panel; one of those methods is *Perturb and Observe Algorithm,* and it is the most economic than any other.

The V(n) and I(n) are the value of voltage and current, respectively, of nth sample, and V($n - 1$) and I($n - 1$) are the voltage and current of the previous samples. After extracting the maximum possible power (Elgendy et al., 2012), it has to be converted into AC by using a power inverter.

2.9.1 What Is an Inverter?

An inverter is the device which changes over DC into AC. The greater part of the business, mechanical, and private loads require alternating current (AC) sources. One of the fundamental issues with alternating current sources is that they cannot be put away in batteries where the capacity is significant for backup power.

The abovementioned defect can be overwhelmed by direct current sources. Alternating current is changed over into DC for capacity purposes. The extremity of direct current sources does not change with time like AC sources along these lines; direct current can be put away in batteries and ultra-capacitors. At whatever point AC is needed to run AC machines, at that point DC is changed over once more into AC to run AC apparatuses.

3 Different Types of Inverters

Inverters are classified into various classifications dependent on the applied input source, connection wise, output voltage wise, and so on.

3.1 Input Source Wise Classification

The inverter can be characterized as the device which changes over DC input supply into AC output where information might be a voltage source or current source. Inverters are fundamentally grouped into two principle classifications.

3.2 Voltage Source Inverter (VSI)

The inverter is known as voltage source inverter when the contribution of the inverter is a consistent DC voltage source. The contribution to the voltage source inverter has a hardened DC voltage source. Hardened DC voltage source implies that the impedance of DC voltage source is zero. Basically, input sources have some insignificant impedance. Ideal voltage sources are supplied with VSI. The AC output voltage is totally dictated by the conditions of exchanging devices in the inverter and the applied DC source.

3.3 Current Source Inverter (CSI)

The inverter is known as current source inverter when the contribution of the inverter is a steady DC current source. Fixed current is supplied to the CSI from the input supply where the input has high impedance. Typically, a huge passive device current is utilized to give solid current. Hence the subsequent current wave is solid that is not affected by the heap. The alternating current output is totally dictated by the conditions of switching devices.

3.4 *Phase Wise Classification Output*

Inverters are isolated to two major classes:

- Single-phase inverters.
- Three-phase inverters.

These classifications are momentarily examined here.

3.4.1 Single-Phase Inverters

This inverter changes over DC contribution to one-phase output. The output voltage/current of one-phase inverter has precisely one phase which has a small recurrence of 50HZ or 60 Hz nominal voltage. The supposed voltage is characterized as the voltage level at which electrical framework works. For example, 120 V, 220 V, 440 V, 690 V, 3.3 KV, 6.6 KV, 11 kV, 33 kV, 66 kV, 132 kV, 220 kV, 400 kV, and 765 kV are distinctive nominal voltages.

The nominal voltages can be straightforwardly accomplished by inverter utilizing an inside transformer or buck-boost circuitry utilizing high nominal voltages, and outer step-up transformers are employed. One-phase inverters are utilized for stumpy loads. Several misfortunes are present in single phase just as the effectiveness of single phase is low as for three-phase inverter. Accordingly, for greater loads the three-phase inverters can be employed.

3.5 *Three-Phase Inverters*

Three-phase inverters convert DC into three-phase power. This power gives three alternating current outputs which are consistently isolated in phase point. Amplitudes and frequencies of every one of the three phases which are generated are the same with small varieties because of the burden, while each phase has a 120° phase difference between one another.

The other converter is DC-AC inverter (Vijetha Inti & Vakula, 2017; Hameed et al., 2016; Inti & Vakula, 2017; Rodriguez et al., 2007; PrakashGautam et al., 2015), and the rating of the inverter is having a prominent role in the cost of solar PV system. Mainly these inverters come with multiple number of strings, and maybe all of those strings have separate MPPT system or with the single MPPT. To make system more reliable, 80%–90% of strings should always be loaded, and string voltage capability will define the module layout, which determines how many modules to be in series and how many modules to be in parallel. These strings are always parallel connected, which increases the current rating of the inverter. The switches and protection circuit ratings are mainly depending on the string voltage and current handling capability.

Apart of the single-phase and three-phase inverter, the inverter which is connected to the PV system has to produce the pure sinusoidal wave. In general, two-level inverters are used in the very first stage of this PV system, but the basic problem with the output is square wave. The electrical loads are operated at the sinusoidal wave, so they always need a filter to minimize the harmonic content in the waveform. In this section of power converters, the filter cost is very high, and it will impact the overall performance of the power system. The other alternative for these types of inverters is multilevel inverter (MLI). These types of inverters will produce stepped waveform, and the shape is almost similar to the sin nature. So there is no need of placing an extra filter to tune the final value.

3.6 Pulse Width Modulation

The technique of PWM is to switching on and off high-power semiconductor switches. These high-voltage and high-current switches will convert AC into DC or vice versa, and these switches also change the voltage and current levels as required. In most of the industrial applications, the PWM-based inverter and DC converter are widely used.

The most powerful PWM method is multicarrier sinusoidal PWM, and it includes two different signals. The first signal is the reference and the second will be the carrier. By examining those signals, pulse needed for switching the method of the inverter can be generated. There are three pulse width modulation techniques which are explained below.

3.7 Single-Pulse Width Modulation (SPWM)

For each half cycle, for controlling, in this technique, we need only single pulse for each cycle. Here there are two signals observed, i.e., square wave signal for reference and a triangular signal as carrier. The gate pulse created will be the effect of the examination of the carrier and the reference signals. The main drawback in this method is the production of higher harmonics.

3.8 Multiple-Pulse Width Modulation (MPWM)

MPWM method is utilized to beat the disadvantage of single-pulse width modulation. Here, various pulses are employed for every half cycle of the voltage at the output. By controlling the frequency of carrier, we can constrain the output.

3.9 Sinusoidal Pulse Width Modulation

In sinusoidal PWM method, a sine wave is employed as a source of perspective and the triangular wave acts as a carrier. The sine wave will be the output and its RMS value of voltage is constrained by the modulation index.

3.10 Modified Sinusoidal Pulse Width Modulation

This change is acquainted to improve the harmonic characteristics. It diminishes the loss because of switching and expands the key component.

An inverter whose usefulness relies on the PWM method is alluded to as pulse width modulation inverters. MSPWM is equipped for keeping up the output voltages as the evaluated voltages relying upon the state regardless of the sort of output associated. By altering the switching frequency width at the oscillator, this pulse width modulation can be done.

Demonstration of PV installation: Electrical layout for 100.8 kWp solar power plants:

How PVSYST helps to design a solar PV power plant in software platform:

Before the discussion of practical methods to install a solar PV system, the most important thing is to analyze the site and electrical structure with a PVSyst software tool. This is the most popular computer tool to design a wide variety of solar systems with a real-time analysis.

Figure 2b shows the complete electrical circuit diagram for 100.8 kWp solar power plants located at the rooftop of Vishnu Institute of Technology (16.5449° N, 81.5212° E). Here the modules are rated with 315 Wp, the number of modules required is 320 (100.8 kWp/315 Wp), and the inverter is taken with 25 KW rating of 4 number. In the market, a wide range of inverter ratings are available, and those may start from 5 KW to even up to 100 KW for solar power stations, and some are available from 0.5 KW for the residential installation. As we said earlier, we have taken a 6-string inverter with individual MPPT, and it is costlier; the same can be taken with only one MPPT for all the 6 strings and it costs also less compared to the above, but the performance is low (varies from 3% to 5%) compared to the previous one.

Out of six strings, only four strings are loaded to increase the durability, and the system will be more reliable, because any failure of string will not affect the system working as we have an additional one. In the distribution generation, power quality (Chakravarthi & Siva Krishna Rao, 2020b; Chakravarthi & Siva Krishna Rao, 2020c) plays a key role, and the grid connected system has to transmit a proper energy to the load centers.

4 Calculation of Modules in Series and Parallel

As only 4 strings to be loaded for 25 KW inverter, each string has a capability of 6.25 KW power. From the nameplate details of solar module, we can see the on-load voltage, open circuit voltage, and short circuit current. By using on-load voltage, panels in series need to be arranged by considering the voltage band of the inverter (see the nameplate details of the inverter), and usually it is between 600 V and 1000 V DC.

Let us consider that the on-load voltage of solar panel is 36 V, the short circuit current rating is 8.4 amps, and the string voltage rating is between 600 V and 900 V DC, so the number of modules to be in series is 20 number (36*20 = 720 V), and the power rating of these 20 panels together will produce a maximum value of 6.04 KW. And as per earlier discussion, four strings are connected in parallel with a total of 80 numbers of panels.

So the total power connected to the single inverter is about 24KW to 25 KW with 80 modules, and the same for the remaining three inverters. After inverter calculations, the output AC power should be connected to the AC distribution box (ACDB), and later the system connections depend on the type of system, i.e., whether on-grid or off-grid.

If the system to be connected is off-grid, there is a battery connection in addition to the AC load point; otherwise the system outpoint wire from the ACDB is connected to the local grid (LV panel) via the molded case circuit breaker (MCCB). The off-grid system is costlier than the on-grid due to its battery bank, and it is not economically suitable for large-scale generation. In the systems, the load and line should be integrated with net meter, which shows the two-way power flow. In the absence or light load conditions, the power will be exported to the line and the net meter shows negative reading. In the case of higher load or night hours, the load will draw power from the main line, so the net meter will read the positive value. At the end of billing cycle, the resultant or net value of meter will be billed.

4.1 On-Grid and off-Grid Systems

The basic difference between these two is that on-grid works with the grid system, while off-grid works independently of it. In the on-grid system, the solar panels convert the solar energy into DC electrical power and the inverter will convert into AC to power the load, and any excess current will be fed back to the line by net meter, and it can be drawn back in based on the requirement. However, the on-grid system will not work during power outage; it will not support any backup power. It is most cost effective and more useful where the power cuts are less. In the places where high power outages, the second system, i.e., off-grid, is more effective. The technology difference between these two is that the off-grid system has a backup source. This backup system consists of a bunch of batteries to store the excess amount of electrical power in DC with less load conditions. And it supplies back to the load whenever the power fails or load increases.

4.2 Cables for DC and AC Power Transmission

Generally, two types of cables are required for the entire system: one is DC cable and the other is AC cable. DC cables (Desai et al., 2020; Mustafa et al., 2020) are used to interconnect solar panels, between panels and inverter through DC distribution box (DCDB). AC cables are connected between the outpoint of inverter and grid or load through ACDB.

For DC connections, single-core cables of 4 or 6 Sq.mm are used based on the current drawn by a single string of an inverter, and for AC (Modi & Singh, 2019; Campanhol et al., 2017), the choice of cable depends on whether the inverter is 1-phase or 3-phase. Usually, for more than 10Kw, the 3-phase inverter is preferred due to its current capability. In the case of 1-phase inverter, 2-core cable is preferable, and for 3-phase inverter, a 3-core or 4-core cable is chosen based on whether the system is delta or star connected.

4.3 Protection System

For the on-grid and off-grid systems, the load is connected to the solar inverters with an AC protection, and the strings and inverters are to be merged with the DC protection. In case of high DC currents, the inverter is protected with the direct current distribution board (DCDB) chamber, which contains high-current DC fuses, and AC side high currents will be isolated with two level protection schemes, which include an alternating current distribution board (ACDB) at the output of the inverter and a molded case circuit breaker (MCCB) before the load side of low voltage (LV) panel. By connecting a dedicated energy meter, the amount energy consumed can be found at the load side in addition to the inverter ratings.

4.4 Protection of Solar Plants against Lighting

During the formation of cloud, a negative charge density will increase in that area, and at the same time, the positive charge density increases at the ground. The normal phenomena of charges are to get settle in stable state, because the charges (Srinivasan & Gu, 2006) will make a common path to collide each other. This collusion can be visually observed during the heavy rains, and it carries a huge amount of electrical energy. While the electrical wave is travelling from cloud to earth, whatever the thing in that path has an enormous effect on the working and it will cause the serious damage.

The process of lightning is as follows:

- Development of thunderstorm.
- Charge separation.
- Clouds bottom includes earth and structures.
- Stepped leader from cloud to ground.
- Streamer from earth to sky.
- Lightning stroke.

Why Lightning Matters?

It causes damage to occupants (humans and animals), structures (wood, brick, concrete, reinforced concrete, and steel frame structure), and electrical/electronic systems. As the solar power plants are more sensible to these types of issues, a protection system is required against the lightning caused by the thunderstorms.

4.5 Lightning Arrestors

It is the one which protects the electrical system or tall buildings against the high-voltage beams emitted (Salman et al., 2018; Yang et al., 2014) during unusual weather conditions. The IEC 62305:2013 is the standard to follow for the design of this type of protection system.

The different types of lightning arrestors (Islam et al., 2017) are:

- Road gap.
- Sphere gap.
- Horn gap.
- Multiple gap.
- Spike rod.

The functions of all these types of arrester are the same mostly. These arrester circuits will provide a low resistance path between the lightning strike point and ground. A lightning arrester is to be placed at the site to protect the entire system (Podporkin et al., 2019) against high-power electric storms caused by clouds. These arresters will provide the safe operation of equipment without any physical or electrical damage.

5 Conclusion

After this chapter, the reader can understand the possible ways of converting the solar energy into other forms with different techniques and also how the solar electrical energy takes place with all the supporting equipment. This chapter mainly covers the different aspects of the installation of solar power plant and easily understands the technical parameters included in the design process of a solar PV power plant as the complete concept has been explained with an example of 100.8 kWp solar power plants. And the other concept covered is the different types of converters available for converting one form of energy into other form. And a considerable discussion is made regarding the types of converters and the controlling methods like PWM techniques. At different levels of solar energy conversion principle, various power converters are required to connect PV system to the AC grid or DC grid. So this discussion will help the readers for easy understanding of solar technology, which is helpful in their research basics. This chapter also includes the need of protection equipment for solar power plants against the unusual weather conditions and faults caused by the malfunction of any electronic circuit or in normal power system fault equipments.

References

Ali, W., Farooq, H., Rehman, A. U., Awais, Q., Jamil, M. and Noman, A., "Design considerations of stand-alone solar photovoltaic systems," *2018 International Conference on Computing, Electronic and Electrical Engineering (ICE Cube)*, Quetta, Pakistan, 2018, pp. 1–6, doi: https://doi.org/10.1109/ICECUBE.2018.8610970.

Ashfaq, A., & Mi, A. (2011). Applications of solar power as a means of renewable energy. *Journal of Industrial Pollution Control, 27*(1), 65–67.

L. B. G. Campanhol, S. A. O. da Silva, A. A. de Oliveira and V. D. Bacon, "Single-stage three-phase grid-tied PV system with universal filtering capability applied to DG systems and AC microgrids," in IEEE Transactions on Power Electronics, vol. 32, no. 12, pp. 9131–9142, Dec. 2017, doi: https://doi.org/10.1109/TPEL.2017.2659381.

Chakravarthi, B. N. C. V., & Siva Krishna Rao, G. V. (2020a). A high gain novel double-boost converter for DC microgrid applications. *Journal of Circuits, Systems and Computers*. https://doi.org/10.1142/S0218126620502461

Chakravarthi, B. N. Ch.V. and Siva Krishna Rao, G. V. (2020b). "Impact of Power Quality Issues in Grid Connected Photovoltaic System," 2020 4th International Conference on Electronics, Communication and Aerospace Technology (ICECA*)*, Coimbatore, pp. 155–158, doi: https://doi.org/10.1109/ICECA49313.2020.9297618.

Chakravarthi, B. N. C. V., & Siva Krishna Rao, G. V. (2020c). Optimal real power penetration to solar PV fed double boost integrated multilevel converter with improved power quality. *Journal of Circuits, Systems and Computers*. https://doi.org/10.1142/S0218126620502564

Desai, A., Pandya, V., Mukhopadhyay, I. and Ray, A., "Temperature effects on DC cable voltage drop in utility scale rooftop solar PV plant based on empirical model," *2020 47th IEEE Photovoltaic Specialists Conference (PVSC)*, 2020, pp. 2397–2402, doi: https://doi.org/10.1109/PVSC45281.2020.9300709.

Elgendy, M. A., Zahawi, B. and Atkinson, D. J. (2012). "Evaluation of perturb and observe MPPT algorithm implementation techniques".

Esram, T., & Chapman, P. L. (2007). Comparison of photovoltaic array maximum power point tracking techniques. *IEEE Trans. Energy Conversion, 22*, 439–449.

C. Etukudor *et al.*, "Optimum tilt and azimuth angles for solar photovoltaic systems in south-west Nigeria," *2018 IEEE PES/IAS PowerAfrica*, Cape Town, South Africa, 2018, pp. 348–353, doi: https://doi.org/10.1109/PowerAfrica.2018.8521047.

Gupta, K. K., Ranjan, A., Bhatnagar, P., Sahu, L. K., & Jain, S. (2016). Multilevel inverter topologies with reduce device count: A review. *IEEE Transactions on Power Electronics, 31*, 1.

Hameed, D., Hamayoon, S., Malik, A. A. and Ansari, O. A. (2016), "Solar grid-tied inverter, with battery back-up, for efficient solar energy harvesting," *2016 IEEE Smart Energy Grid Engineering (SEGE)*, Oshawa, ON, Canada, pp. 95–99, doi: https://doi.org/10.1109/SEGE.2016.7589507.

Inti, V. V. V., & Vakula, V. S. (2017). Design and Matlab/Simulink implementation of four switch inverter for microgrid utilities. *Energy Procedia 117, 6*(1), 615–625.

H. R. Iskandar, E. Darmawan, Y. B. Zainal, G. Angga Setia, N. Winanti and F. Haz (2019) "Design of Solar Power Plant for Electrical Engineering Department Laboratory," *2019 2nd International Conference on High Voltage Engineering and Power Systems (ICHVEPS)*, Denpasar, Indonesia, 145–150, doi: https://doi.org/10.1109/ICHVEPS47643.2019.9011041.

Islam, M. Z., Rashed, M. R. and Yusuf, M. S. U. (2017). "ATP-EMTP modeling and performance test of different type lightning arrester on 132kv overhead transmission tower," *2017 3rd International Conference on Electrical Information and Communication Technology (EICT)*, pp. 1–6, doi: https://doi.org/10.1109/EICT.2017.8275172.

Kalagotla Chenchireddy ,V. Jegathesan, L. Ashok Kumar" Different Topologies of Inverter: A Literature Survey" Lecture Notes in Electrical Engineering.

Kannaiyan, S., Bokde, N. D., & Geem, Z. W. (2020). Solar collectors modeling and controller design for solar thermal power plant. *IEEE Access, 8*, 81425–81446. https://doi.org/10.1109/ACCESS.2020.2989003

Khanfh, T. L. M. (2007). A multilevel modular capacitor clamped DC-DC converter. *IEEE Trans on Industry Applications, 43*(6), 1628–1638.

Liang, B. Y. and Li, Y. S., "A review of DC/DC converter based on MMC," *2017 7th International Conference on Power Electronics Systems and Applications - Smart Mobility, Power Transfer & Security (PESA)*, Hong Kong, China, 2017, pp. 1–6, doi: https://doi.org/10.1109/PESA.2017.8277749.

H. Liu, X. Xu, Y. Meng, D. Yu, H. Liu and K. Shi (2018). "A research for the influence of tilt angles of the solar panel on photovoltaic power generation," *2018 International Conference on Smart Grid and Clean Energy Technologies (ICSGCE)*, Kajang, Malaysia, 95–100, doi: https://doi.org/10.1109/ICSGCE.2018.8556655.

Luque, A., & Hegedus, S. (2010). *Handbook of photovoltaic science and engineering* (2nd ed.). Wiley. ISBN 978-0-470-72169-8.

Modi, G. and Singh, B. (2019), "Solar PV-battery microgrid for three phase four wire AC distribution system," *2019 International Conference on Computing, Power and Communication Technologies (GUCON)*, 1008–1014.

Mustafa, E., Afia, R. S. A. and Tamus, Z. Á., "Investigation of photovoltaic DC cable insulation integrity under Thermal Stress," *2020 IEEE 3rd International Conference on Dielectrics (ICD)*, 2020, 13–16, doi: https://doi.org/10.1109/ICD46958.2020.9341899.

Ortiz, G., Biela, J., Bortis D. and Kolar, J. W., "1 Megawatt 20 kHz isolated bidirectional 12kV to 1.2kV DC-DC converter for renewable energy applications", *Power Electronics Conference (IPEC) 2010 International*, 3212–3219, 2010.

Podporkin, G. V., Enkin, E. Y., Dmitriy, B. O. and V. E. Pilschikov (2019). "Multi-chamber disc-type lightning arrester for 13.8 kV overhead lines protection," *2019 11th Asia-Pacific International Conference on Lightning (APL)*, pp. 1–5, doi: https://doi.org/10.1109/APL.2019.8816039.

Shivam PrakashGautam, Lalit Kumar; Shubhrata Gupta, Department of Electrical Engineering, National Institute of Technology, Raipur 492010, India , "Hybrid topology of symmetrical multilevel inverter using less number of devices" IET Power Electron., 2015, Vol. 8, Iss. 11, pp. 2125–2135&The Institution of Engineering and Technology 2015.

Ramakumar, R., Allison, H. J., & Hughes, W. L. (1975). Solar energy conversion and storage systems for the future. *IEEE Transactions on Power Apparatus and Systems, 94*(6), 1926–1934. https://doi.org/10.1109/T-PAS.1975.32038

Rodriguez, J., Bernet, S., Wu, B., Pontt, J. O., & Kouro, S. (2007). Multilevel voltage-source-converter topologies for industrial medium-voltage drives. *IEEE Transactions on Industrial Electronics, 54*(6), 2930–2945.

Rojas Aldana, A. A., Barrera Beltrán, O. E. and Trujillo, R. C. L. (2015), "Design and implementation of a DC-DC converter for photovoltaic applications," *2015 IEEE PES Innovative Smart Grid Technologies Latin America (ISGT LATAM)*, Montevideo, Uruguay, pp. 321–325, doi: https://doi.org/10.1109/ISGT-LA.2015.7381175.

Salman, S. *et al.* (2018). "Design and implementation of surge protective device for solar panels," *2018 2nd IEEE Conference on Energy Internet and Energy System Integration (EI2)*, pp. 1–6, doi: https://doi.org/10.1109/EI2.2018.8582097.

Sohag, H. A., Hasan, M., Khatun, M. and Ahmad, M. (2015). "An accurate and efficient solar tracking system using image processing and LDR sensor," *2015 2nd International Conference on Electrical Information and Communication Technologies (EICT)*, Khulna, Bangladesh, pp. 522–527, doi: https://doi.org/10.1109/EICT.2015.7392008.

Srinivasan, K. and Gu, J. (2006). "Lightning as atmospheric electricity," *2006 Canadian Conference on Electrical and Computer Engineering*, 2258–2261, doi: https://doi.org/10.1109/CCECE.2006.277659.

Turcek, J., Hrasko, M., Altus, J., " Photovoltaics in present days and their coexistence with power system, " Communications-Scientific letters of the University of Zilina, 2A/2011, Vol. 13, 2011, p.: 109.-113., 1335–4205.

Vijetha Inti, V. V., & Vakula, V. S. (2017). Simulation model of a fuel cell based three phase inverter with reduced number of switches. *International Journal of Pure and Applied MathematicsVolume, 115*(8), 609–614.

Whavale S. and Dhavalikar M. (2018). "A review of Adaptive solar tracking for performance enhancement of solar power plant," *2018 International Conference on Smart City and Emerging Technology (ICSCET)*, Mumbai, India, 1–8, doi: https://doi.org/10.1109/ICSCET.2018.8537352.

Yang, J., Fu, L. and Yu, Y.-H. (2014). "Research on experiment of islanding protection device of grid-connected photovoltaic system base on RTDS," *2014 China International Conference on Electricity Distribution (CICED)*, pp. 114–118, doi: https://doi.org/10.1109/CICED.2014.6991676.

Recent Advancement in Preparation Strategy of Efficient Graphene-Cds-Based Photocatalysts for H_2 Production by Water Splitting

Bachcha Lal, Chandradhwaj Nayak, Sanjay Kumar, Amit Kumar Gupta, and Arvind Singh

Abstract The most promising and lucrative alternative to nonrenewable fossil fuels has been identified as hydrogen (H_2). Photocatalytic water splitting is regarded as one of the most ecologically benign ways to long-term H_2 generation. Because of its bandgap and band placement, CdS is regarded as the finest photocatalyst. Because of its unique structure and characteristics, such as its hierarchical network, enormous specific surface area, and good electrical conductivity, graphene has been largely used as a cocatalyst and support for CdS. These characteristics allow active areas to be exposed and facilitate electron transport. CdS photocatalysts based on graphene showed tremendous potential as high-performance catalysts for photocatalytic H_2 generation. Recent preparation approaches for synthesizing graphene-based CdS are described and addressed in this study.

Keywords Hydrogen · Water splitting · CdS · Graphene · Preparation techniques

B. Lal
TD College, Jaunpur, India

C. Nayak
India Indira Gandhi Institute of Technology, Sarang, India

S. Kumar
Department of Chemical Engineering, Marwadi University, Rajkot, India

A. K. Gupta
Chemical Engineering Department, BIT Sindri, Sindri, India

A. Singh (✉)
Department of Chemical Engineering and Biochemical Engineering, Rajiv Gandhi Institute of Petroleum Technology, Amethi, India

D. B. Pal, J. M. Jha (eds.), *Sustainable and Clean Energy Production Technologies*, Clean Energy Production Technologies,
https://doi.org/10.1007/978-981-16-9135-5_9

1 Introduction

As a renewable source of energy, solar energy can be harvested in the form of heat, electricity, and hydrogen. Solar energy stored in the form of hydrogen (H_2) may become possible solution to a long-term supply from the renewable energy sources (Sherif et al., 2005). Moreover, in the creation of new sources of energy, hydrogen is one of the most appealing fuels for the twenty-first century. In addition, the combustion of hydrogen produces water without GHG emission and air pollutants (Barbir, 2009). As a result, sustainable hydrogen generation is an important goal for the development of alternative future energy systems that provide a clean and inexpensive energy source. H_2 has the highest energy content per unit mass. It has heating value of 140 $MJ.kg^{-1}$ that is 3 times superior to liquid hydrocarbon-based fuels (Mazloomi & Gomes, 2012). This could be fed to a wide range of area, such as internal combustion engines, turbines, fuel cells, and cooking gas in kitchen.

Further hydrogen is abundant on earth but not available in free form on earth. It is found in the form of water (combines with other elements oxygen) or hydrocarbons and coal (combines with carbon). It is not a primary source, and hence, it needs to be produced.

At present, commercial processes are based on:

- Processing of carbonaceous materials mainly natural gas or naphtha.
- Electrolysis of water.

Essentially any carbonaceous matter can be processed by the following processes for hydrogen production by (a) steam reforming (SR), (b) partial oxidation (PO), and (c) oxidative steam reforming (OSR).

These can be portrayed as commercially mature technologies that can be deployed at affordable cost while achieving high efficiency. The hydrogen generation process, particularly when employing the SR route of CH_4, may achieve the competence range of 66 to 75%. The efficiency of the PO route of CH_4, on the other hand, is estimated to be around 50%. Hydrogen gas may also be generated from H_2O using the H_2O-electrolysis technique, which accounts for around 95% of the total amount of H_2 production.

SR is normally carried out at the temperature range of 700–900 °C and pressure is about 0.1–0.3 MPa. The major drawbacks are generation of greenhouse gases (Docekal, 1986) and also the requirement of pure oxygen and generation of more CO_2 than steam reforming. A well-known electrochemical method to produce H_2 using electricity is the water electrolysis (Zeng & Zhang, 2010). It permits the splitting of H_2O molecule into O_2 and H_2. In this process, a chemical reaction, i.e., water electrolysis, is taking place by external electric energy in an electrochemical cell. The reaction, however, is very endothermic; thus the required energy input is provided by electricity.

Sustainable H_2 generation is a crucial goal for the development of alternative future energy systems which will provide a clean and inexpensive source of energy. Moreover, it is a carbon-free and renewable hydrogen generation method. H_2 can be produced using a renewable source, i.e., solar energy, and it dissociates from water, which is abundant on earth. It has more attraction and this can be termed renewable hydrogen.

The reduction of water into H_2 under solar irradiation is referred to as photocatalytic water splitting. The simplicity is seemed in photocatalytic water splitting. To get H_2 from water, it just requires catalyst powder in solution and solar light. There is no greenhouse gas generation or any negative consequences on the climate. In practice, photonic energy (e.g., sunlight), water, and a photocatalyst (a semiconductor) are required in this process.

A PEC process consists of electrolyzer having a semiconductor electrode of desired bandgap. When light falls on this semiconductor electrode, it generates electron–hole pairs. These electrons are forced to move by an external bias to another electrode, where it produces H_2 by dissociation of water. Holes are utilized by the sacrificial agents present in the electrolyte. For example, if sulfide and sulfite ions are presented in electrolyte (Liu et al., 2013), then:

$$CdS \xrightarrow{h\upsilon} CdS^* + h^+ + e^- \quad (1)$$

$$2H_2O + 2e^- \rightarrow 2H_2 + 2OH^- \quad (2)$$

$$2S^{2-} + SO_3^{2-} + 2h^+ \rightarrow S_2O_3^{2-} \quad (3)$$

Overall, H_2 is a great energy carrier with a variety of distinct characteristics. It is an efficient and cleanest fuel. Hydrogen can be produced by dissociation of water and on combustion it reforms water. Thus, it is completely renewable and does not produce any pollutant. It potentially offers the cleanest and sustainable way to produce hydrogen. However, the exploration for suitable semiconductor to be employed for the water dissociation into hydrogen is still an open challenge.

2 Photocatalytic H_2 Production System by Water Splitting

A semiconductor having narrow bandgap where electron–hole pairs are formed due to activation by photons of wavelength less than 1100 nm can act as a catalyst. In addition, the conduction band (CB) should be more negative compared to reduction potential of water for electron to carry out the reaction. An electrolyte (typically $SO_3^{2-} + S^{2-}$) gets oxidized by the hole left in the valence band.

The entire water dissociation reaction process consists of three major steps [as shown in Fig. 1]: The first one is the production of charge carriers (electron/hole) after photon absorption. Next is the separation of charge carriers and migration of electron to conduction band, and the last one is interaction between charge carriers to species present onto surface. The first two stages are photo-physical in nature, while the last one is a chemical step. In addition, recombination of electrons and holes may also take place spontaneously. Backward reaction in the liquid phase is

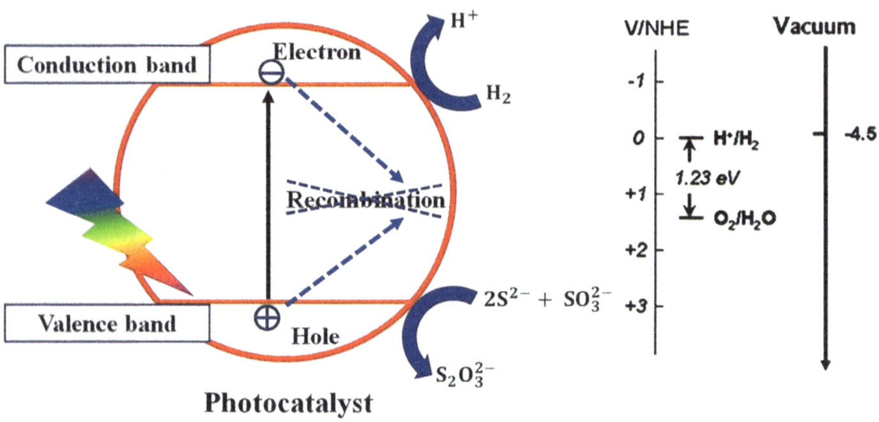

Fig. 1 Fundamental steps of photocatalytic water splitting for hydrogen production

also possible. Therefore, research is being carried out for (a) enhancement in solar light harvesting by developing semiconductors and (b) rapid charge separation system and elimination of electron hole recombination. Apart from these, changes to the particle shape of the material and changes to the surface structure are necessary to improve the reaction rate of redox couples.

Mostly commercial hydrogen is produced by steam forming of natural gas and it is sold at ~Rs. 45 per kg. At the current state of development of processes for solar hydrogen, the price would be approximately 10 times higher.

2.1 *Metal Sulfide Photocatalysts*

Photocatalysts based upon metal sulfide have been studied more towards H_2 production. Conduction bands of these semiconductors are more negative which are favorable for dissociation of H_2O into H_2 (K. Zhang & Guo, 2013). Moreover, these sulfide-based photocatalysts are more responsive to solar light compared to oxide-based photocatalysts to accomplish photocatalytic hydrogen generation; sacrificial reagents are always required. Metal sulfides, for example, usually display good photocatalytic activity in aqueous solution when Na_2S is used along with Na_2SO_3 as sacrificial reagents (Priya & Kanmani, 2009). In the past few decades, various metal sulfide–based photoactive catalysts were reported for H_2O dissociation in the presence of several sacrificial reagents. Most metal sulfide photocatalysts are composed of metal cations with a d10 structure (e.g., group IB, Cu, Ag; group IIB, Zn, Cd; group IIIA, Ga, In; group IVA, Ge, Sn) (Maeda & Domen, 2007; Zhang & Guo, 2013). Therefore, the CBs of sulfide semiconductor are composed of d orbital and sp-orbital with the valence band (VB)s consisting of S3p-orbital. These S3p-orbitals are comparatively highly negative than O2p orbitals ensuing more negative positions of CB that is enough to reduce H_2O into H_2 and narrow bandgaps that enable active solar spectrum response. ZnS (Kudo & Sekizawa, 2000; Tsuji & Kudo, 2003), CdS, and CdSe (J. Huang et al., 2012; Tongying et al., 2012), i.e., group IIB chalcogenides, are the mostly studied photocatalyst.

Because of its small bandgap, CdS is perhaps the most investigated metal-sulfide photocatalyst for H_2 evolution (2.4 eV) (Meissner et al., 1988). This provides very good suitability for the visible solar spectrum and better band positions with regard to the dissociation of H_2O into H_2. So, this fulfils the requirement of ideal photocatalyst except one major drawback, which is separation of photogenerated charges (Hu et al., 2013; Yan et al., 2009).

2.2 *Charge Recombination Phenomena and its Prevention*

The photo-assisted generation of electrons–holes needs to be separated and be available for chemical reaction. Else they will recombine and produce heat or radiation only. The recombination of photoinduced electron is too rapid compared to its migration to surface and water splitting. In general, the time for recombination $\sim 1 \times 10^{-9}$ s (Serpone et al., 1995) is ten times faster to the time taken for chemical reaction. It has been reported that 70% of photogenerated carriers recombine after excitation, and due to this major problem, photocatalytic efficiency is not achieved as required. Clearly, the photo-assisted generated charge carriers must be efficiently separated in order to enhance the utilization rate of the charge carriers by transferring to the surface active sites of photocatalysts and achieve high H_2 generation rate by photocatalytic water splitting.

The reports in the literature [Li et al., 2015; Xiang & Yu, 2013; Xie et al., 2013] imposed the use of cocatalysts, often a noble metal or metal oxide or a mix of the two, to produce heterojunctions (HJN) which are embedded on the surface of the host photocatalyst to improve photogenerated charge separation.

Photocatalysts loaded with the cocatalyst Pt are the most efficient and have higher H_2 production photocatalytic activity by dissociation of water utilizing visible light irradiation. However, Pt is costly as well as scarce metal, so attention is shifting to find some alternative having the same or a better efficiency. In this context, graphene has been widely studied. It behaves like metals with a high work function (Song et al., 2012). The value of graphene's work function is 4.42 eV. Moreover, at $pH = 0$, graphene's reduction potential is stated to be -0.08 eV vs. NHE, which is lesser negative to the CB level of CdS (-0.52 V vs. NHE) (Yong Xu & Schoonen, 2000), thus resulting to the thermodynamically easier transport of the photoinduced electrons from the main catalyst CdS to support/cocatalyst graphene. It has high electron conductivity which may facilitate a quick transfer of electron to the solid-liquid interface. Therefore, graphene seems to be a promising substrate that can replace noble metals to present cost-effective cocatalyst option.

2.3 Graphene: Background and Current Status

Graphene, a novel two-dimensional carbon-based substance, is a promising substrate having zero bandgap. Basically, the CB (π^*-state) and the VB (π-state) of graphene meet at six points which is also called Dirac points (Avouris, 2010), as shown in Fig. 2. Graphene's band structure has symmetry around the Dirac point, along with position of the Fermi level which is situated in between CB and VB. Because of this, graphene behaves as zero-bandgap material or semi-metal. With this exclusive band structure, graphene shows high conductivity that is essential for high electron mobility. Furthermore, graphene's band structure may be modified through hetero-atom doping resulting in the shift of Fermi level and also a shift in Dirac point. Subsequently, graphene shows *p*-type or *n*-type semiconductivity having a narrow bandgap [Fig. 2] (Guo et al., 2010; Guo et al., 2011).

The most frequent type of doping is with oxygen, and the resultant substance is known as graphene oxide (GO). Doping with oxygen causes structural deformation due to the creation of sp^3 C–O bonds, which disrupts the conducting π-network. This doping may be obtained through chemical exfoliation of graphite through oxidization [Fig. 3]. It renders graphene with a variety of oxygen containing functional groups (Chen et al., 2012; Xu et al., 2008). These groups are epoxy, carboxyl, or hydroxyl attached on edges as well as basal plane. GO shows insulator properties due to the creation of sp^3 hybridized carbon atoms which interrupt the delocalized π-conjugation. By reducing GO, the inherent electronic and structural characteristics of graphene may be modified. Experiments have revealed, however, that a number of oxygen groups and defects remain also after reduction (Chen et al., 2012; Xu et al., 2008). Due to this, the product is referred to as reduced graphene oxide (rGO). Furthermore, this rGO behaves like a semiconductor with high electron conductivity.

The chemical functionalization can also be applied on GO to tune its electronic properties (Chen et al., 2012). The carbon–oxygen bonds on graphene cause

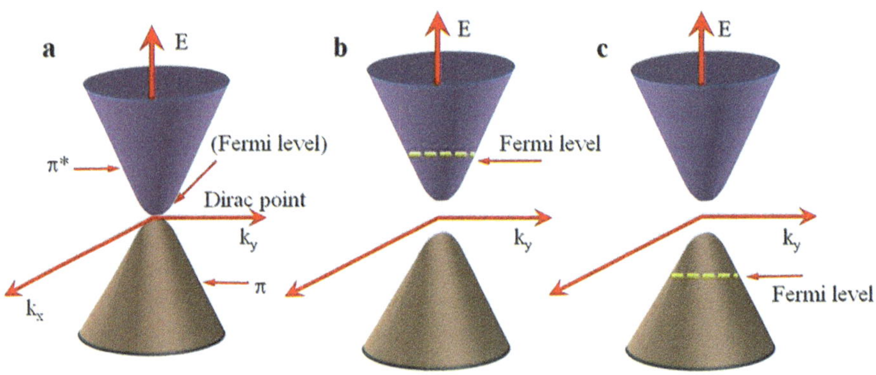

Fig. 2 (**a**) Energy band and structure of graphene, (**b**) n-type graphene, (**c**) p-type graphene by suitable heteroatom doping (Avouris, 2010)

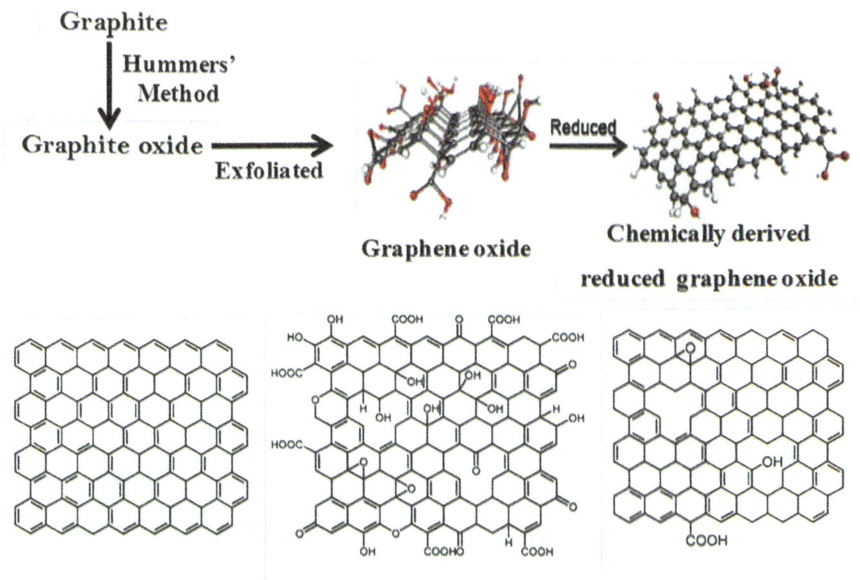

Fig. 3 Schematic diagram of the synthesis of the graphite oxide, GO, rGO

electronic states to shift and a wider bandgap to develop. Carbon–oxygen covalent linkages disrupt the extended sp^2 entangled structure, confining the source π–electrons within solitary sp^2 regions. Despite this, the $CB_{minimum}$ is still made up of the anti-bonding π^*–orbital, but the $VB_{maximum}$ is made up of the O2p orbital rather of the π–orbital of graphene. As a result, it is well understood that GO/rGO with extra oxygenated components is a p-type semiconductor (Gómez-Navarro et al., 2007). The existence of various oxygen-containing functional groups in rGO results in an abundance of active sites present onto the surface, those take the form of floating oxygen bonds, vacancy defects, and localized orbital sites. These active sites where functional groups may be attached to chemical interaction may take place as a precursor in graphene-based composite.

As previously discussed, that heterojunction (HJN) plays a key role in photocatalysis. A HJN is described as the boundary of two various semiconductors with inequitable bandgaps; these can lead in band alignments. Generally, heterojunction in photocatalysts is classified into three categories: having a stroking gap, referred to as type-I HJN; next having a staggered gap, referred to as type-II HJN; and the last having the breached gap, referred to as type-III HJN. CB and VB of A-semiconductor are positioned higher and lower than the respective bands of B-semiconductor under type-I HJN photocatalysts [Fig. 4a]. As a result, when exposed to solar light, electrons and holes accrue at the CB and VB levels of B semiconductor, respectively. Figure 4b shows that the VB and CB positions of A-semiconductor are superior to B-semiconductor in type-II HJN photocatalysts. When light irradiation emissions take place, photoinduced electron will move to

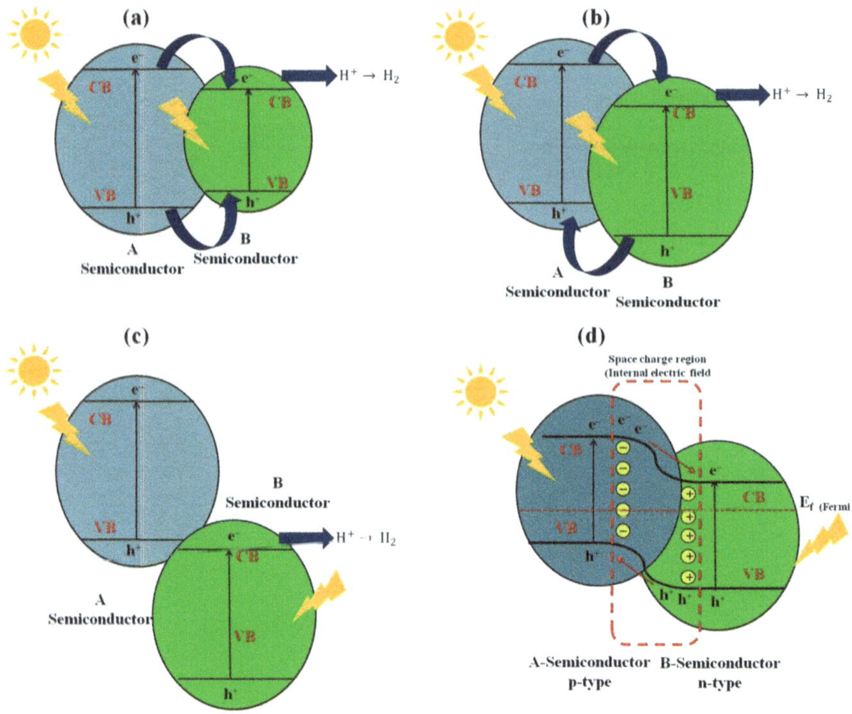

Fig. 4 Schematic representation of various types of heterojunction in photocatalysis: : (**a**) Type-I heterojunction; (**b**) Type-II heterojunction; (**c**) Type-III heterojunction; (**d**) p–n heterojunction

B-semiconductor, whereas photoinduced holes will move to A-semiconductor, ensuing in a distinction of electrons–holes. The redox potential of the type-II HJN photocatalyst will be lowered, as with the type-I HJN. The type-III HJN photocatalyst is akin to the type-II HJN photocatalyst, as seen in Fig. 4c. The spaced gap is so intense, resulting inhibition of overlapping of the bandgaps. Therefore, movement of electrons–holes for best separation may not arise well for the type-III HJN.

However a type-II HJN could indeed preferably segregate electron–hole pairs in regime, and hence, improvement in mitigation of electron–hole recombination is not up to type-II HJN, and it is not sufficiently able to prevent high throughput electron–hole recombination on the semiconductor. As a result, a p–n HJN photocatalysts were first proposed, capable of accelerating electron–hole transfer throughout the HJN and therefore enhancing photoactivity by introducing an appropriate electric field. Integrating p- and n-type semiconductors yields an efficacious p–n HJN photocatalyst. Before being exposed to solar light, electrons on the n-type semiconductor close the p–n interface which are likely to diffuse into the p-type semiconductor, resulting in a positive charge system (see Fig. 4d). Similarly, the holes on the p-type semiconductor close the p–n interface which are likely to diffuse into the n-type semiconductor, resulting in system with negative charge. The diffusion phenomenon of electrons–holes will prolong until the system reaches Fermi level equilibrium. As a result, the region near the p–n interface becomes charged, resulting in a "charged" space, also known as the internal electric field. When light having energy equal to or greater than the bandgap of both semiconductors is irradiated, excitation process is taken place in both semiconductors, resulting in photoinduced generation of electron–hole pairs. These will migrate to the CB of the n-type semiconductor and the VB of the p-type semiconductor, respectively, under the influence of the internal electric field, resulting in the distinction of the electron–hole pairs. Moreover, thermodynamically feasibility is also present due to higher position of VB and CB of the p-type semiconductor with respect to the n-type semiconductor. As a result of the synergistic relation between the internal electric field and alignment of band, the electron–hole separation efficiency in p–n heterojunction photocatalysts is faster than that of type II HJN photocatalyst system [Low et al., 2017].

These HJNs were revealed by several authors. DP Pal et al. (2021) used X-ray photoelectron spectroscopy (XPS) and Transmission electron microscopy (TEM) to demonstrate heterojunction in CdS/graphene photocatalyst. Alam et al. (2020a, 2020b) also prepared N, S-rGO, and multi-walled carbon nanotube (MWCNT) supported CdS photocatalyst. They postulated HJN in these photocatalyst with HRTEM, electrochemical impedance spectroscopy, Raman spectra, and Mott-Schottky analysis. It may be concluded that these techniques are useful to demonstrate the type of HJN formation in graphene-CdS photocatalyst along with all the composites in the field of photocatalysis.

Various review articles (Li et al., 2015; Xiang & Yu, 2013; Xie et al., 2013) have been published to demonstrate the roles of graphene in photocatalysis. These give a thorough summary of current graphene-based material research developments. There has been no such review available that focuses specifically on the preparation technique of these composites.

3 GO/rGO-CdS-Based Photocatalysts

In the following section, a review based upon different techniques for the preparation of CdS/rGO was done. Moreover, a comparative study of H_2 production is depicted in Table 1. In situ growth, solution mixing, hydrothermal, and solvothermal techniques are the most often utilized preparation methods. The sections that follow will provide a more thorough overview of the different synthesis pathways.

3.1 In Situ Growth Strategy

The direct growth technique is frequently utilized in the preparation of CdS/graphene and its composites. Functionalized GO and Cd salts are the most frequent reagents of graphene and metal compounds, respectively. Typically, the salt

Table 1 GO/rGO-CdS-based visible light active ($\lambda \geq 420$ nm) photocatalysts for H_2 production by dissociation of water

Sl.	Photocatalyst	Preparation route	Sacrificial agents	Light source	H_2 production ($\mu mol\ h^{-1}$)	Reference
1.	rGO/CdS	In situ growth	–	–	–	Cao et al. (2010)
2.	rGO/CdS-Pt	Solvothermal	Lactic acid	350 W, Xe-lamp	1120	Li et al. (2011)
3.	CdS/N-rGO	Sol-gel	Na_2SO_3, Na_2S	350 W, Xe-lamp	50	Jia et al. (2011)
4.	CdS/rGO	Solvothermal	Na_2SO_3, Na_2S	300 W Xe-lamp	480	Zeng et al. (2011)
5.	rGO/CdS	Two-phase	Methanol	400 W high-press. Hg lamp	110	Gao et al. (2012)
6.	rGO/CdS	Precipitation	Na_2SO_3, Na_2S	300 W Xe-lamp	314	Peng et al. (2012)
7.	CdS/rGO	Hydrothermal	Na_2SO_3, Na_2S	200 W Xe-lamp	70	Ye et al. (2012)
8.	CdS@TaOH-RGO-Pt	Hydrothermal	Na_2SO_3, Na_2S	–	633	Hou et al. (2012)
9.	CdS/Al_2O_3 /RGO	Solid state	Na_2SO_3, Na_2S	500 W, tungsten halogen lamp	350	Khan et al. (2012)
10.	CdS/ZnO /RGO	Solid state	Na_2SO_3, Na_2S	500 W, tungsten halogen lamp	751	Khan et al. (2012)
11.	CdS/graphene	Sol-gel	Na_2SO_3, Na_2S	300 W Xe-lamp	500	Lv et al. (2012)
12.	CdS QDs–$ZnIn_2S_4$/ GO	Hydrothermal	Na_2SO_3, Na_2S	300 W Xe-lamp	2700	Hou et al. (2013)
13.	CdS-MoS_2 / graphene	Hydrothermal	Lactic acid	500 W, tungsten halogen lamp	1800	Jia et al. (2014)
14.	CdS-ZnS-GO	Precipitation	Na_2SO_3, Na_2S	300 W, Xe lamp	750	Wang et al. (2015)
15.	CdS- graphene	Hydrothermal	Na_2SO_3, Na_2S	350 W, Xe Arc lamp	800	Li et al. (2015)
16.	Graphene-CdS		Na_2SO_3, Na_2S	300 W, Xe lamp	1344	Cao et al. (2015)

		Solvothermal method				
17.	rGO-CdS	Solvothermal method	$(NH_4)_2SO_3$	300 W, Xe-lamp	1400	Hong et al. (2015)
18.	Graphene–CdS	Solvothermal process	Na_2SO_3, Na_2S	300 W, Xe lamp	175	Xu et al. (2016)
19.	CdS-MoS_2 / graphene	Solvothermal method	Lactic acid	300 W, xenon lamp	1913	Yu et al. (2016)
20.	rGO-CdS	Solvothermal method	Na_2SO_3, Na_2S	150 W, Xe arc lamp	3.85	Murillo Leo et al. (2017)
21.	CdS/graphene	Solvothermal method	Lactic acid	350 W Xenon	1890	Xia et al. (2019)
22.	CdS@BNG	Calcination method	TEA	PLS-SXE 300	6465.33 μmol h^{-1} g^{-1}	Li et al. (2021)
23.	CdS/rGO/MoS_2	Hydrothermal	Lactic acid	300 W Xe Lamp	4460 μmol h^{-1} g^{-1}	Ben Ali (2017)
24.	NiO@Ni-ZnO/ rGO/CdS	Solution method	Na_2SO_3, Na_2S and triethanolamine	Xe lamp (PLX-300)	5640 μmol·h^{-1}·g^{-1}	Chen et al. (2017)
25.	NS-RGO/CdS	Calcination method	Lactic acid	300 W Xe lamp	1701 μmol h^{-1} g^{-1}	Han et al. (2018)
26.	rGO/CdS	In situ method	Ethanol	4 LED lamps 4 W each,	2477 μmol/h g	Quiroz-Cardoso et al. (2019)
27.	rGO/CdS/Pt	In situ method	Na_2SO_3, Na_2S	300 W Xe lamp	10,480 μmol g^{-1} h^{-1}	Dai et al. (2020)
28.	rGO/CdS	Solid-gas reaction method	Na_2SO_3, Na_2S	100 W Xe lamp	312 μmol g^{-1} h^{-1}	DB Pal et al. (2021)
29.	N,S-rGO/CdS/ MWCNT	Solid-gas reaction method	Na_2SO_3, Na_2S	100 W Xe lamp	455 μmol g^{-1} h^{-1}	Alam et al. (2020b)
30.	N,S-rGO/CdS	Solid-gas reaction method	Na_2SO_3, Na_2S	100 W Xe lamp	373 μmol g^{-1} h^{-1}	Alam et al. (2020a)

is combined with GO before being converted to the appropriate sulfide, resulting in a GO/metal sulfide composite. After reducing GO during composite production, graphene-based metal composites were formed.

Cao and his coworker synthesized graphene/CdS composite by a one-step technique. In this process, GO and $Cd(CH_3COO)_2.2H_2O$ were dissolved into dimethyl sulfoxide (DMSO) where DMSO released H_2S during this one-step reaction. It acted both the reducing agent of GO as well as the sulfide source. Thus, simultaneously deposition of CdS and the reduction of GO occurred on the graphene. It produced a single-layer graphene/CdS quantum dot nanocomposite and a uniform distribution of CdS nanoparticles having a size of 10 nm on graphene sheets [Cao et al., 2010].

Quiroz-Cardoso et al. (2019) prepared CdS-GO by in situ method. First, GO was disseminated in butanol for 90 min using sonication. The precipitation technique was used to create CdS nanofibers after the GO sheets were distributed. $Cd(NO_3)_2{\cdot}4H_2O$ was dissolved under constant stirring in the aforementioned GO-butanol at room temperature for 15 min, and then ethylenediamine was added dropwise to this solution. Finally, dropwise additions of carbon disulfide were made. The resulting solution was heated to 100 °C for 1 h while being magnetically stirred and then allowed to cool at ambient temperature while still being magnetically stirred. The yellow precipitate was collected by filtering, washed with DD water and ethanol solution, and lastly dried for 1 h at 80 °C. When compared to the bare semiconductor CdS, H_2 generation activity was increased by 1.8 times. This finding is mostly due to the distribution of CdS fibers on the GO sheets achieved by this synthesis technique, in which the fibers were grown on the carbon network, increasing the interfacial interaction between both materials and hence enhancing electron transmission from CdS to the GO sheets.

3.2 *Two-Phase Method*

Gao et al. (2012) reported GO–CdS synthesis by the two-phase method: first, CdS was prepared by mixing solution of cadmium chloride hydrate and sulfur into oleylamine and then heated at 433 K in N_2 atmosphere under constant stirring. Finally CdS was dispersed in toluene along with as-prepared GO. The morphology and structure of nanocomposite was investigated by Atomic force microscopy (AFM), TEM, and X-ray powder diffraction (XRD). The results of Fourier-transform infrared spectroscopy (FTIR) and XPS spectroscopy confirmed the excellent deposition of CdS on GO sheets. Because of its strong light absorbing capabilities and efficient charge transport, it aided in the successful evolution of hydrogen. GO–CdS has much more hydrogen evolution activity than CdS, i.e., four times higher.

Jia and his coworkers (Jia et al., 2014) prepared CdS–graphene samples by combining CdS nanoparticles and graphene oxide (GO) in ethanol, followed by 2 h of sonication and 24 h of stirring at room temperature. An enhanced Hummers' technique was used to prepare graphene in the beginning. CdS was produced independently. CdS–graphene had the greatest H_2 evolution rate of 3.067 mL h^{-1}, which was almost 5 times greater than CdS. XRD and TEM analysis ruled out the development of any new phases in this combination during preparation. A high degree of light-harvesting ability and effective charge transfer between these two nanomaterials via the 2D graphene structure were linked to improved hydrogen generation.

Khan et al. (2012) prepared GO-loaded CdS/Al_2O_3 and CdS/ZnO by solid-state route. In brief, a mortar and pestle were used to grind CdS/oxide for 0.5 h. The CdS/metal oxide and GO were ground to ensure that they were in tightly contact with each other. To investigate the influence of GO on CdS/metal oxide photocatalysis, the final powder was employed in photocatalytic activity. The enhanced hydrogen activity of ternary composites was attributed to graphene oxide's sheetlike structure, which offers a wide surface area for efficient mass transfer and easy charge transport, enhancing disintegration of photogenerated charge carriers.

3.3 *Sol-Gel Method*

Jia and his research group (Jia et al., 2011) studied N-graphene-CdS photocatalyst. It was prepared by adding cadmium chloride into GO solution of ethanol. Solution of sulfur reagent Na_2S was added gradually into the suspension with stirring, washed, and finally heated under N_2 environment at 673 K for 2 h. Loading N-doped graphene substantially increased the H_2 evolution rate of CdS. The rate of H_2 production was 210 mmol h^{-1}, which was more than 5 times higher than the rate on pure CdS. The greater H_2 evolution was ascribed to N-higher graphene's electrical conductivity, suggesting that photogenerated electrons migrate to the composite's surface considerably in more easy way, enhancing the suppression of photoinduced electron–hole charge carriers. It was also postulated that crystallinity, surface area, morphology, and bandgap of N-graphene/CdS and CdS were not to ascribe for the significant increase in H_2 generation.

Lv et al. (2012) prepared CdS–graphene oxide. In this process, first sulfonated graphene was dissolved in distilled water under ultrasonic. Then aqueous solution Cd-precursor ($CdCl_2$) was added dropwise under stirring followed by dropwise addition of aqueous solution of Na_2S. This resultant was stirred for 3 h. It was then retrieved by filtration, washed several times with deionized water, and dried in a vacuum oven at 70 °C. Morphology analysis confirmed that the obtained graphene-CdS photocatalyst was nanocomposite. Higher photocatalytic activity of this composite compared to bare CdS attributed to highly dispersion of graphene on CdS and co-catalyst-semiconductor interfacial interaction's enhancement. Therefore, graphene efficiently accepted and transported electrons from the excited CdS and suppressed charge recombination resulting improved interfacial charge transportation.

3.4 Precipitation/Solution Method

Li and his coworkers reported the preparation of GO–CdS by precipitation method. Graphene oxide (GO) was first synthesize by following modified Hummers' technique. Following that, the requisite quantity of cadmium acetate and exfoliated GO were added to water to form 0.1 M solution. Then, gradually, 100 mL of Na_2S solution (0.11 M) was added while stirring. For 1 h, the resulting mixture was kept in stirring condition. The result was rinsed with water and acetone after centrifugation and then put for drying at 313 K. The highest H_2 generation rate of GO/CdS composite was 1.5 times that of bare CdS. Enhanced activity attributed to coupling of GO with the CdS nanoparticles. It was efficiently able to absorb and transfer the electrons from the excited semiconductor, which reduced charge recombination and enhanced interfacial charge transportation (Peng et al., 2012).

Wang et al. (2015) prepared GO doped ZnS–CdS by precipitation method. First, GO was prepared by following the modified Hummers' method. The appropriate quantity of GO solution was then added to 50 mL of DI water while constantly stirring in a usual procedure. Under continual stirring, specified quantities of Zn $(CH_3COO)_2$ and $Cd(CH_3COO)_2$ were added to an aqueous solution of GO. Following that, 50 mL of aqueous Na2S solution was added dropwise with steady stirring for 1 h. The combined solution was rinsed with DI water and vacuum dried for 24 h at 333 K. The surface of GO nanosheets was loaded with ZnS and CdS nanoparticles, resulting in a sheetlike structure. The photocatalytic H_2 production rate of ZnS–CdS/GO was almost double that of ZnS–CdS. Improved hydrogen production rate was attributed to high carrier transport property of GO nanosheets that constructed transport carrier pathway between ZnS and CdS nanoparticles. It also enhanced the cooperative effects in the heterostructure.

Chen et al. (2017) synthesized ZnO-rGO-CdS photocatalyst. During the synthesis process, zinc powder was initially disseminated in distilled water while being stirred magnetically. Following that, for 10 min, GO aqueous solution was added to the aforementioned solution while being constantly magnetically stirred. In the aqueous solution, a dark precipitate of Zn/rGO developed. Under continual stirring, a specified quantity of Ni- acetate aqueous solution was progressively put into the aqueous solution having Zn/rGO. The solution was kept in a 60 °C oven for 24 h to allow Ni-ZnO/rGO to develop. Washing of the Ni-ZnO/rGO was repeated. Then, for 1 h, an aqueous solution of $CdCl_2$ was added to the abovementioned mixed solution under continual stirring and solar irradiation. After that, aqueous solution containing Na_2S was added gradually with continual stirring for 1 h under light irradiation. The precipitate was rinsed with water before being vacuum dried at 80 °C for 24 h. Finally, heterostructures of NiO@Ni-ZnO/rGO/CdS were created. Chemical connections generated between the semiconductor phases and rGO nanosheets provide seamless photoexcited transfer, improving the synergistic impact between photocatalysts.

3.5 *Calcination Method*

Li et al. (2021) used a calcination technique to create a boron nitride rGO CdS ternary photocatalyst. Initially, GO nanosheets were evenly distributed in distilled water, followed by the addition of boric acid and urea to produce a clear solution. At 80 °C, the water evaporated. This combination was then calcined for 5 h at 900 °C with N_2 gas flowing over it. This boron nitride rGO and CdS Nano particles combination was distributed evenly in deionized water before being transferred to a closed stainless steel reactor and reacted for 8 h at 220 C. The rate of H_2 generation was increased 11 times over that of pure CdS nanoparticles. This improvement can be ascribed to the creation of the boron nitride/rGO interface. As a result, photoinduced electrons may be transported quickly, effectively inhibiting electron–hole recombination.

Han and his coworkers (Han et al., 2018) synthesized a CdS/NS-rGO nanocomposite. $Cd(CH_3COO)_2.2H_2O$ was initially dissolved in DI water in the preparation process. The Na_2S solution was then added to the solution gradually while vigorously stirring. Continued stirring was performed for the next 24 h. The orange CdS precipitate was then centrifuged and washed several times with DI water and ethanol to eliminate unreacted reactants. This preprepared CdS was combined with NS-rGO in ethanol before being evaporated at 60 °C. The sample was then annealed in argon flow at 400 °C for 2 h. The rate of H_2 generation was found to be 5.8 times greater than that of pure CdS. It was hypothesized that co-doped graphene with the appropriate amount of nitrogen and sulfur atoms can modify the electron distribution and enhance the charge density of a spot on the surface. Furthermore, by loading CdS on NS-rGO, higher charge separation efficiency, a broader light absorption range, a narrower bandgap, and a lower H* absorption free energy could be obtained, resulting in increased photoactivity.

3.6 *Hydrothermal/Solvothermal Method*

Many research groups have reported preparation of GO/rGO supported CdS by solvothermal or hydrothermal techniques.

Li and his research group followed solvothermal method to synthesize graphene-CdS photocatalyst. In the preparation method, graphene oxide (GO), $Cd(Ac)_2$ as a Cd precursor, and dimethyl sulfoxide (DMSO) as a solvent and sulfur source were used. The activity for hydrogen production increased 4.87 times rather than CdS alone. They demonstrated that graphene enhanced the activity due to its properties as an excellent supporting material as well as collector and transporter of electron to separate photogenerated charge carriers (Li et al., 2011).

P. Zeng and his group (Zeng et al., 2011) also prepared rGO–CdS by following hydrothermal method. In detail, Cd precursor, $Cd(Ac)_2.2H_2O$, was added with DMSO solvent containing GO. These were first ultrasonically dispersed in water and then transferred to a Teflon-lined stainless steel autoclave (TLSSAC) for further hydrothermally treatment at 453 K. For comparison, CdS and rGO–CdS were also prepared by precipitation. The activity for hydrogen production increased two times rather than CdS alone. Physical mixtures of rGO-CdS did not show enhanced photoactivity for H_2 evolution compared to CdS. Enhanced activity was reported due to implantation of rGO. Electron-transportation medium and acceptor, as well as CdS support, were some of the roles played by rGO.

Ye et al. (2012) also prepared CdS–graphene nanocomposite by hydrothermal treatment. First, GO was dissolved in water under ultrasonic treatment for 1 h. Then a cadmium precursor ($Cd(Ac)_2$) was added. Subsequently, the aqueous solution of Na_2S was added dropwise into this solution under constant stirring. Then it was put into a TLSSAC at 453 K hydrothermal treatment. Hydrogen production rate was reported to be 4.8 times higher than CdS. They reported that CdS nanoparticles having the size of~35 nm were dispersed on the graphene sheets in the nanocomposites. The incorporation of graphene into CdS also resulted in a significant narrowing of the bandgap, demonstrating the strong interactions between CdS and graphene.

Hong et al. (2015) also used the solvothermal technique to create rGO-CdS. GO was ultrasonically disseminated in an aqueous solution of DMF in a conventional method. Then Cd-precursor $Cd(Ac)_2.4H_2O$ was added, followed by TAA as a sulfur source, and stirred for 0.5 h. This clear solution was placed in a TLSSAC at 433 K for 24 h. This technique produced hexagonal and cubical phases of CdS, according to the researchers. The development of a linkage between CdS and GO, which demonstrated a more effective transfer of photoinduced electrons than mechanically loaded CdS, was responsible for the increased H_2 generation.

Hou et al. (2013) also prepared the ternary composite CdS–$ZnIn_2S_4$–GO by hydrothermal treatment. In which commercially available GO was ultrasonically dispersed in an anhydrous ethanol solution. Then pre-prepared CdS–$ZnIn_2S_4$ was added. The constant stirring was done for to get homogeneous suspension. Then, it was transferred into TLSSAC at 393 K. CdS–$ZnIn_2S_4$-GO showed about fourfold enhancement compared to $ZnIn_2S_4$ alone. They claimed that adopting this preparation approach resulted in three-dimensional architectures of CdS quantum dots integrated in the porous assembly of marigold-like $ZnIn_2S_4$ heterostructures. It led to enhanced hydrogen production with incorporation of GO. Again, Hou et al. (2012) prepared CdS@TaON core shell composite embedded with GO as a cocatalyst. In the preparation of CdS@TaOn, an aqueous solution of Na_2S, nanoparticles of as-prepared TaON were added cadmium acetate solution. The yellow liquid that resulted was agitated for 24 h before being filtered. The wet solid was suspended in water and transported to a TLSSAC for hydrothermal treatment at 473 K for 72 h. Then it was dried at 368 K for 24 h under vacuum. This as-prepared CdS@TaON core shell was mixed with GO in ethanol solution. This mixing solution was put for ageing purpose under constant stirring to get homogeneous suspension. Then this suspension was put to a TLSSAC at 393 K.

Murillo Leo et al. (2017) prepared CdS–rGO hybrid by following solvothermal method. Commercially GO was reduced by solvothermal method. Then, Cd precursor (cadmium acetate) and sulfur precursor (thiourea) were mixed with as-prepared rGO in the synthesis of hybrid. The TLSSAC was filled with this mixture and ethylenediamine. It was then put at 393 K for 12 h. Wrinkle and rippled shape structures were formed by this method. Hydrogen production was enhanced by 1.5 times compared to CdS. The extremely crystalline structure, crystal size, and the nature of the interaction of CdS with rGO were attributed to the migration and transport of charges in the CdS-rGO hybrids.

Yu and his coworkers (Yu et al., 2016) also prepared CdS/MoS_2/graphene hallows spheres. First GO was prepared by following modified Hummers' method. Then CdS/MoS_2/graphene hybrids were prepared by a one-step hydrothermal method. In this process, first, cadmium acetate (Cd precursor) with GO was added into DI water, next sonication was done, and then cysteine (sulfur precursor) and sodium molybdate (Mo precursor) were added under constant stirring. This homogenous solution was then transferred into TLSSAC and put at 573 K for 24 h. The unique hollow structures of composite and 2D graphene with high specific surface area were attributed to the high H_2 production activity as reported by them.

Cao et al. (2015) used a solvothermal technique to synthesize graphene-CdS nanowires. It resulted in a high hydrogen evolution rate that was about twice as fast as that of bare CdS nanowires. The graphite oxide was originally synthesized using a modified Hummers' technique. The GO was then dispersed ultrasonically in ethylenediamine (EDA), followed by the addition of $Cd(NO_3)_2.4H_2O$ and thiourea. This homogenous solution was then transferred to a TLSSAC, which was then filled to 80% capacity with EDA and placed at 433 K for 48 h. Increased H_2 generation rate was due to graphene's excellent electron conductivity as well as surface area.

Xu and his research group prepared graphene–CdS composite by solvothermal process. rGO and cadmium acetate were added with the solution of DMSO. The suspension was put into a 100 mL TLSSAC. The temperature was then raised to 453 K and it was held there for 12 h. This group reported the occurrence of aggregation of graphene sheets by using this preparation process. They claimed that a heterojunction developed between CdS and graphene, allowing electrons to move from CdS to graphene and react with protons at graphene's active spots. As a result, charge carrier recombination was prevented. The composite catalyst produced H_2 at a rate 19.3 times faster than the physical mixed sample (Xu et al., 2016).

Xia and his coworkers (Xia et al., 2019) used the solvothermal technique to create CdS/graphene nanorods. Graphene nanorods (NR) were first created on multi-walled carbon nanotubes (MWCNT). The graphene solution produced in N, N-dimethylformamide was then dissolved in acetate precursor of Cd. Furthermore, thiourea was added while stirring. This combination was then placed in a 100 mL TLSSAC at 160 °C for 12 h to undergo hydrothermal treatment. The filtered precipitate was rinsed with distilled water and ethanol before being dried in a vacuum oven at 70 °C for 12 h. Graphene NRs with a high specific surface area, good light absorption, and active edge sites were shown to have a 3.5-fold increase in H_2 generation activity when compared to bare CdS. This improvement can be ascribed to enhanced charge separation efficiency, longer carrier lifetime, and faster surface transfer reaction of electron–hole pairs.

Ben Ali et al. (2017) developed a noble metal-free ternary $CdS/MoS_2/rGO$ photocatalyst. In the preparation procedure, NH_2CSNH_2 solution was introduced gradually into $Cd(Ac)_2.2H_2O$ solution under continual stirring. The aforementioned mixture was then sonicated with aqueous MoS_2 sheets/graphene oxide and mixed for 30 min. The resultant solution was put in a TLSSAC and maintained the temperature at 200 °C for 24 h. The finished product was rinsed many times with water and dried overnight at 80 °C. The increase in H_2 production was attributed to the intimate contact between them, which resulted in an obstruction of electron–hole recombination.

Dai et al. (2020) synthesized MWCNT exploited rGO supported CdS photocatalyst. First, rGO/MWCNT was constructed by utilizing MWCNT as a graphitic carbon precursor. To synthesis MWCNT using rGO, a modified Hummers' technique was used. The rGO-CdS was then produced using the solvothermal technique. $CdCl_2$•$2.5H_2O$ and S (Sulfur) were dissolved and agitated in a solution comprising DETA (diethylenetriamine) and purified water. The prepared rGO was then added to the aforementioned mixture and mixed well for 30 min. The mixture was then placed in an autoclave reactor and hydrothermally treated at 353 K for 48 h. The as-prepared products were then cleaned multiple times with distilled water and dried overnight in a vacuum freeze drying oven. The 2D platform structure of rGO offers good supporting material and improved electron collecting for the facilitated H_2 production activity and large S_{BET}, resulting in 4 times higher H2 generation (10.48 mmol/g-h) compared to bare CdS. Second, because of its 2D planar conjugation structure in rGO/CdS-DETA composites, rGO's high electrical conductivity reduces the recombination rate of photoexcited carriers. Third, the CdS-DETA has a unique inorganic-organic structure that enables active sites for it to easily form a strong interaction with rGO.

Commonly, precursors of "Cd" and "S" and GO/rGO were dispersed ultrasonically and kept at elevated temperatures (150–200 °C) in TLSSAC for a few hours.

On the basis of above discussion, it is concluded that GO/rGO can accept photogenerated electrons from the CBs of CdS and further will efficiently prevent the recombination of photogenerated electron–hole resulting into improved photocatalytic H_2 production activity. One major responsible factor is its high work function (4.42 eV). Moreover, GO/rGO has high electronic conductivity, and as a result the accepted electrons can easily move across its 2D plane to the surface for H_2 evolution. Therefore, GO/rGO plays an efficient role of an electron acceptor and transporter. Further, for an efficient transfer of electron from CdS to GO/rGO to take place, the nature of interface plays a major role.

4 Conclusion

Solar H_2 production from water is a critical method for alleviating the world's energy deficit by directly using abundant solar energy in the form of molecular hydrogen, a green fuel. Although photocatalytic technology has advanced and has been applied to the use of solar energy to produce H_2 and other fuels, its efficiency remains a difficulty. In order to improve the efficiency of this technology and excellence in photocatalytic H_2 production, these two steps of the process, i.e., charge creation and charge survival, should be substantially improved. CdS is essentially a photocatalyst that responds to visible light; its absorption characteristics may be tuned up to the whole visible range or the infrared area. As evidenced by several research studies, combining two photocatalysts, particularly CdS, with graphene has been proven to be extremely effective in increasing photocatalyst light absorption capabilities. This combination also enhances charge recombination. This synergetic phenomenon has resulted in excellent H_2 production by water splitting. which results in higher H_2 production by water splitting.

References

Alam, Z., Verma, B., & Sinha, A. S. K. (2020a). Creation of heterojunction in CdS supported on N, S-rGO for efficient charge separation for photo-reduction of water to hydrogen. *International Journal of Hydrogen Energy, 45*, 4095–4112. https://doi.org/10.1016/j.ijhydene.2019.12.056

Alam, Z., Verma, B., & Sinha, A. S. K. (2020b). Synthesis and characterization of MWCNT incorporated N, S-rGO supported CdS photocatalyst for the dissociation of water to hydrogen by visible light. *International Journal of Hydrogen Energy, 45*, 175–189. https://doi.org/10.1016/j.ijhydene.2019.10.183

Avouris, P. (2010). Graphene: Electronic and photonic properties and devices. *Nano Letters, 10*(11), 4285–4294. https://doi.org/10.1021/nl102824h

Barbir, F. (2009). Transition to renewable energy systems with hydrogen as an energy carrier. *Energy, 34*(3), 308–312. https://doi.org/10.1016/j.energy.2008.07.007

Ben Ali, M., Jo, W.-K., Elhouichet, H., & Boukherroub, R. (2017). Reduced graphene oxide as an efficient support for CdS-MoS2 heterostructures for enhanced photocatalytic H2 evolution. *International Journal of Hydrogen Energy, 42*(26), 16449–16458. https://doi.org/10.1016/j.ijhydene.2017.05.225

Cao, A., Liu, Z., Chu, S., Wu, M., Ye, Z., Cai, Z., . . . Liu, Y. (2010). A facile one-step method to produce graphene–CdS quantum dot nanocomposites as promising optoelectronic materials. Advanced Materials, 22(1), 103–106. doi: https://doi.org/10.1002/adma.200901920.

Cao, M., Wang, P., Ao, Y., Wang, C., Hou, J., & Qian, J. (2015). Investigation on graphene and Pt co-modified CdS nanowires with enhanced photocatalytic hydrogen evolution activity under visible light irradiation. *Dalton Transactions, 44*(37), 16372–16382. https://doi.org/10.1039/C5DT02266E

Chen, D., Feng, H., & Li, J. (2012). Graphene oxide: Preparation, functionalization, and electrochemical applications. *Chemical Reviews, 112*(11), 6027–6053. https://doi.org/10.1021/cr300115g

Chen, F., Zhang, L., Wang, X., & Zhang, R. (2017). Noble-metal-free NiO@Ni-ZnO/reduced graphene oxide/CdS heterostructure for efficient photocatalytic hydrogen generation. *Applied Surface Science, 422*, 962–969. https://doi.org/10.1016/j.apsusc.2017.05.214

Dai, K., Hu, T., Zhang, J., & Lu, L. (2020). Carbon nanotube exfoliated porous reduced graphene oxide/CdS- diethylenetriamine heterojunction for efficient photocatalytic H2 production. *Applied Surface Science, 512*, 144783. https://doi.org/10.1016/j.apsusc.2019.144783

Docekal, J. (1986). Hydrogen production from hydrocarbons. *International Journal of Hydrogen Energy, 11*(11), 709–714. https://doi.org/10.1016/0360-3199(86)90139-4

Gao, P., Liu, J., Lee, S., Zhang, T., & Sun, D. D. (2012). High quality graphene oxide-CdS-Pt nanocomposites for efficient photocatalytic hydrogen evolution. *Journal of Materials Chemistry, 22*(5), 2292–2298. https://doi.org/10.1039/C2JM15624E

Gómez-Navarro, C., Weitz, R. T., Bittner, A. M., Scolari, M., Mews, A., Burghard, M., & Kern, K. (2007). Electronic transport properties of individual chemically reduced graphene oxide sheets. *Nano Letters, 7*(11), 3499–3503. https://doi.org/10.1021/nl072090c

Guo, B., Fang, L., Zhang, B., & Gong, J. R. (2011). Graphene doping: A review. *Insciences J, 1*(2), 80–89.

Guo, B., Liu, Q., Chen, E., Zhu, H., Fang, L., & Gong, J. R. (2010). Controllable N-doping of graphene. *Nano Letters, 10*(12), 4975–4980. https://doi.org/10.1021/nl103079j

Han, W., Chen, L., Song, W., Wang, S., Fan, X., Li, Y., . . . Peng, W. (2018). Synthesis of nitrogen and sulfur co-doped reduced graphene oxide as efficient metal-free cocatalyst for the photo-activity enhancement of CdS. Applied Catalysis B: Environmental, 236, 212–221. doi: https://doi.org/10.1016/j.apcatb.2018.05.021.

Hong, Y., Shi, P., Wang, P., & Yao, W. (2015). Improved photocatalytic activity of CdS/reduced graphene oxide (RGO) for H2 evolution by strengthening the connection between CdS and RGO sheets. *International Journal of Hydrogen Energy, 40*(22), 7045–7051. https://doi.org/10.1016/j.ijhydene.2015.04.005

Hou, J., Wang, Z., Kan, W., Jiao, S., Zhu, H., & Kumar, R. V. (2012). Efficient visible-light-driven photocatalytic hydrogen production using CdS@TaON core-shell composites coupled with graphene oxide nanosheets. *Journal of Materials Chemistry, 22*(15), 7291–7299. https://doi.org/10.1039/C2JM15791H

Hou, J., Yang, C., Cheng, H., Wang, Z., Jiao, S., & Zhu, H. (2013). Ternary 3D architectures of CdS QDs/graphene/ZnIn2S4 heterostructures for efficient photocatalytic H2 production. *Physical Chemistry Chemical Physics, 15*(37), 15660–15668. https://doi.org/10.1039/C3CP51857D

Hu, Y., Gao, X., Yu, L., Wang, Y., Ning, J., Xu, S., & Lou, X. W. (2013). Carbon-coated CdS Petalous nanostructures with enhanced photostability and photocatalytic activity. *Angewandte Chemie, 125*(21), 5746–5749. https://doi.org/10.1002/ange.201301709

Huang, J., Mulfort, K. L., Du, P., & Chen, L. X. (2012). Photodriven charge separation dynamics in CdSe/ZnS Core/Shell quantum dot/cobaloxime hybrid for efficient hydrogen production. *Journal of the American Chemical Society, 134*(40), 16472–16475. https://doi.org/10.1021/ja3062584

Jia, L., Wang, D.-H., Huang, Y.-X., Xu, A.-W., & Yu, H.-Q. (2011). Highly durable N-doped graphene/CdS nanocomposites with enhanced photocatalytic hydrogen evolution from water under visible light irradiation. *The Journal of Physical Chemistry C, 115*(23), 11466–11473. https://doi.org/10.1021/jp2023617

Jia, T., Kolpin, A., Ma, C., Chan, R. C.-T., Kwok, W.-M., & Tsang, S. C. E. (2014). A graphene dispersed CdS-MoS2 nanocrystal ensemble for cooperative photocatalytic hydrogen production from water. *Chemical Communications, 50*(10), 1185–1188. https://doi.org/10.1039/C3CC47301E

Khan, Z., Chetia, T. R., Vardhaman, A. K., Barpuzary, D., Sastri, C. V., & Qureshi, M. (2012). Visible light assisted photocatalytic hydrogen generation and organic dye degradation by CdS-metal oxide hybrids in presence of graphene oxide. *RSC Advances, 2*(32), 12122–12128. https://doi.org/10.1039/C2RA21596A

Kudo, A., & Sekizawa, M. (2000). Photocatalytic H2 evolution under visible light irradiation on Ni-doped ZnS photocatalyst. *Chemical Communications, 15*, 1371–1372. https://doi.org/10.1039/B003297M

Li, Q., Guo, B., Yu, J., Ran, J., Zhang, B., Yan, H., & Gong, J. R. (2011). Highly efficient visible-light-driven photocatalytic hydrogen production of CdS-cluster-decorated graphene nanosheets. *Journal of the American Chemical Society, 133*(28), 10878–10884. https://doi.org/10.1021/ja2025454

Li, Q., Li, X., Wageh, S., Al-Ghamdi, A. A., & Yu, J. (2015). CdS/graphene nanocomposite photocatalysts. *Advanced Energy Materials, 5*(14), 1500010-n/a. https://doi.org/10.1002/aenm.201500010

Li, W., Wang, X., Ma, Q., Wang, F., Chu, X.-S., Wang, X.-C., & Wang, C.-Y. (2021). CdS@h-BN heterointerface construction on reduced graphene oxide nanosheets for hydrogen production. *Applied Catalysis B: Environmental, 284*, 119688. https://doi.org/10.1016/j.apcatb.2020.119688

Liu, Y., Yu, Y.-X., & Zhang, W.-D. (2013). MoS2/CdS heterojunction with high photoelectrochemical activity for H2 evolution under visible light: The role of MoS2. *The Journal of Physical Chemistry C, 117*(25), 12949–12957. https://doi.org/10.1021/jp4009652

Low, J. X., Yu, J. G., Jaroniec, M., Wageh, S., & Al-Ghamdi, A. A. (2017). Heterojunction Photocatalysts. *Advanced Materials, 29*, 1601694. https://doi.org/10.1002/adma.201601694

Lv, X.-J., Fu, W.-F., Chang, H.-X., Zhang, H., Cheng, J.-S., Zhang, G.-J., . . . Li, J.-H. (2012). Hydrogen evolution from water using semiconductor nanoparticle/graphene composite photocatalysts without noble metals. Journal of Materials Chemistry, 22(4), 1539–1546. doi: https://doi.org/10.1039/C1JM14502A.

Maeda, K., & Domen, K. (2007). New non-oxide photocatalysts designed for overall water splitting under visible light. *The Journal of Physical Chemistry C, 111*(22), 7851–7861. https://doi.org/10.1021/jp070911w

Mazloomi, K., & Gomes, C. (2012). Hydrogen as an energy carrier: Prospects and challenges. *Renewable and Sustainable Energy Reviews, 16*(5), 3024–3033. https://doi.org/10.1016/j.rser.2012.02.028

Meissner, D., Memming, R., & Kastening, B. (1988). Photoelectrochemistry of cadmium sulfide. 1. Reanalysis of photocorrosion and flat-band potential. *The Journal of Physical Chemistry, 92*(12), 3476–3483. https://doi.org/10.1021/j100323a032

Murillo Leo, I., Soto, E., Vaquero, F., Mota, N., Navarro, R. M., & Fierro, J. L. G. (2017). Influence of the reduction of graphene oxide (rGO) on the structure and photoactivity of CdS-rGO hybrid systems. *International Journal of Hydrogen Energy, 42*(19), 13691–13703. https://doi.org/10.1016/j.ijhydene.2016.11.154

Pal, D. B., Rathoure, A. K., & Singh, A. (2021). Investigation of surface interaction in rGO-CdS photocatalyst for hydrogen production: An insight from XPS studies. *International Journal of Hydrogen Energy, 46*, 26757–26769. https://doi.org/10.1016/j.ijhydene.2021.05.173

Peng, T., Li, K., Zeng, P., Zhang, Q., & Zhang, X. (2012). Enhanced photocatalytic hydrogen production over graphene oxide–cadmium sulfide nanocomposite under visible light irradiation. *The Journal of Physical Chemistry C, 116*(43), 22720–22726. https://doi.org/10.1021/jp306947d

Priya, R., & Kanmani, S. (2009). Batch slurry photocatalytic reactors for the generation of hydrogen from sulfide and sulfite waste streams under solar irradiation. *Solar Energy, 83*(10), 1802–1805. https://doi.org/10.1016/j.solener.2009.06.012

Quiroz-Cardoso, O., Oros-Ruiz, S., Solís-Gómez, A., López, R., & Gómez, R. (2019). Enhanced photocatalytic hydrogen production by CdS nanofibers modified with graphene oxide and nickel nanoparticles under visible light. *Fuel, 237*, 227–235. https://doi.org/10.1016/j.fuel.2018.10.013

Serpone, N., Lawless, D., Khairutdinov, R., & Pelizzetti, E. (1995). Subnanosecond relaxation dynamics in TiO2 colloidal sols (particle sizes Rp = 1.0-13.4 nm). Relevance to heterogeneous Photocatalysis. *The Journal of Physical Chemistry, 99*(45), 16655–16661. https://doi.org/10.1021/j100045a027

Sherif, S. A., Barbir, F., & Veziroglu, T. N. (2005). Towards a hydrogen economy. *The Electricity Journal, 18*(6), 62–76. https://doi.org/10.1016/j.tej.2005.06.003

Song, S. M., Park, J. K., Sul, O. J., & Cho, B. J. (2012). Determination of work function of graphene under a metal electrode and its role in contact resistance. *Nano Letters, 12*(8), 3887–3892. https://doi.org/10.1021/nl300266p

Tongying, P., Plashnitsa, V. V., Petchsang, N., Vietmeyer, F., Ferraudi, G. J., Krylova, G., & Kuno, M. (2012). Photocatalytic hydrogen generation efficiencies in one-dimensional CdSe Heterostructures. *The Journal of Physical Chemistry Letters, 3*(21), 3234–3240. https://doi.org/10.1021/jz301628b

Tsuji, I., & Kudo, A. (2003). H2 evolution from aqueous sulfite solutions under visible-light irradiation over Pb and halogen-codoped ZnS photocatalysts. *Journal of Photochemistry and Photobiology A: Chemistry, 156*(1), 249–252. https://doi.org/10.1016/S1010-6030(02)00433-1

Wang, X., Yuan, B., Xie, Z., Wang, D., & Zhang, R. (2015). ZnS–CdS/graphene oxide heterostructures prepared by a light irradiation-assisted method for effective photocatalytic hydrogen generation. *Journal of Colloid and Interface Science, 446*, 150–154. https://doi.org/10.1016/j.jcis.2015.01.051

Xia, Y., Cheng, B., Fan, J., Yu, J., & Liu, G. (2019). Unraveling Photoexcited charge transfer pathway and process of CdS/graphene nanoribbon composites toward visible-light photocatalytic hydrogen evolution. *Small, 15*(34), 1902459. https://doi.org/10.1002/smll.201902459

Xiang, Q., & Yu, J. (2013). Graphene-based photocatalysts for hydrogen generation. *The Journal of Physical Chemistry Letters, 4*(5), 753–759. https://doi.org/10.1021/jz302048d

Xie, G., Zhang, K., Guo, B., Liu, Q., Fang, L., & Gong, J. R. (2013). Graphene-based materials for hydrogen generation from light-driven water splitting. *Advanced Materials, 25*(28), 3820–3839. https://doi.org/10.1002/adma.201301207

Xu, J., Wang, L., & Cao, X. (2016). Polymer supported graphene–CdS composite catalyst with enhanced photocatalytic hydrogen production from water splitting under visible light. *Chemical Engineering Journal, 283*, 816–825. https://doi.org/10.1016/j.cej.2015.08.018

Xu, Y., & Schoonen, M. A. (2000). The absolute energy positions of conduction and valence bands of selected semiconducting minerals. *American Mineralogist, 85*(3–4), 543–556.

Xu, Y., Bai, H., Lu, G., Li, C., & Shi, G. (2008). Flexible graphene films via the filtration of water-soluble noncovalent functionalized graphene sheets. *Journal of the American Chemical Society, 130*(18), 5856–5857. https://doi.org/10.1021/ja800745y

Yan, H., Yang, J., Ma, G., Wu, G., Zong, X., Lei, Z., . . . Li, C. (2009). Visible-light-driven hydrogen production with extremely high quantum efficiency on Pt–PdS/CdS photocatalyst. Journal of Catalysis, 266(2), 165–168. https://doi.org/10.1016/j.jcat.2009.06.024.

Ye, A., Fan, W., Zhang, Q., Deng, W., & Wang, Y. (2012). CdS-graphene and CdS-CNT nanocomposites as visible-light photocatalysts for hydrogen evolution and organic dye degradation. *Catalysis Science & Technology, 2*(5), 969–978. https://doi.org/10.1039/C2CY20027A

Yu, X., Du, R., Li, B., Zhang, Y., Liu, H., Qu, J., & An, X. (2016). Biomolecule-assisted self-assembly of CdS/MoS2/graphene hollow spheres as high-efficiency photocatalysts for hydrogen evolution without noble metals. *Applied Catalysis B: Environmental, 182*, 504–512. https://doi.org/10.1016/j.apcatb.2015.09.003

Zeng, K., & Zhang, D. (2010). Recent progress in alkaline water electrolysis for hydrogen production and applications. *Progress in Energy and Combustion Science, 36*(3), 307–326. https://doi.org/10.1016/j.pecs.2009.11.002

Zeng, P., Zhang, Q., Peng, T., & Zhang, X. (2011). One-pot synthesis of reduced graphene oxide-cadmium sulfide nanocomposite and its photocatalytic hydrogen production. *Physical Chemistry Chemical Physics, 13*(48), 21496–21502. https://doi.org/10.1039/C1CP22059D

Zhang, K., & Guo, L. (2013). Metal sulphide semiconductors for photocatalytic hydrogen production. *Catalysis Science & Technology, 3*(7), 1672–1690. https://doi.org/10.1039/C3CY00018D

Design and Storage of Solar Thermal Energy Production

B. Kumar, M. K. Das, and J. N. Roy

Abstract Available solar energy is in diluted form; therefore, we need a reflector to collect solar thermal energy. Parabolic dish collector is a good source for medium- and high-temperature ranges. It is used to increase the concentrated heat flux at the receiver surface. Due to the high concentration ratio, the temperature at the localized surface of the receiver is very high. Therefore, to reduce the localized point heat flux on the flat receiver surface, we disperse the concentrated flux over the entire surface using the optimization of the receiver placement. In the existing design criteria, the receiver is placed at a focal point called the focal plane, but with the help of numerical simulation of the parabolic dish collector, we can disperse the concentrated flux up to periphery. A new approach has been used to design a parabolic dish collector to increase the life of the system without sacrificing efficiency. The dish collector system has been designed for 1 kW output for 6 h a day. The projected area of the dish collector is 7.08 m^2. Rim angle is varied from 15° to 90°. The concentration ratio is varied for 60, 80, 100, and 120. It is established that the rim angle of 45° is 8% more efficient than the rim angle 15°. The surface energy per unit time (4.45 kJ/s) is maximum at 45° rim angle, and surface energy per unit time (4.08 kJ/s) is minimum at rim angle 15°. When the receiver is placed at the optimal plane, the dispersion of energy density is up to the periphery, and the maximum loss is less than 2% in the case of 45° rim angle. When the rim angle is beyond 75°, the energy density is not distributed up to the periphery, and energy density is high at the center.

B. Kumar (✉)
School of Energy Science and Engineering, Indian Institute of Technology Kharagpur, Kharagpur, West Bengal, India

Department of Mechanical Engineering, Vignan's Institute of Information Technology, Duvvada, Visakhapatnam, Andhra Pradesh, India

M. K. Das
Department of Mechanical Engineering, Indian Institute of Technology Kharagpur, Kharagpur, West Bengal, India

J. N. Roy
School of Energy Science and Engineering, Indian Institute of Technology Kharagpur, Kharagpur, West Bengal, India

D. B. Pal, J. M. Jha (eds.), *Sustainable and Clean Energy Production Technologies*, Clean Energy Production Technologies,
https://doi.org/10.1007/978-981-16-9135-5_10

Solar thermal energy is stored by phase change material (PCM) of medium- and high- temperature range. The bulk temperature of NaCl is higher than that of LiBr and NaOH, but the total energy stored by NaCl is less than the LiBr and NaOH. Energy stored in NaOH is more than 90% from NaCl and 13.5% from LiBr. PCM should be selected so that it should melt completely but not reach up to the vaporization phase. Therefore, for latent heat storage, NaOH is a better option than LiBr and NaCl.

Keywords Parabolic dish collector · Heat flux · Beam radiation · Rim angle · Phase change material · Numerical simulation

1 Introduction

Energy is an essential need for every system. All forms of energy existing in nature are solar in origin. The most significant advantages of solar energy are that it is clean and environmentally friendly. One major disadvantage is that it is a diluted and intermittent form of energy; even in a hot environment, available radiation flux is about 1 kW/m^2. Total radiation over a day is about 4–7 kWh/m^2. The average annual solar radiation available in India is a minimum of 3.4 kWh/m^2 to a maximum of 5.9 kWh/m^2 per day throughout the year. To convert solar energy, we need some special kind of device like heat exchange called a collector. Since solar energy is intermittent energy, we need to store the solar thermal energy using phase change material (PCM) to mitigate the demand and supply.

1.1 Solar Thermal Energy

People from the prehistoric era are using solar energy for different applications. In 212 BC, the Greek physician Archimedes used concave metallic mirrors to focus at a single point to burn the ship (Anderson, 1977). Athanasius Kircher carried out some experiments to validate Archimedes story in 1601–1680 and put fire on a woodpile (Meinel & Meinel, 1977). Kreider and Kreith (1975) reported that August Monchot tried to convert solar energy into a low-pressure steam engine during 1864–1878. In 1928, Pasteur used a solar concentrator as solar distillation, reported by Malik et al. (1982).

There are different types of collector available such as flat-plate collector (Ravi et al., 2021), evacuated tube collector (Morrison, 2001), linear Fresnel reflector (Nelson et al., 1975; Kritchman et al., 1979; Zhu et al., 2017; Cheng et al., 2018; Ghodbane et al., 2019; Oscar et al., 2020), parabolic trough collectors (Bellos & Tzivanidis, 2019; Li et al., 2020; Reddy & Ananthsornaraj, 2020) and parabolic dish collector (Anderson, 1977; Morrison, 2001; Reddy & Veershetty, 2013; Tian & Zhao, 2013; Sup et al., 2015; Senthil & Cheralathan, 2017; Senthil & Cheralathan, 2019; Bopche et al., 2020; Sinha & Gulhane, 2020), etc. In a solar dish/Stirling power plant, Li et al. (2011) focused on the concentrator-receiver system's radiation flux and temperature distributions. Pavlovic et al. (2017) did experimental work with a spiral absorber of a solar thermal dish collector. A preliminary assessment of the solar flux distribution on the flat receiver of a heliostat device was performed by Gadhe et al. (2018). Tsekouras et al. (2018) has tested a linear Fresnel collector with a trapezoidal cavity receiver optically and thermally. For a wide range of available offset dish solar concentrators, Ruelas et al. (2018) examined thermal and concentration efficiency. The impacts of circumsolar radiation on the optimal performance of a Stirling heat engine coupled with a parabolic dish solar collector were investigated by Malali et al. (2019). An analysis of the solar flux distribution in concentrated solar energy was performed by He et al. (2019). Bahrami et al. (2019) focused on the experimental and theoretical analysis of a solar dish collector system. The solar concentrator device for energy supply and management is also being researched by various authors (Manchanda & Kumar, 2018; Alayi et al., 2019).

1.1.1 Challenges of Solar Energy

Solar thermal energy is available in diluted form; therefore, it requires a large collector area to convert solar energy to practical applications. The concentration ratio of the dish collector system is very high, and flux distribution on the receiver surface is pinpointed. Pinpointed flux concentration on flat-plate receiver will damage the receiver. A detailed investigation is needed to study the method of dispersing the concentrated flux into the entire receiver surface without hampering the system's performance.

It is an intermittent form of energy, so, we need to store the energy to mitigate the demand and supply. Many industrial processes require a constant supply of energy; thus, we need to keep the energy in latent heat energy storage. However, the main disadvantage of latent heat storage is its low thermal conductivity, which mainly falls into the range of 0.2–0.7 W/(m-K). Therefore, nanoparticle with high thermal conductivity material is mixed with the base material to enhance the material's thermal conductivity.

1.1.2 Solar Thermal Applications

Water Heating: Water heating is the most common and attractive solar thermal application. Flat-plate collectors and evacuated tube collectors are used for domestic water heating applications. Solar water heating systems with a natural circulation system or thermosyphon and forced circulation system (Sukhatme & Nayak, 2011). In the natural circulation system, cold water flows in the collector through the gravity effect. It receives heat energy from the collector and gets evaporated and collected in a storage tank for hot water applications. In forced circulation, cold water is pumped into the collector at a constant rate, and it gets heat from the collector and is stored in the storage tank for further applications. Ravi et al. (2021) reviewed the flat-plate collector process heating system for industry and power generation.

Distillation: In the entire world, distillation and desalination are performed in the long run. Brine water transforms into vapor form in solar still system and gets condensate which is collected via the vent, and pure water is separated for further use.

Drying: One of the significant preservation methods for grain, fruits, and vegetables is drying. Drying is a dynamic process involving unstable heat and mass transfer or physical or chemical transformations that can affect the consistency of the product. Many products such as fruit, vegetables, grains, and herbs are dried using a solar drier (Prasad & Mullick, 1983). Norton (1992) gave detailed information on the solar drier. Thermal drying methods account for 10–20% of overall industrial energy consumption (Belessiotis & Delyannis, 2011).

Cooking: One of the most important solar energy applications is the solar cooker. Approximately 36% of energy is consumed in cooking (Bansal et al., 2014). Therefore, in the domestic market, solar cookers have great potential. The solar cooker is used for cooking food, pasteurization, and sterilization (Cuce & Cuce, 2013; Yettou et al., 2014).

Refrigeration and cooling: For cooling and refrigeration applications worldwide, solar energy can provide a cheap and renewable energy source. Mediterranean countries can save about 50% of their energy costs by installing a solar-powered cooling system (Balaras et al., 2007). Due to refrigerant and toxic Freon gas, solar refrigeration has become more desirable for cooling purposes today (Thirugnanasambandam et al., 2010). Recent statistics show that most absorption cooling systems are manufactured with a LiBr/water pair using single-effect absorption technology, where these systems are implemented by a solar flat-plate collector or an evacuated-tubular collector with hot water (Zhai et al., 2011). The recent development in the absorption cooling cycle using concentrating solar collector is described by Alsagri et al. (2020).

Power generation: One of the most significant applications of solar energy sources is the production of electric power. Solar thermal power generation is based on the collection of energy from the sun through mirrors or reflectors guided to a receiver. The energy obtained is transferred to a heat transfer fluid in the receiver that supplies heat via the traditional thermodynamic cycle to produce electricity. Flat-plate collectors or solar ponds are used for low-temperature power generation systems. Parabolic trough technology is used in medium-temperature range power generation systems. The high-temperature power generation system needs a central receiver tower or paraboloidal dish collector. Qin et al. (2020) did a thorough analysis of solar power generation and the progress of solar-aided power generation.

1.2 Design of Parabolic Dish Collector System

A schematic of parabolic solar dish reflector and the geometrical relationship between reflector and receiver is shown in Fig. 1. Solar beam radiation enters from the top and is reflected by the parabolic dish reflector. The reflected ray converges towards a small area on the focal plane where the receiver is mounted. Rim angle (φ) (Sup et al., 2015) is the angle subtended by the perpendicular line drawn through the center of the receiver to the edge of the reflector.

The parabolic dish collector system has been designed for 1 kW output for 6 h a day to investigate the effect of rim angle and focal plane distance. This can be scaled up for the larger system and various industrial applications. The assumption for the design is based on the availability of local beam radiation. The average beam radiation intensity in India is about 700 W/m^2, reported by Funk (2000).

Collector geometric concentration (C_r) is defined in Eq. 1.

$$C_r = \frac{A_p}{A_b} \quad (1)$$

where C_r is the geometric concentration ratio, A_p is the projected aperture area of the reflector (m^2), and A_b is the receiver area or absorber area (m^2). We have taken C_r as 60, 80, 100, and 120 for further design. Various losses such as optical loss, pumping loss, thermal loss, and reflector loss have been considered to estimate system output. The optical efficiency (η_o) (ratio of the energy absorbed by the absorber to the energy incident on the reflector) of most solar concentrators lies between 0.6 and 0.7 (Mohammed, 2012). In this solar dish collector, a combination of a working fluid and PCM substance is used to store the absorbed energy from the absorber. The thermal efficiency (η_{th}) (ratio of the useful energy delivered to the energy incident on the reflector) of most solar concentrators lies between 0.4 and 0.6 (Mohammed, 2012). The efficiency of the pump (η_P) (ratio of water horsepower output to the shaft

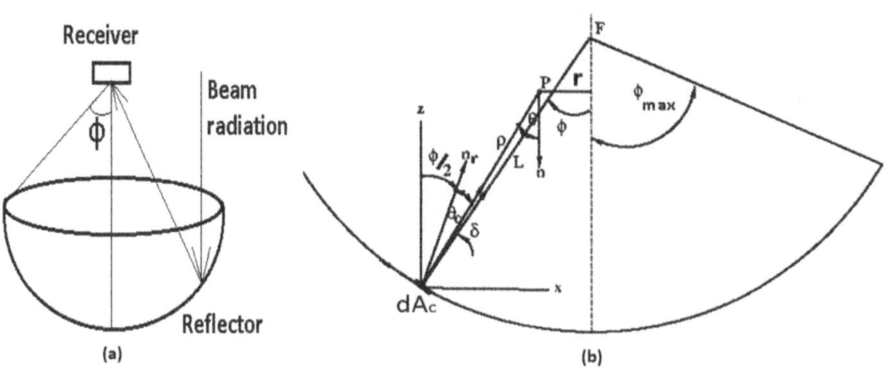

Fig. 1 (**a**) Schematic of parabolic dish reflector. (**b**) Geometrical relationships between reflecting source and receiver

horsepower input) varies from 0.2 to 0.8 (Ulanicki et al., 2008). The reflector efficiency (η_{ref}) (the energy incident reflected on the reflector) of a good solar concentrator lies between 0.8 and 0.95. For designing parabolic dish collector system, we have taken various parameters: $\eta_o = 0.7$, $\eta_{th} = 0.4$, $\eta_p = 0.8$, $\eta_{ref} = 0.9$, and $G_{bn} = 700$ W/m^2 for 6 h operating time in a day. The estimated area of the reflector, absorber, and focal point is given in Eqs. 2–5 (Mohammed, 2012).

$$A_p = Q/(\eta_o \times \eta_{th} \times \eta_P \times \eta_{ref} \times G_{bn}) \tag{2}$$

$$A_b = \frac{A_p}{C_r} \tag{3}$$

$$f = \left(\sqrt{A_p \times \frac{4}{\pi}}\right) / \left(4\left(\frac{1}{\sin\phi} - \frac{1}{\tan\phi}\right)\right) \tag{4}$$

$$A_{ap} = 4\pi f^2 \frac{\sin^2\phi}{1 + \cos\phi} \tag{5}$$

where f is the focal length of the parabolic reflector (m) and A_{ap} is the total surface area of the reflector (m^2). From Eqs. 2–4, the aperture area, receiver area, and focal length are calculated. The projected area of the reflector is 7.08 m^2. The receiver diameter for different concentration ratios is calculated using Eq. 3. The other concentration ratio gives different receiver radius such as 19.4, 16.7, 15, and 13.7 cm of concentration ratios of 60, 80, 100, and 120, respectively.

1.2.1 Governing Equations

The geometrical relationship between reflector and receiver is shown in Fig. 1(b). The differential area elements on the surface of the reflector are dA_c, and P is the point on the receiver at a distance of r from the center of the receiver. The surface normal at the reflector and focal planes are n_s and n, respectively, and F is the focal point. The following angles are defined:

$$\delta = \angle \mathrm{F\,dA_c\,P} \tag{6}$$

$$\theta_c = \angle \mathrm{P\,dA_c}\,\mathrm{n}_s \tag{7}$$

$$\theta = \angle \mathrm{dA_c\,P\,n} \tag{8}$$

The concentration ratio (C_r) is also a function of the radial position in the receiving plane. Integrating over all azimuthal angles, the concentration ratio at the receiver is given in Eq. 9.

$$C_r(\mathrm{r}) = \frac{1}{G_{bn}} \int \frac{I \cos(\theta) \cos(\theta_c)}{|\rho|^2} \mathrm{d}A_c \tag{9}$$

$$I = \left\{ \begin{matrix} G_{bn}/\left(\pi \sin^2(s)\right) & \delta \leq s \\ 0 & \delta > s \end{matrix} \right\} \tag{10}$$

$$q = C_r(r) G_{bn} \tag{11}$$

where I is the radiant intensity (W/m^2 steradian), G_{bn} is the incident normal beam radiation (W/m^2), q is the heat flux (W/m^2), s is the maximum solar half-angle (4.65 mrad) (Howell et al., 2015), and dA_c is a differential area element on the surface of the reflector (m 2).

The incident flux on the surface of the receiver is computed using the Monte Carlo ray-tracing (MCRT) method. The illuminated surface boundary condition is used to release rays directly from the surface of the reflector to the receiver. For numerical simulation, some assumptions are made:

- All incoming rays are parallel to each other.
- Ray propagation is computed uniformly throughout the material.
- The collector has zero pointing errors.

1.2.2 Validation

Computational models are validated with the published result of Jeter (1986) and Shuai et al. (2008). The distribution of concentration ratio in a radial direction is done by Jeter (1986) and flux distribution with radial direction by Shuai et al. (2008). The current computational work, which has a focal length of 3 m, a receiver radius of 0.03 m, and an absorption coefficient of 0, has been validated with Jeter (1986).

2 Results and Discussion

2.1 Flux Distribution on the Focal Plane and Optimal Plane

The available average beam radiation (G_{bn}) is 700 W/m^2. Heat flux density at the receiver is simulated for different rim angles and different focal plane distances. The projected area of the reflector is kept constant, i.e., 7.08 m^2 (Eq. 2). The radius of the receiver (r_r) is varied for different concentration ratios (60, 80, 100, and 120) and the different rim angle (φ) (15°, 30°, 45°, 60°, and 90°). The absorption coefficient (α) is kept at 0.1. The focal length of different rim angles is calculated using Eq. 4, which gives a 30° rim angle of 2.8 m. For 30° rim angle, Fig. 2 shows (a) simulation domain of solar dish reflector, (b) surface flux distribution on the focal plane, (c) radial flux distribution on the focal plane, (d) surface flux distribution on an

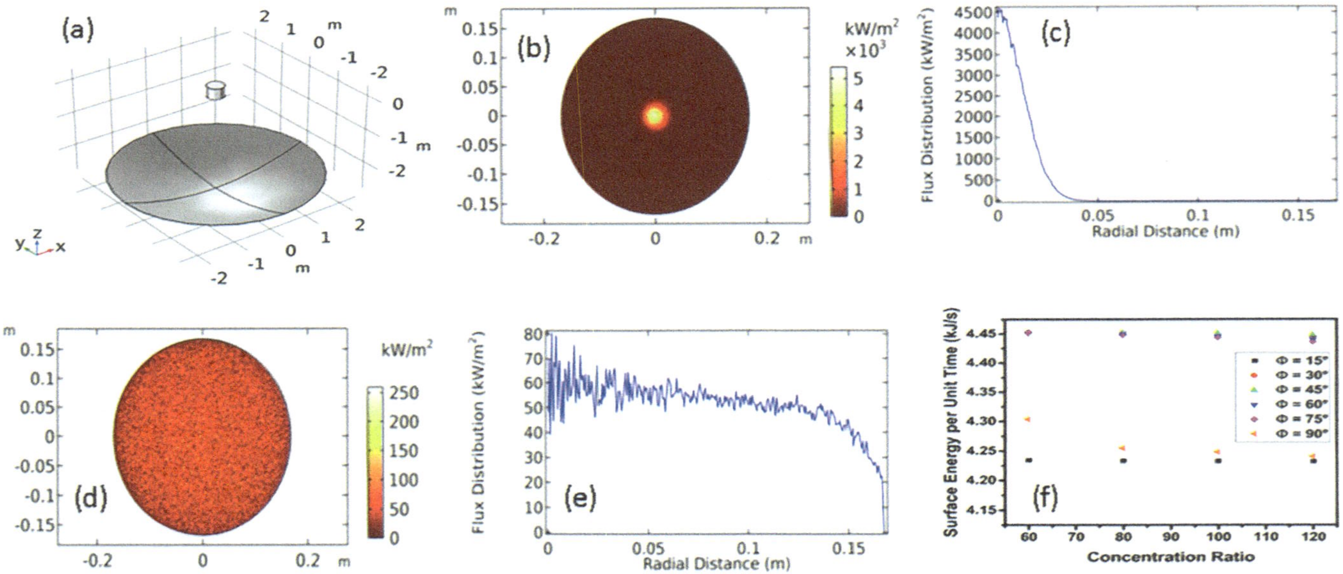

Fig. 2 For 30° rim angle, (**a**) simulation domain of solar dish reflector, (**b**) surface flux distribution on the focal plane, (**c**) radial flux distribution on the focal plane, (**d**) surface flux distribution on an optimal plane, (**e**) radial flux distribution on an optimal plane, (**f**) comparison of surface energy per unit time when the receiver is placed at the focal plane

optimal plane, (e) radial flux distribution on an optimal plane, and (f) comparison of surface energy per unit time when the receiver is placed at the focal plane.

2.2 *The Relation between the Optimal Plane and the Focal Plane*

The receiver is placed at the focal point where all reflected rays from the dish reflector are collected at the receiver's center, which is considered the focal plane (f_p). If the receiver is placed so that all reflected rays collected up to the periphery of the receiver, this is regarded as an optimal plane (o_p). To produce a relationship between the optical plane and the focal plane, we derive in Eq. 12.

$$o_p = f - \frac{d}{2\sqrt{C_r}} \times \frac{\cos(\phi)}{\sin(\phi)} \tag{12}$$

where o_p (m) is an optimal plane, f (m) is the focal point, d (m) is the diameter and φ (°) is the rim angle, and C_r is the concentration ratio. The optimal plane for any focal length (f) and projected diameter (d) of the dish reflector can be directly calculated from Eq. 12 for different rim angles ($15° \leq \varphi \leq 90°$) and the concentration ratio ($60 \leq C\,r \leq 120$).

2.3 *Solar Thermal Energy Storage Using Phase Change Material*

Solar thermal energy can be stored using sensible heating storage, listed in Table 1 and latent heat storage material listed in Table 2. In the process in which working temperature is very strict, we require a latent heat thermal storage system. Many researchers work on solar thermal energy storage using phase change material. Solomon (1979) worked on 1D enthalpy-based numerical study on phase change material (PCM) using the Crank-Nicolson scheme. Lecomte and Mayer (1985) did latent heat analysis in a thermal system using the finite difference method. Geometric design with phase change material for solar-aided energy storage using 2D fixed grid conjugate enthalpy model was numerically studied by Esen et al. (1998). Ismail et al. (2003) did a numerical study on 2D conjugate heat transfer enthalpy-based model for melt phase of natural convection of circular cylinder. A review of free cooling of a building using phase change material was done by Adeel and Zia (2013). Dynamic cycling analysis of phase change material for a battery pack system using experimental analysis was done by Yan et al. (2016). Experimental analysis of phase change material with nano enhancement for thermal management system was done by Sharma et al. (2017). Elias and Stathopoulos (2019) reviewed phase change

Table 1 Sensible heat storage materials (Tian & Zhao, 2013)

Materials	Working temperature (°C)	Density (kg/m^3)	Specific heat (kJ/(kg K))	Thermal conductivity (W/(m-K))
Reinforced	200–400	2200	0.85	1.5
Sand-rock	200–300	1700	1.30	1
Cast iron	200–400	7200	0.56	37
Cast steel	200–700	7800	0.6	40
Magnesia bricks	200–1200	3000	1.15	5
Silica bricks	200–700	1820	1	1.5
Mineral oil	200–300	770	2.6	0.12
Synthetic oil	250–350	900	2.3	0.11
Silicone oil	300–400	900	2.1	0.1
Liquid sodium	270–530	850	1.3	71

Table 2 Latent heat storage materials

Thermophysical property of material	Density ρ_s: ρ_l (kg/m 3)	Specific heat cp_s: cp_l (kJ/(kg K))	Thermal conductivity k_s: k_l (W/(m K))	Heat of fusion λ (kJ/kg)	Melting temperature T_m (°C)
Paraffin (RT100) (Tian & Zhao, 2013)	880	–	0.2	124	100
$LiNO_3$ (Cardenas & Leon, 2013)	2380–1910	1.2–1.84	0.518–0.505	372	253
NaOH (Wei et al., 2018)	2130–1780	2.01–2.09	0.92–0.7	160	318
KOH (Wei et al., 2018)	2044–1470	1.47	0.5	149.7	380
LiBr (Agyenim et al., 2010)	3460–2528	1.45–1.68	0.42–0.4	203	550
NaCl (Li et al., 2019)	2160–1556	0.987–1.19	0.52–0.48	482	802

material for thermal energy storage with material property and its applications. Experimental analysis of heat exchanger with sodium acetate trihydrate as a phase change material for energy storage is carried by Lin et al. (2020).

2.4 Mathematical Model

The heat transfers in the container and PCM are assumed to be transient and three dimensional. Body forces and convection are neglected. The container and PCM interface are uniform. Purely conductive and radiative heat transfer is allowed for all domains. For modelling, a three-dimensional (3D) heat transfer enthalpy-based model has been used. Enthalpy-based model is the best model for the study of the phase-field model. Many authors have used enthalpy-based model in 1D or 2D. In this work, the 3D model is used; all three dimensions are considered for simulation. It gives more accuracy, but its cost of simulation is very high.

The thermophysical property of the container and PCM is listed in Table 3.

2.5 Governing Equation

The governing equations are the following:

$$\rho \frac{\partial h}{\partial t} + \nabla.\mathbf{q} = \ldots Q \tag{13}$$

where

$$\mathbf{q} = -k\nabla T$$

Table 3 Thermophysical property of the container and PCM

Thermophysical property of material	Density (ρ_s: ρ_l) (kg/m^3)	Specific heat (cp_s: cp_l) (kJ/(kg K))	Thermal conductivity (k_s: k_l) ($W/(m^2$-K))	Heat of fusion λ (kJ/kg)	Melting temperature T_m (K)
Stainless steel (Polished) (Mills, 2011)	7609	0.59	23	–	1740
NaOH (Wei et al., 2018)	2130–1780	2.01–2.09	0.92–0.7	160	591
LiBr (Agyenim et al., 2010)	3460–2528	1.45–1.68	0.42–0.4	203	823
NaCl (Li et al., 2019)	2160–1556	0.987–1.19	0.52–0.48	482	1075
$LiNO_3$ (Cardenas & Leon, 2013)	2380–1910	1.2–1.84	0.518–0.505	372	526
KOH (Wei et al., 2018)	2044–1470	1.47	0.5	149.7	653

$$\nabla = \frac{\partial}{\partial r}\widehat{r} + \frac{1}{r}\frac{\partial}{\partial \theta}\widehat{\theta} + \frac{\partial}{\partial z}\widehat{z}$$

$$\nabla^2 = \frac{1}{r}\frac{\partial}{\partial r}\left(r\frac{\partial}{\partial r}\right) + \frac{1}{r^2}\frac{\partial}{\partial \theta}\left(\frac{\partial}{\partial \theta}\right) + \frac{\partial}{\partial z}\left(\frac{\partial}{\partial z}\right)$$

For solid container,

$$h = c_p T \tag{14}$$

For solid-phase PCM,

$$h = c_p T : (T \leq T_m - \varepsilon) \tag{15}$$

For the transition phase PCM,

$$h = c_p T + \frac{\lambda}{2\varepsilon}(T - T_m + \varepsilon) : (T_m - \varepsilon \leq T \leq T_m + \varepsilon) \tag{16}$$

For the liquid phase PCM,

$$h = c_p T + \lambda : (T \geq T_m + \varepsilon) \tag{17}$$

Nondimensional Stefan number (Ste):

$$\text{Ste} = \frac{c_p(T_m - \varepsilon - T_b)}{\lambda} \tag{18}$$

Melt fraction (Φ):

$$\Phi = (1 - \text{Ste}) \tag{19}$$

Effective thermophysical property during the transition phase:

$$\rho_e = \Phi\rho_s + (1 - \Phi)\rho_l : (T_m - \varepsilon \leq T \leq T_m + \varepsilon) \tag{20}$$

$$k_e = \Phi k_s + (1 - \Phi)k_l : (T_m - \varepsilon \leq T \leq T_m + \varepsilon) \tag{21}$$

$$\alpha_m = \frac{1}{2}\frac{(1 - \Phi)\rho_l - \Phi\rho_s}{\Phi\rho_s + (1 - \Phi)\rho_l} \tag{22}$$

$$c_{pe} = \frac{1}{\rho_e}\left(\Phi\rho_s c_{ps} + (1 - \Phi)\rho_l c_{pl}\right) + \lambda\frac{\partial \alpha_m}{\partial T} : (T_m - \varepsilon \leq T \leq T_m + \varepsilon) \tag{23}$$

The bulk temperature and total energy storage by PCM is calculated using volumetric integration at each time interval shown in Eq. 24.

$$< \psi > = \frac{1}{\nabla V} \sum \int_{n}^{k=1} \psi(x, y, z, t)\, dV \quad (24)$$

where ρ (kg/m^3) is the density, c_p (kJ/(kg-K)) is the specific heat capacity, h (kJ/kg) is the specific enthalpy, k ($W/(m^2$-K)) is the thermal conductivity, ... Q (W/m^3) is the volumetric heat generation, T_b (K) is the bulk temperature of PCM, T_m (K) is the melting temperature, α_m is the amount of the PCM which is under melting process, Ste. is Stefan number, ε (K) is the transition interval between solid and liquid phase (Sar & Kaygusuz, 2002), Φ is the solid-phase PCM, λ (kJ/kg) is the heat of fusion, subscripts (s) is for solid-phase PCM and (l) is for liquid phase PCM, ψ is a function, <ψ> is the average of function, dV is the control volume, $k = 1$ to n is the time interval, (x, y, z) is the spatial coordinate, and t is the time coordinate.

Various assumptions have been taken for the simulation setup:

1. The solid container and PCM interface is uniform.
2. The solid-liquid interface of PCM is progressed uniformly.
3. Super-heating and subcooling effects are neglected.
4. Fully implicit backward differentiation formula (BDF) is used for numerical simulation (COMSOL Multiphysics®).

2.6 *Validation*

The present computational model is validated with the experimental work of Zivkovic and Fujii (2001) and the experimental work of Siyabi et al. (2018). Zivkovic and Fujii (2001) has used a closed cuboidal box filled with $CaCl_2.6H_2O$ PCM for their experiment. They recorded the center temperature of PCM with the help of a thermocouple. All faces were insulated except two opposite ends having a constant wall temperature of 60 °C. Siyabi et al. (2018) used a cuboidal box filled with Rubitherm (RT 47) PCM for experimental work. All faces were insulated, and the constant power of 2 W is supplied at the bottom surface and the inside temperature was recorded with the help of the thermocouple. It is noticed that the maximum deviation at the center temperature is 4.7% and 2.1% with respect to numerical solution. The variation in the result is less than 5%, which is within the permissible limits.

3 Results and Discussion

A container made of polished stainless steel whose outer radius is 16.7 cm having a height of 16.2 cm and a thickness of 10 mm is considered [taken from ray optics design for concentration ratio 80 and rim angle 45°]. It is filled with a different mass of phase change material (PCM). Mass of the PCM is selected based on receiver volumetric capacity and density of PCM. Radially variable inward heat flux is taken for the bottom surface, and all surfaces are exposed at a constant surface emissivity of 0.1 for radiation heat transfer. Heat flux on the receiver surface at the center is about 78.7 kW/m^2 and near the periphery approximately 18.02 kW/m^2.

Initial and boundary conditions:when $t = 0$

$$T(\overline{S}, 0) = 295\,K$$

when $t > 0$

$$\mathbf{q} = \begin{cases} q_0 & (At\,the\,bottom\,surface,\,Inward\,direction) \\ \epsilon\sigma\left(T^4 - T_{amb}^4\right) & (At\,the\,periphery\,and\,top\,surface,\,Outward\,direction) \end{cases}$$

where $\overline{S}$ is the position vector and q is the surface heat flux, ϵ is the surface emissivity (0.1), and σ is Stefan–Boltzmann constant (5.670373×10^{-8} $W/(m^2\text{-}K^{-4})$) and $T_{\text{amb}} = 295$ K.

3.1 Different PCM at Constant Volume

Three different PCM, viz., NaCl, LiBr, and NaOH, with the same volume of 0.01 m 3 are used for analysis [stainless steel container with outer radius is 16.7 cm, with a height of 16.2 cm and a thickness of 10 mm, taken from ray optics design for concentration ratio 80 and rim angle 45°].

The transverse temperature contour of NaOH at the sixth hour is shown in Fig. 3 (a). The temperature at the bottom surface gradually increases from 850 K to 1025 K, and the temperature at the top surface increases from 370 to 460 K. The temperature distribution profile of NaOH along the diameter at the middle of PCM is shown in Fig. 3(b). Figure 3(b) shows that as time increases, temperature gradient near the periphery reduces, and the temperature of PCM increases. Figure 3(c) shows the temperature contour on a horizontal plane at the top, middle, and bottom surface of NaOH at the sixth hour. At the bottom surface, the temperature at the center is high with respect to the periphery. Still, in the case of the middle and top surfaces, the temperature at the center is low with respect to the periphery. This is because the heat flux on the bottom surface at the center is very high and gradually reduces towards the periphery, but in the case of middle and top surfaces, temperature is penetrating inward direction due to the high thermal conductivity of the container.

Comparison of different PCM is shown in Fig. 3 (d), (e), and (f). Figure 3(d) shows melt fraction vs time of different PCM. From Fig. 3(d), it is clear that NaOH gets 75% melted, LiBr gets 15% melted, and NaCl gets about 2% melted after the sixth hour. PCM can completely melt if the flux at the bottom surface is higher than the available flux. Figure 3(e) shows bulk temperature vs time. The bulk temperature of PCM is calculated using Eq. 24. Figure 3(e) shows that the bulk temperature of

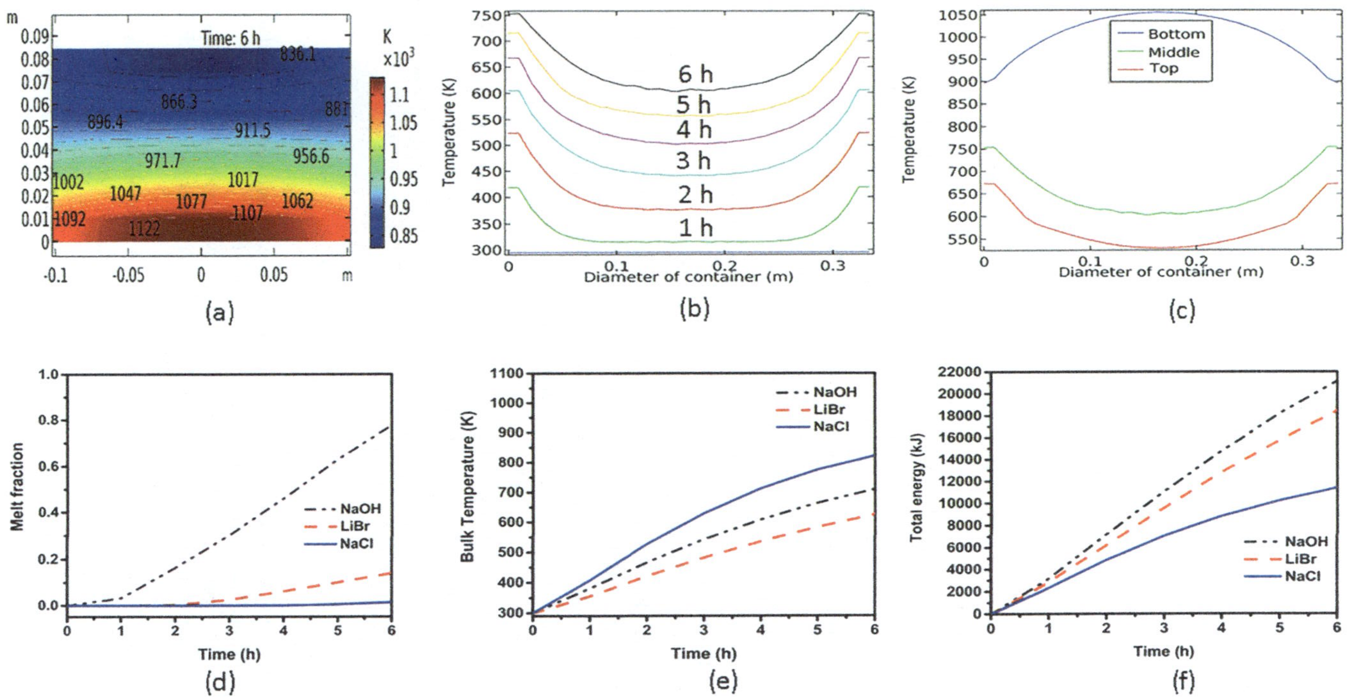

Fig. 3 (**a**) The temperature contour of NaOH PCM along a vertical plane at the sixth hour. (**b**) Temperature distribution profile of NaOH along the diameter at a different time interval. (**c**) Temperature distribution profile of NaOH at the sixth hour along the diameter at the bottom, middle, and top surfaces. (**d**) Comparison of melt fraction vs time. (**e**) Comparison of bulk temperature vs time. (**f**) Comparison of total energy vs time

NaCl is about 820 K, NaOH is about 710 K, and LiBr is about 600 K. Figure 3(f) represents total stored energy to the system at different time intervals. From Fig. 3(f), the total energy stored at the sixth hour by NaOH is about 21 MJ, LiBr is about 18.5 MJ, and NaCl is about 11 MJ. Energy stored in NaOH is more than 13.5% from LiBr and 90% from NaCl. The bulk temperature of NaCl is higher than that of LiBr and NaOH, but the total energy stored by NaOH is less than that of LiBr and NaCl.

4 Conclusions

A new approach has been used for the design of a dish collector to increase the life of the system without sacrificing efficiency. The dish collector system has been designed for 1 kW output for 6 hours a day. The projected area of the dish collector is 7.08 m^2. Rim angle is varied from 15° to 90°. The concentration ratio is varied for 60, 80, 100, and 120. It is established that the rim angle of 45° is 8% more efficient than the rim angle of 15°. The surface energy per unit time (4.45 kJ/s) is maximum at 45° rim angle, and surface energy per unit time (4.08 kJ/s) is minimum at rim angle 15°. When the receiver is placed at the optimal plane, the dispersion of energy density is up to the periphery, and the maximum loss is less than 2% in the case of a 45° rim angle. When the rim angle is beyond 75°, the energy density is not distributed up to the periphery, and energy density is high at the center.

The bulk temperature of NaCl is higher than that of LiBr and NaOH, but the total energy stored by NaOH is less than the LiBr and NaCl. NaOH gets 75% melted, LiBr gets 15% melted, and NaCl gets about 2% melted after the sixth hour. The total energy stored by NaOH is about 21 MJ, LiBr is about 18.5 MJ, and NaCl is about 11 MJ at the sixth hour. Energy stored in NaOH is more than 13.5% from LiBr and 90% from NaCl. Therefore, NaOH is a better option for latent heat storage at the same heat flux than LiBr and NaCl. PCM should be selected so that it should melt completely but not reach up to the vaporization phase.

References

Adeel, W., & Zia, U. (2013). Phase change material (PCM) storage for free cooling of buildings: A review. *Renewable and Sustainable Energy Reviews, 18*, 607–625.

Agyenim, F., Hewitt, N., Eames, P., & Smyth, M. (2010). A review of materials, heat transfer and phase change problem formulation for latent heat thermal energy storage systems (LHTESS). *Renewable and Sustainable Energy Reviews, 14*, 615–628.

Alayi, R., Kasaeian, A., & Atabi, F. (2019). Thermal analysis of parabolic trough concentration photovoltaic/thermal system for using in buildings. *Environmental Progress and Sustainable Energy, 38*, 1–12.

Alsagri, A. S., Alrobaian, A. A., & Almohaimeed, S. A. (2020). Concentrating solar collectors in absorption and adsorption cooling cycles: An overview. *Energy Conversion and Management, 223*, 113420.

Anderson, B. (1977). *Solar energy: Fundamentals in building design*. McGraw-Hill.

Bahrami, M., Avargani, V. M., & Bonyadi, M. (2019). Comprehensive experimental and theoretical study of a novel still coupled to a solar dish concentrator. *Applied Thermal Engineering, 151*, 77–89.

Balaras, C. A., Grossman, G., Henning, H. M., Ferreira, C. A. I., Podesser, E., Wang, L., & Wiemken, E. (2007). Solar air conditioning in Europe: An overview. *Renewable and Sustainable Energy Reviews, 11*, 299–314.

Bansal, M., Saini, R., & Khatod, D. (2014). Optimal sizing of a solar-biogas-based cooking system for a cluster of villages. *International Journal of Sustainable Energy, 33*, 1017–1032.

Belessiotis, V., & Delyannis, E. (2011). Solar drying. *Solar Energy, 85*, 1665–1691.

Bellos, E., & Tzivanidis, C. (2019). Alternative designs of parabolic trough solar collectors. *Progress in Energy and Combustion Science, 71*, 81–117.

Bopche, S. B., Kumar, R., Singh, I. (2020). Development of a novel two-stage parabolic dish collector-receiver system for efficiency improvement. Energy Sources, Part A: Recovery, Utilization, and Environmental Effects, 1–32.

Cardenas, B., & Leon, N. (2013). High temperature latent heat thermal energy storage: Phase change materials, design considerations and performance enhancement techniques. *Renewable and Sustainable Energy Reviews, 27*, 724–737.

Cheng, Z. D., Zhao, X. R., He, Y. L., & Qiu, Y. (2018). A novel optical optimization model for linear Fresnel reflector concentrators. *Renewable Energy, 129*, 486–499.

Cuce, E., & Cuce, P. M. (2013). A comprehensive review on solar cookers. *Applied Energy, 102*, 1399–1421.

Elias, C. N., & Stathopoulos, V. N. (2019). A comprehensive review of recent advances in materials aspects of phase change materials in thermal energy storage. *Energy Procedia, 161*, 385–394.

Esen, M., Durmuay, A., & Durmuay, A. (1998). Geometric design of solar-aided latent heat store depending on various parameters and phase change materials. *Solar Energy, 62*, 19–28.

Funk, P. A. (2000). Evaluating the international standard procedure for testing solar cookers and reporting performance. *Solar Energy, 68*, 1–7.

Gadhe, P. M., Sapali, S. N., & Kulkarni, G. N. (2018). Experimental evaluation of the solar flux distribution on the flat receiver of a model heliostat system. *International Journal of Renewable Energy Research-IJRER, 8*, 878–887.

Ghodbane, M., Boumeddane, B., Said, Z., & Bellos, E. (2019). A numerical simulation of a linear Fresnel solar reflector directed to produce steam for the power plant. *Journal of Cleaner Productionl, 231*, 494–508.

He, Y. L., Wang, K., Qiu, Y., Du, B. C., Liang, Q., & Du, S. (2019). Review of the solar flux distribution in concentrated solar power: Non-uniform features, challenges, and solutions. *Applied Thermal Engineering, 149*, 448–474.

Howell, J., Menguc, M., & Siegel, R. (2015). *Thermal radiation heat transfer* (6th ed.). CRC Press.

Ismail, K. A. R., Silva, D., & Maria, D. G. E. (2003). Numerical solution of the phase change problem around a horizontal cylinder in the presence of natural convection in the melt region. *International Journal of Heat and Mass Transfer, 46*, 1791–1799.

Jeter, S. M. (1986). The distribution of concentrated solar radiation in paraboloidal collectors. *Journal of Solar Energy Engineering, 108*, 219–225.

Kreider, J. F., & Kreith, F. (1975). *Solar heating and cooling: Engineering, practical design, and economics*. N. P..

Kritchman, E., Friesem, A., & Yekutieli, G. (1979). Efficient Fresnel lens for solar concentration. *Solar Energy, 22*, 119–123.

Lecomte, D., & Mayer, D. (1985). Design method for sizing a latent heat store/heat exchanger in a thermal system. *Applied Energy, 21*, 55–78.

Li, Q., Li, C., Du, Z., Jiang, F., & Ding, Y. (2019). A review of performance investigation and enhancement of shell and tube thermal energy storage device containing molten salt based phase change materials for medium and high temperature applications. *Applied Energy, 255*, 1–37.

Li, Q., Zhang, Y., Wen, Z. X., & Qiu, Y. (2020). An evacuated receiver partially insulated by a solar transparent aerogel for parabolic trough collector. *Energy Conversion and Management, 214*, 112911.

Li, Z., Tang, D., Du, J., & Li, T. (2011). Study on the radiation flux and temperature distributions of the concentrator-receiver system in a solar dish/Stirling power facility. *Applied Thermal Engineering, 31*, 1780–1789.

Lin, W., Zhang, W., Ling, Z., Fang, X., & Zhang, Z. (2020). Experimental study of the thermal performance of a novel plate type heat exchanger with phase change material. *Applied Thermal Engineering, 178*, 115630.

Malali, P. D., Chaturvedi, S. K., & Agarwala, R. (2019). Effects of circumsolar radiation on the optimal performance of a Stirling heat engine coupled with a parabolic dish solar collector. *Applied Thermal Engineering, 159*, 113961.

Malik, M., Tiwari, G. N., Kumar, A., & Sodha, M. (1982). *Solar distillation: A practical study of a wide range of stills and their optimum design, construction, and performance*. Pergamon Press.

Manchanda, H., & Kumar, M. (2018). Study of water desalination techniques and a review on active solar distillation methods. *Environmental Progress and Sustainable Energy, 37*, 444–464.

Meinel, A. B., & Meinel, M. P. (1977). *Applied solar energy: An introduction* (Vol. 77). Addison-Wesley Publishing Company.

Mills, K. C. (2011). *Thermophysical properties of selected commercial alloys*. Woodhead Publishing Limited Cambridge England.

Mohammed, I. L. (2012). Design and development of a parabolic dish solar water heater. *International Journal of Engineering Research and Applications, 2*, 822–830.

Morrison, G. L. (2001). *Solar collectors. Solar energy: The state of the art* (pp. 145–221). ISES.

Nelson, D., Evans, D., & Bansal, R. (1975). Linear Fresnel lens concentrators. *Solar Energy, 17*, 285–289.

Norton, B. (1992). *Solar energy thermal technology*. Springer.

Oscar, A. L. N., Arturo, J. A. A., Jaramillo, O., Ramirez-Minguela, J., Castro, J. C., Cesar, E. D. A., & Sergio, C. A. (2020). A numerical analysis of the energy and entropy generation rate in a linear Fresnel reflector using computational fluid dynamics. *Renewable Energy, 146*, 1083–1100.

Pavlovic, S., Bellos, E., Le Roux, W. G., Stefanovic, V., & Tzivanidis, C. (2017). Experimental investigation and parametric analysis of a solar thermal dish collector with spiral absorber. *Applied Thermal Engineering, 121*, 126–135.

Prasad, K., & Mullick, S. (1983). Heat transfer characteristics of a solar air heater used for drying purposes. *Applied Energy, 13*, 83–93.

Qin, J., Hu, E., & Li, X. (2020). Solar aided power generation: A review. *Energy and Built Environment, 1*, 11–26.

Ravi, K. K., Krishna, C. N. V. V., & Sendhil, K. N. (2021). Solar thermal energy technologies and its applications for process heating and power generation a review. *Journal of Cleaner Production, 282*, 125296.

Reddy, K. S., & Ananthsornaraj, S. (2020). Design, development and performance investigation of solar parabolic trough collector for large-scale solar power plants. *Renewable Energy, 146*, 1943–1957.

Reddy, K. S., & Veershetty, G. (2013). Viability analysis of solar parabolic dish stand- alone power plant for indian conditions. *Applied Energy, 102*, 908–922.

Ruelas, J., Sauceda, D., Vargas, J., & Garcia, R. (2018). Thermal and concentration performance for a wide range of available offset dish solar concentrators. *Applied Thermal Engineering, 144*, 13–20.

Sar, A., & Kaygusuz, K. (2002). Thermal and heat transfer characteristics in a latent heat storage system using lauric acid. *Energy Conversion and Management, 43*, 2493–2507.

Senthil, R., & Cheralathan, M. (2017). Effect of non-uniform temperature distribution on surface absorption receiver in parabolic dish solar concentrator. *Thermal Science, 21*, 11–19.

Senthil, R., & Cheralathan, M. (2019). Enhancement of the thermal energy storage capacity of a parabolic dish concentrated solar receiver using phase change materials. *Journal of Energy Storage, 25*, 100841.

Sharma, S., Micheli, L., Chang, W., Tahir, A., Reddy, K., & Mallick, T. (2017). Nano- enhanced phase change material for thermal management of BICPV. *Applied Energy, 208*, 719–733.

Shuai, Y., Xia, X. L., & Tan, H. P. (2008). Radiation performance of dish solar concentrator/cavity receiver systems. *Solar Energy, 82*, 13–21.

Sinha, R., & Gulhane, N. P. (2020). Numerical study of radiation heat loss from solar cavity receiver of parabolic dish collector. *Numerical Heat Transfer, Part A: Applications, 77*, 743–759.

Siyabi, A. I., Khanna, S., Mallick, T., & Sundaram, S. (2018). Multiple phase change material (PCM) configuration for PCM-based heat sinks-an experimental study. *Energies, 11*, 1629.

Solomon, A. D. (1979). Melt time and heat flux for a simple PCM body. *Solar Energy, 22*, 251–257.

Sukhatme, S. P., & Nayak, J. K. (2011). *Solar energy principles of thermal collection and storage* (3rd ed.). Tata McGraw Hill.

Sup, B. A., Zainudin, M. F., Ali, T. Z. S., Bakar, R. A., Ming, G. L. (2015). Effect of rim angle to the flux distribution diameter in solar parabolic dish collector. Energy Procedia 68, 45–52. 2nd International Conference on Sustainable Energy Engineering and Application (ICSEEA) 2014 Sustainable Energy for Green Mobility.

Thirugnanasambandam, M., Iniyan, S., & Goic, R. (2010). A review of solar thermal technologies. *Renewable and Sustainable Energy Reviews, 14*, 312–322.

Tian, Y., & Zhao, C. (2013). A review of solar collectors and thermal energy storage in solar thermal applications. *Applied Energy, 104*, 538–553.

Tsekouras, P., Tzivanidis, C., & Antonopoulos, K. (2018). Optical and thermal investigation of a linear Fresnel collector with trapezoidal cavity receiver. *Applied Thermal Engineering, 135*, 379–388.

Ulanicki, B., Kahler, J., & Coulbeck, B. (2008). Modeling the efficiency and power characteristics of a pump group. *Journal of Water Resources Planning and Management, 134*, 88–93.

Wei, G., Wang, G., Xu, C., Ju, X., Xing, L., Du, X., & Yang, Y. (2018). Selection principles and thermophysical properties of high temperature phase change materials for thermal energy storage: A review. *Renewable and Sustainable Energy Reviews, 81*, 1771–1786.

Yan, J., Li, K., Chen, H., Wang, Q., & Sun, J. (2016). Experimental study on the application of phase change material in the dynamic cycling of battery pack system. *Energy Conversion and Management, 128*, 12–19.

Yettou, F., Azoui, B., Malek, A., Gama, A., & Panwar, N. (2014). Solar cooker realizations in actual use: An overview. *Renewable and Sustainable Energy Reviews, 37*, 288–306.

Zhai, X. Q., Qu, M., Li, Y., & Wang, R. Z. (2011). A review for research and new design options of solar absorption cooling systems. *Renewable and Sustainable Energy Reviews, 15*, 4416–4423.

Zhu, Y., Shi, J., Li, Y., Wang, L., Huang, Q., & Xu, G. (2017). Design and thermal performances of a scalable linear Fresnel reflector solar system. *Energy Conversion and Management, 146*, 174–181.

Zivkovic, B., & Fujii, I. (2001). An analysis of isothermal phase change of phase change material within rectangular and cylindrical containers. *Solar Energy, 70*, 51–61.

Clean Energy Using Hydroelectric Generation from Rivers

Rahul Rajawat, Anuj Kumar, Gaurav Dwivedi, and Anoop Kumar Shukla

Abstract In today's world, many countries, especially developing countries, are facing energy crisis, an increase in industrialization for development programs being the apparent reason. Fossil fuels being the most prominent source for the energy production are having an adverse effect on our environment. Also, with the rate, we are exploiting the energy resources; fossils fuels may be a principal source of energy that may see its end in the near future. If we continue to meet this demand using conventional methods like thermal power plants, environmental pollution is the most prominent aspect that we will have to continue to compromise. While pollution and climate change are the biggest challenges in the modern era, switching our attention to renewable energy sources to meet our energy demands can be the best feasible solution. Out of different renewable energy sources available, hydropower is the most readily available and clean sources of renewable energy worldwide. It can be considered as the leading source of renewable energy across the world. In this chapter, an attempt has been made to study the different ways to harness hydropower such as using static head, kinetic head, or any other disruptive technology. Environmental aspect of hydropower has also been studied. A review of different technologies such as hydrokinetic, vortex flow turbine, and water wheels have been carried out along with conventional hydropower. Thus, our effort has emphasized hydropower's overall scenario as the most reliable renewable energy source with more emphasis on small hydropower. This book chapter aims to discuss the hydropower production scenario and its all-around aspects and efforts, which are made to develop it as the most significant factor for the sustainable future.

R. Rajawat · G. Dwivedi (✉)
Energy Centre, Maulana Azad National Institute of Technology Bhopal, Bhopal, India

A. Kumar
School of Mechanical Engineering, VIT University, Vellore, India

A. K. Shukla
Department of Mechanical Engineering, Amity University Noida, Noida, Uttar Pradesh, India

D. B. Pal, J. M. Jha (eds.), *Sustainable and Clean Energy Production Technologies*, Clean Energy Production Technologies,
https://doi.org/10.1007/978-981-16-9135-5_11

Keywords Clean energy · Hydroelectric generation · Energy · Rivers · Hydrokinetic · Head

1 Introduction

Energy is considered as one of the essential aspects of wealth, social infrastructure development, and improved quality of life for both developed and developing countries. When we use energy resources, many factors are considered, such as geographical distribution, availability, and pricing. One of the most important is the environmental aspect. Thus, renewable energy can be regarded as the most feasible solution for the coming future (Capik et al., 2012). Potential resources of renewable energy are solar, wind, hydropower, biomass, and geothermal. These are the primary sources of energy and, most importantly, are clean and inexhaustible. Out of the available resources of renewable energy, hydropower seems to be the most promising. Hydropower holds tremendous potential as the most potential source for clean energy. With the currently installed potential of around 1308 GW globally, hydropower had seen significant development in recent years. In the Indian context, the installed hydropower stands at about 50 GW (International Hydropower Association (Status Report), 2020). In general, hydropower can be classified in terms of size (generating capacity) and the type of scheme (canal based, run of river, pumped storage, dam toe based). However, there is no fixed definition, and the following can be the typical classification bands, as in Table 1 (International Renewable Energy Agency, 2012).

However, different countries have different definition of hydropower as shown in Table 2 (International Renewable Energy Agency, 2012).

Table 1 Classification of hydropower according to capacity (International Renewable Energy Agency, 2012)

Type	Capacity (MW)
Large hydropower	100 MW or more
Medium hydropower	20 MW to 100 MW
Small hydropower	1 MW to 20 MW
Mini hydropower	100 KW to 1 MW
Micro hydropower	5 KW to 100 KW
Pico hydropower	Up to 5 KW

Table 2 Classification of small hydropower (International Renewable Energy Agency, 2012)

Country	Small hydropower definition(MW)
Brazil	Less than or equal to 30 MW
Canada	Less than 50 MW
China	Less than or equal to 50 MW
European nation	Less than or equal to 20 MW
India	Less than or equal to 25 MW
Norway	Less than or equal to 10 MW
Sweden	Less than or equal to 1.5 MW
United States	5–100 MW

Though hydropower has got many advantages, building large dams has seen to cause environmental issues affecting the river flow; thus switching to small hydropower as compared to large hydropower is a more sought-out option for many countries including India.

Small hydropower plays a very crucial role in strategies related to rural electrification and mini-grid (Kougias et al., 2019). Small hydropower provides high conversion efficiency and is equipped with operational and economic superiority. It also provides spectacular flexibility. Identification of the potential sites and installing hydropower projects can save earth from the emissions of the fossils fuels. Quantitative knowledge of the topographical setting and stream network is the key to identify potential sites for hydropower projects. Hydropower potential locations for installation of small hydropower are shown in Fig. 1. Besides conventional hydropower, different technologies such as vortex hydroturbine, hydrokinetic technology, etc. can also be utilized to generate hydropower. More developments in this field can help in exploring greater potential of hydropower sector. Thus, there is a lot need to be developed in the field of hydropower if we want it as the long-serving energy resource for the future and can help in accomplishing the goal of sustainable development.

This book chapter provides a comprehensive study in the field of hydropower development both globally and in India. The technological aspect is also analyzed and briefly discussed the recent developments on small hydropower.

1.1 Global Scenario

Hydropower is one of the largest renewable electricity production sources and plays a vital role in moving towards clean energy and a sustainable future. Hydropower as the source of electricity was first seen in the nineteenth century. The first hydroelectric power plant was built in 1882 across the Fox River in the United States. Germany was the first country to produce a three-phase hydroelectric system in 1891. By 1900 many small hydropower projects were in operation as the technology developed across the globe. The twentieth century also witnessed rapid innovations in the field of hydropower. Towards the end of the twentieth century, China and Brazil became world leaders in the hydropower sector. Low-cost hydropower began to be the most viable source for meeting growing energy demands. Towards the end of the twentieth century, the world also started considering hydropower's environmental and social impact and reassessed hydropower's value. A report published by World Commission on Dams added a challenge to the existing practices and tried to shift focus for sustainable development. Thus the focus shifted towards how to best plan and execute the hydropower projects (Brief History of Hydropower, 2018). With more development in the field and the focus on sustainable development, a 500 GW increase in hydropower installed worldwide capacity from 2000 to 2017, around a 65% increase compared to the late twentieth century.

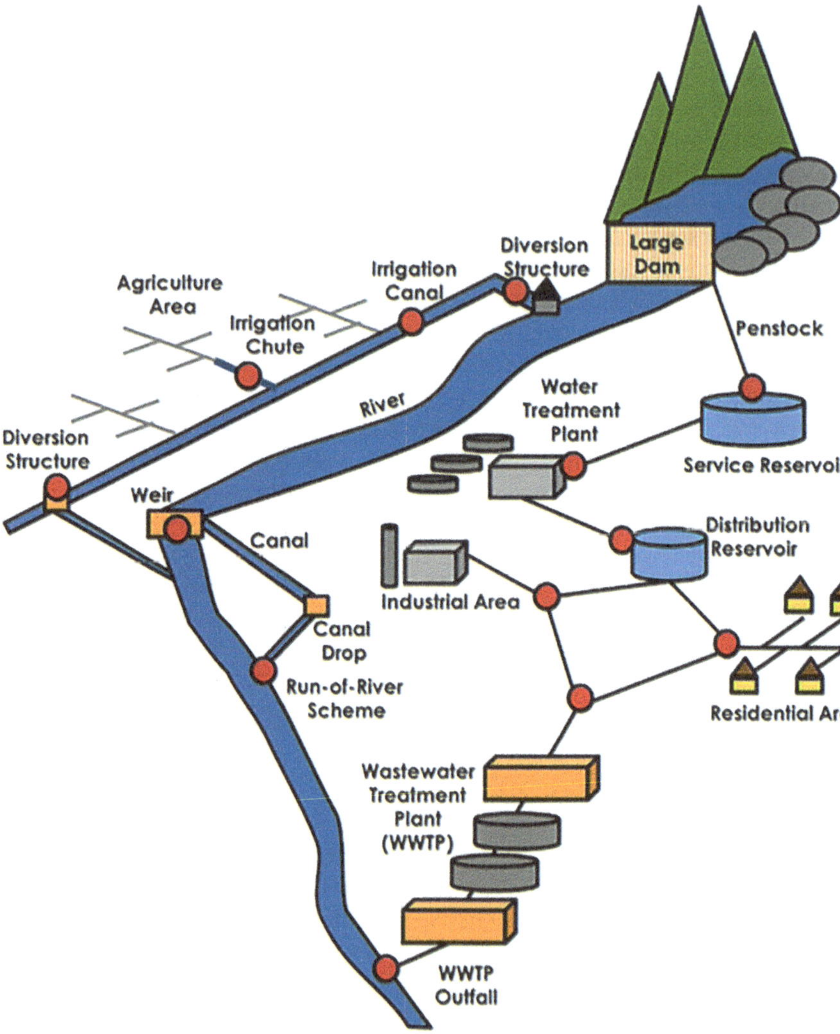

Fig. 1 Potential locations for installation of small hydropower (Sari et al., 2018)

Currently, the global hydropower installed potential is around 1308 GW. Electricity generation from hydropower is approximately 4306 TWh, a record of the most generous renewable energy contribution in history. China is the leading country with an installed potential of around 356.4 GW, followed by Brazil 109.06 GW, the United States with 102.75 GW, Canada 81.39 GW, and India 50.07 GW. The installed potential of the top 10 countries of the world is shown in Table 3 (International Hydropower Association (Status Report), 2020).

The world's most giant hydroelectric power plant has set up in China, named Three Gorges with a potential of 22.5 GW; Itaipu, which is in Brazil and Paraguay

Table 3 Top 10 countries in hydropower installed capacity (International Hydropower Association (Status Report), 2020)

Country	Installed capacity (GW)	Capacity added in 2019 (MW)
China	356.40	4170
Brazil	109.06	4919.18
United States	102.75	7.6
Canada	81.39	–
India	50.07	154.1
Japan	49.91	–
Russia	49.86	462.5
Norway	32.67	134
Turkey	28.50	219
Italy	22.59	95

Table 4 Top 10 hydropower projects in the world (The 10 biggest hydroelectric power plants in the world, 2020)

Project	Potential (GW)	Hydraulic head (m)	Major problems
Three Gorges, China	22.5	80.6	Caused ecological changes including increased risk of landslides, also flooded cultural sites
Itaipu, Brazil and Paraguay	14	118	Over 10,000 families were displaced, caused ecological changes
Xiluodu, China	13.8	–	Relocated 7300 people and can displace 50,000 people in the long run
Guri, Venezuela	10.2	128	Due to overdependence on this dam, it caused blockouts in 2010 and 2016
Belo Monte, Brazil	9.39	89.3	Caused loss of vegetation and natural spaces, changed quality of water and migration routes of fishes, temporary disruption of water supply, and loss of biodiversity
Tucurui, Brazil	8.37	–	Relocation of 14,000 people and caused migration which resulted in increase in cases of AIDS and malaria
Grand Coulee, USA	6.8	116	The environmental effects of the dam negatively impacted the life of native inhabitants
Xianhjiaba, China	6.4	–	Led to biodiversity loss, food insecurities, loss of landscape, and other ecological imbalance
Sayona, Russia	6.4	194	Met certain accidents due to powerful spring floods and suffered a catastrophic accident
Longtan, China	6.3	179	Being another mega dam, it caused ecological imbalance

with 14 GW; and Guri in Venezuela with the prospect of 10.2 GW. Three out of the ten most significant hydroelectric projects are in China which is also evident from the fact that it is the leading project in the potential of hydropower. The top

10 hydropower projects are shown in Table 4 (The 10 biggest hydroelectric power plants in the world, 2020).

With more future developments, hydropower can continue to be the most important source of renewable energy worldwide and help the world fight the challenge of a sustainable future.

1.2 Indian Scenario

India is continuously emerging in the field of hydropower being the fifth largest producer of hydropower all over the world. With growing demand and carbon emission being a major concern, hydropower development is becoming more favorable. Hydropower projects are divided into two categories, one is large hydro and the other is small hydro. The capacity of around 25 MW is considered as small hydropower in India. Large hydropower development in India faces issue of resistance by environmentalists, local communities, and also NGOs due to the issue of deforestation and resettlement issues. As a result, India's government does not consider large hydropower as a source of renewable energy, while most of the small hydro potential is through the run of a river that uses the natural flow of the river to drive the turbine. Thus any dam built is relatively small, which doesn't face the issue of deforestation and resettlement. Therefore, the India's ministry of power looks into sizeable hydropower projects; the responsibility of small hydropower is given to the ministry of new and renewable energy (Central Electricity Authority, 2021).

The first hydropower station in India was developed in Darjeeling with a capacity of 130 MW. Hydropower development in India commenced over a decade ago. The share of hydropower in India in the total installed capacity has decreased over the years. It was around 50% in 1960–1961, but now it is roughly about 14–15%, including small and large hydropower.

India's hydropower installed capacity is about 50.07GW, being the fifth globally in the hydroelectric power capacity (International Hydropower Association (Status Report*)*, 2020). India's hydropower potential is estimated at 148,700 MW at a 60% plant load factor. In contrast, 56 sites for the pumped storage scheme have also been identified with a possibility of around 94,000 MW. The small hydropower program scheme is currently running, intending to encourage the state government entities and independent private producers to set up new small hydro projects for realizing 21,000 MW potential can be realized in a phased and planned manner. In addition to ongoing projects, the immediate objective shall be to encourage them to start work on new projects with a potential of around 1000 MW, with a target to meet a cumulative capacity of 6000 MW by 2022.

According to the reassessment study, the Brahmaputra river basin is the largest capacity of 66,065 MW, out of which only 3974 MW have been developed. In terms of developed capacity, the Indus river basin is at the top with a developed capacity of

14,294.3 MW out of the identified potential of 33,832 MW (Ministry of Power (Government of India), 2019).

2 Estimation of Hydropower Potential

Hydropower can be harnessed basically in two ways as shown in Fig. 2. One is the conventional method which uses the static head and other is using kinetic head. The estimation of hydropower potential is based on the type of head which are discussed in the subsequent sections.

2.1 Conventional Hydropower (Using Static Head)

The governing equation for the calculation of power using static head will be given by Eq. 1:

$$\text{Power }(P) = \rho \times Q \times g \times H \tag{1}$$

The power output is directly proportional to the discharge rate and available head. This method is based upon building dams across rivers and producing electricity with the help of static head or potential energy. The discharge rate can be calculated using different direct and indirect methods. Velocity area method, dilution method, ultrasonic method, etc. are direct methods, while slope area method and stage discharge method are indirect methods. Head is the vertical change in elevation between the reservoir water level and the tailrace water level. Therefore, power from water increases with head and discharge but is limited by pipe friction.

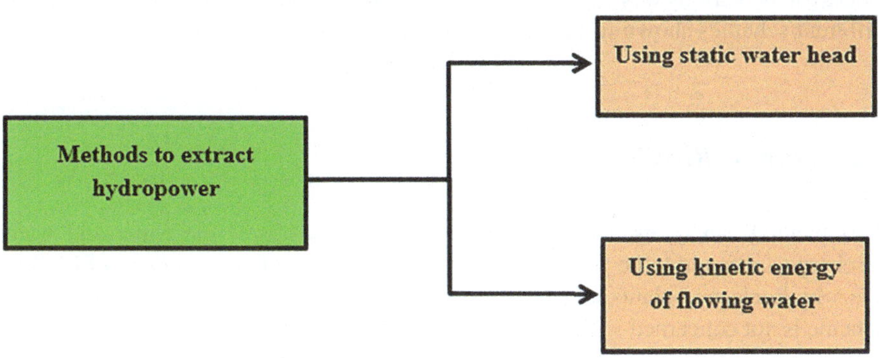

Fig. 2 Methods to extract hydropower

2.2 Using Kinetic Head (Hydrokinetic)

Flowing water stream within itself holds a large amount of kinetic energy which can be tapped using hydrokinetic energy technologies. In the modern era, apart from conventional methods, novel and innovative methods are being developed to harness the hydropower; thus hydrokinetic has emerged as the potential solution to exploit the vast hydropower potential available in flowing rivers/canals (Saini et al., 2020).

The governing equation for the available hydro kinetic energy will be given by Eq. 2:

$$P_{\text{th}} = \frac{1}{2} \times \rho \times A \times V^3 \times C_{\text{P}} \tag{2}$$

where ρ is the density of the fluid which is passing through the turbine (kg/m^3), A is the rotor swept area (m^2), and V is the free water flow velocity (m/s). The measure of dynamic fluid efficiency of the turbine is done through power coefficient C_P, and it depends on the manufacturer (Bernad et al., 2008). On the basis of configuration, hydrokinetic turbines can be basically classified into two types, i.e., axial flow turbines and cross-flow turbines. All the categories of hydrokinetic turbines consist of blades which are mounted on a support structure, which is further connected to a gearbox and generator system. Axial flow turbine consists of a rotor axis which rotates parallel to the incoming water flow. Cross flow turbines are broadly classified into two categories, i.e., vertical axis turbine and in-plane axis turbine. Vertical axis turbines are those in which the rotor axis is on the vertical plan of the water surface, while in the case of in-plane axis turbine, the rotor axis is on the horizontal plane of the water surface.

3 Different Schemes for Conventional Hydropower

There can be various ways in which conventional hydropower can be extracted, with different schemes shown in Fig. 3, which will be discussed further.

3.1 Run of River

This method can be used in areas where elevation drops of the water and water discharge are found to be consistent. The schematic layout is as shown in Fig. 3(a). Thus hydroelectric plants can be constructed directly into the river. Run-of-river scheme is not concerned with large water reservoirs and regulations because streamlines gets diverted continuously and processed through turbines, which thus returns

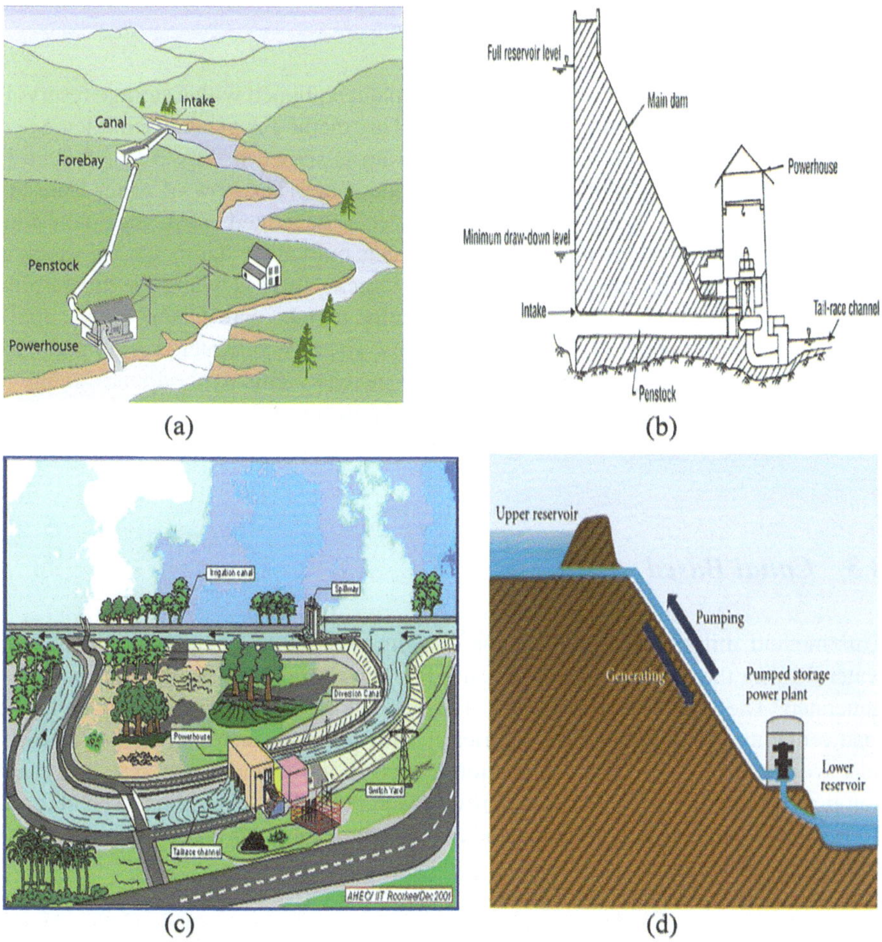

Fig. 3 Different schemes for extracting conventional hydropower: (**a**) run of river, (**b**) dam toe based, (**c**) canal based, (**d**) pumped storage (Varun et al., 2010; Kaunda et al., 2012; Dave et al., 2015; Mejbel Ali et al., 2018)

to rivers (Richards et al., 2017). Run-of-river plant can either be with pondage or without pondage. It can supply base load during good flow conditions and peak load during low flow conditions. Since there are seasonal changes in the river flow, the output may get affected unless interconnected with the grid. Recently a study was done regarding the public's perception of run-of-river projects across Europe. Four perspectives emerged on the importance of run-of-river from the analysis done where it maintains regional control, fights climate change, promotes citizen well-being, and protects natural ecosystem (Venus et al., 2020).

3.2 Dam Toe Based

This is the most common type of hydropower plant equipped with a storage reservoir provided by building a dam across the river. This method is generally implemented in mountainous regions. The schematic diagram is shown in Fig. 3 (b). A dam is constructed across the river in order to store the water. The flow of water from the storage is controlled through which the power is generated. Here the main dam includes the intake system, and the powerhouse is built at the dam's toe. Penstock is installed directly through the dam's body, which conveys water to the turbine (Singal & Saini, 2008). The water storage helps us take care of the water supply fluctuations during flood and draught situations. The flow of water is regulated with a dam's help, and the practical head can also increase. This method's advantage is that it makes the stream flows better utilized for power production and can also serve other water resource development purposes.

3.3 Canal Based

This method utilizes the drops on the existing or proposed irrigation canals. The water flowing through the canal fall is used for energy conversion and thus power is generated. The typical layout is shown in Fig. 3(c). They are planned close to the load center and provide reliable electricity. They provide electricity cost effectively and avoid transmission and distribution losses. In civil works, the system's main components are diversion channel, spillway, and powerhouse building (Singal et al., 2008). The powerhouse can be located at the central canal, or a good bypass can be constructed to locate it.

3.4 Pumped Storage Plant

In this method, hydropower technology is used to store off-peak electricity, which can be used during peak periods. The schematic representation is shown in Fig. 3(d). This plant generally consists of two reservoirs, upper reservoir and the lower reservoir. Pumping of water takes place from low reservoir to high reservoir, and in case of peak in demand, it runs downhill through turbine. In order to pump the water back to the reservoir, a reversible turbine can be used. Otherwise, a separate pump can also be used. On the basis of reservoir volume size, the pumped storage can be classified into pluri-annual pumped storage, seasonal pumped storage, weekly pumped storage, daily pumped storage, and hourly pumped storage. In case of day-night energy arbitrage, daily pumped storage-type plant is generally built. It is one the most prevalent pumped storage applications nowadays. Weekly pumped storage finds its application where the energy from the intermittent sources like solar and wind has to be stored. This technology has emerged over in recent years. In order to store water and energy beyond a yearlong period, pluri-annual pumped storage scheme is used. They are also rare and are estimated to increase further due to increasing energy needs in the coming future. Seasonal pumped storage is also not widely used in the current scenario but is expected to have an enormous potential for the future (Hunt et al., 2020).

Pumped storage plants can have turbines which can be both fixed rotation speed turbine and variable speed turbine. Pumping and generation capacities of fixed-speed turbines are generally expected to be invariable.

4 Large Hydropower

Based on the power generation capacity, hydropower plants can be divided into two categories: one is large hydropower and the other is small hydropower. Different countries have different size criteria to define hydropower projects. Generally, hydropower projects with a capacity of more than 50 MW can be considered as large hydropower.

Large hydropower projects use large water turbines, and large dams are built to store water. Though large hydropower produces a large amount of electricity, its environmental impacts are always a point of debate. The decision to construct large hydropower is based on the country's assessment of the economic development and energy needs (Hennig et al., 2013). Since large hydropower involves constructing big dams, loss of biodiversity, habitats, and aquatic vegetation can be seen in the dam area. It can also lead to accumulation of sediments in the reservoir (Almeida et al., 2005). It also affects the water quality and obstructs migration movements of fishes.

By considering various negative impacts of large hydropower, small hydropower projects are more emphasized by various countries worldwide.

5 Recent Developments in Small Hydropower

Small hydropower, which is considered the clean and renewable energy source, plays a vital role in providing electricity and optimizing energy structure (Zhang et al., 2020). The global installed capacity for small hydropower projects of capacity up to 10 MW is around 78 GW (World Small Hydropower Development Report (UNIDO*)*, 2019).

A study was performed that described the procedure for assessing the fatigue life of penstocks (made up of steel) in hydropower plants based on crack propagation theory. Stress distribution in the material of penstock shells, amplitude, frequency variability, and forecasting of future operation of the hydropower plants were the issues having a significant impact on the fatigue strength. Authentic objects were used for performing the tests and the procedure adopted for the research work was well developed. The results obtained can largely contribute to applying good engineering practices for hydropower plants equipped with steel penstocks (Adamkowski et al., 2021).

Recently, a study of the Bhilangana-III hydropower project located in India was performed, focusing on the high head Francis turbine's sediment erosion. Silt erosion is one among the major problems associated with the hydroelectric power projects installed in the regions of Himalaya. The methodology was performed on the Francis turbine components of the hydroelectric plant. The operation lasted for about 6993.12 h, and a sediment load of around 29,291 MT was noted to be passed through turbine. Analysis of the severely eroded parts of the turbine was performed, observing the wear patterns. CFD analysis was also performed in order to identify erosion-prone zones (Sharma et al., 2021).Corrosion of the components used in hydroelectric power projects can lead to significant technical problems which in turn results into economic losses. Thus, a study was performed to analyze the failure cause of the draft tube casing from a turbine which was used in a hydroelectric power project. Chemical composition of the parts that underwent failure was analyzed and also the microstructural analysis was performed. Light microscopy technique and scanning electron microscopy technique were performed in order to evaluate the present phases. After performing a detailed analysis, it was concluded that the draft tube failure was due to intercrystalline corrosion of stainless sheets. The unsuitable chemical composition of the sheet material and its microstructure were the causes of failure (Horynová et al., 2017).

Another study was performed to design and implement a cross-flow turbine used in Pico hydropower electric project. The study mainly focused on the implementation and design considerations at various nozzle positions, and performance evaluation was also done. To determine the stress and degree of deformation of blade material under the hydraulic jets impact, ANSYS was used. The shaft analysis was also performed under both static and dynamic loading to analyze whether the shaft undergoes plastic or nonplastic deformation at both ends. Response surface methodology was performed, and various simulations/runs (69) were obtained. The factors being considered were distance of nozzle from the shaft, height of the nozzle, and angle of attack in the experimental design. The water head for testing the turbine was 6.4 m and discharged at 0.0042 m^3/s. It was observed that the shaft was safe both in static and dynamic conditions. For turbine blading, carbon steel was found to be suitable (Achebe et al., 2020). Underground pumped storage hydropower (UPSH) electric projects can be one of the potential methods to balance supply-demand electricity crisis. Recently a research was presented which involved a novel method in order to determine the round-trip efficiency regarding the aspect of pumped storage-type hydroelectric projects. Two Francis turbines of 124.9 and 214.7 MW power output while two Francis pump with 114.8 and 199.7 MW power input were selected. 3D CFD numerical simulations were performed for the evaluation. Typical mass water was 45.000 t and the maximum value of gross pressure was 4.41 MP. The pressure value inside the underground reservoir was found to be having a significant impact on the round-trip efficiency, and thus the results obtained were found to be different from conventional pumped storage plant. It was concluded that the reservoir pressure had the negative impacts on the energy balance and also on the financial results (Menéndez et al., 2020).

Kaplan turbines (low head reaction type) are generally preferred for run-of-river type of hydropower projects. Recently a study regarding computational analysis (both static and dynamic) of Kaplan turbine was done by varying profile of the turbine blades. Also a parametric study was performed on a typically complex geometry of Kaplan turbine with complex geometry by varying turbine blade profile at different angles in a CAD model, while CREO was used to develop blade surface geometry and design. Further results showed that by varying the geometry of the blade profile, there can be improvement in CAD models using computational techniques. The study led to 5.43% enhancement in the turbine performance (Janjua et al., 2020).

Micro hydro can be considered as the potential technology which is befitting for irrigation networks installation. It can help in decreasing energy over pressures and can also significantly reduce net energy consumption rate of the irrigation process. Recently a study was performed in which regression models and also the artificial neutral networks computing were utilized in order to predict and analyze potential of energy recovery for micro hydropower across a large spatial scale in on-demand pressure irrigation networks. In the absence of hydraulic models, these modeling techniques were then used to predict the potential of energy recovery across the 164,000 hectare of provinces of Spain (viz., Cordoba and Seville). It was found that if the micro hydro technique could have been adopted in the 2018 irrigation session, it would have led to energy consumption reduction of around 12.8% (Crespo Chacón et al., 2020). Though it is a well-known fact that the small hydropower sector has vast investment potential, the uncertainty issue related to the project cost leads to lack of enthusiasm among the investors. Recently a paper was published which deals with the solution of a trade-off between the investors and the concerned insurance companies. Applied fuzzy logic approach was adopted for risk index assessment. For this research work, thirty-six SHP projects were analyzed by classifying them into different risk classes. The primary objectives of the study were to analyze and identify the risks associated with the investment in the SHP projects of Uttarakhand (India), also to estimate the risk index of SHP projects which have been identified and suggesting various risk mitigation measures which can help investors directly in cost reduction. It was also revealed that in this region, investment loss to the investors was around 35% of the total loss. Thus the authors proposed a variable insurance product which can be an innovative risk mitigator. Various risk resolution strategies were also proposed such as risk acceptance, risk avoidance, risk reduction, research, and risk transfer. For example, risk transfer means shifting the risk to some other person, a group, or an organization. Fixed timescale insurance can help in reducing socio-economic and environmental risks. Power purchase agreements can reduce risk of price uncertainty. In case of geological clearance, construction of projects should be avoided in geographically unstable areas (Roy & Roy, 2020).

It is evident that small hydropower projects interrupt continuity of rivers, causing migration of fishes and causing downstream flow regime to change during its operation. Thus, a study was performed which enumerates the environmental and ecological impacts related to small hydropower projects. Techniques adopted were environmental monitoring data, surveys of different sites, and also the document evidence following an integrated well-developed approach. For case study, ecological impact of *Catalouk* SHP plant (located at the Ceyhan River Basin in Turkey) was assessed. Scorecard evaluating the ecological impact revealed that there were notable gaps relating to the hydro-peaking practice and downstream fish passage and also for management of sediments. Various measures for mitigation were also proposed such as environmental flows enhancement, river basin construction for the purpose of hydro-peaking, and sustainable management of sediments through hydrosuction technology (Alp et al., 2020).

Recently a novel control system to improve the hydro-governor's action was proposed. It mainly focused on allowing more flexibility over the control action and primary frequency control improvement, which would lead to the improvement in the response frequency of overall system. Evolutionary game theory strategy was the basis of the supplementary control. Supplementary control was combined with simulated annealing algorithm. The design aspects and analysis of the proposed methodology was proposed in this work (Chamorro et al., 2019). Another study was done regarding the problem of short-term hydrothermal scheduling (STHS). Different mathematical programming model approaches were presented. Thus for every specific issue regarding the formulation process of STHS, an extensive comparison and discussion were provided. This was regarded as the beginning point in order to find more effective methods which helps in dealing with the challenges associated with unit-based STHS problem (Kong et al., 2020).

Small hydropower projects have the tremendous potential to meet the renewable energy targets globally. But exploitation of rivers in order to meet the increasing energy needs can lead to various ecological and environmental impacts. Thus there is a requirement for water allocation requirement strategies. In this context, a study was proposed which used analytical tools to meet the design considerations of run-of-river hydropower plants. This hypothetical case study was done in a Scottish river catchment. The methodology included the estimation of the hydropower potential along the river flow networks and analysis of environmental aspects related to small hydropower plants. It was concluded that policy and design interaction shall be allowed which will lead to different effective plans and roadmaps for designing run-of-river hydropower plants which can be economically profitable and also does not cause much harm to the ecosystem. One approach can be to build small-sized installations where hydrological connectivity can be sustained through bypass flows that are larger than the capacity. Another strategy can be the construction of relatively larger power plants that can eventually fulfill the ecological needs through the means of flow discharge lower than turbine cut-off value and flow discharge to be minimum (Basso et al., 2020).

6 Hydrokinetic Energy

Hydrokinetic energy conversion technology is one of the novel technologies in the field of hydropower. In order to exploit its entire potential, a lot of research and optimization work are being carried out (Kumar & Saini, 2016). Some of the recent studies regarding hydrokinetic has been discussed further.

Recently a study was done in which a surrogated model (with an oscillating hydrofoil) in a closed water channel, which was following a motion law comparable to the real angle of incidence of a Darrieus turbine blade along its rotation. Tip speed ratio (TSR) and oscillation frequency were the oscillation parameters. The result showed that the flexible blades provide reduction in the higher thrust and also the normal forces value were found to be reduced. Cyclic normal force variations were also reduced resulting in increased turbine lifetime (Hoerner et al., 2019). A study investigated the effect of a ducted augmented system on a straight blade Darrieus hydrokinetic turbine in order to analyze the performance, fluid loads, and stress induced. Real-time computation of stress and loads was done and compared for both non-ducted turbine and ducted turbine. It was found that the power generated and stress experienced in the case of ducted turbine was twice as compared to non-ducted. Thus this study recommended that with careful design considerations taken into aspect, duct augmented system shall be a preferred choice (Tunio et al., 2020). In a recent study, a blocking plate with optimal considerations was located to the upstream side of the retarding valve in order to enhance the performance of Darrieus hydrokinetic turbine. Experimental investigation was carried out to get the specific location of the turbine. The result showed that the coefficient of power of turbine increased considerably (from 0.125 to 0.36) (Patel et al., 2019). With reference to the hydro farm arrangement, an experimental analysis was performed on Darrieus turbine (straight blade) focusing on the application of tidal current and parametric optimization. The result concluded that the minimum distance of 3D and 7D was found to be essential along the span-wise and stream-wise direction, respectively (Patel et al., 2017).

Estuaries have a high potential of exploiting hydrokinetic energy since it is influenced by both the river and the marine flows. In this context a study was performed using hydrodynamic numerical modeling technique evaluating the production of hydrokinetic energy in the Douro estuary. Basically the modeling aimed at analyzing and reproducing the combined effects of tidal and river flows on the velocity patterns of main current. Also locations of estuarine having maximum energy exploitation were identified. Since there is high variability in river flows due to precipitation patterns, different scenarios of discharge were studied, which includes spring tides, neap tides, and low and high flows of river. It was concluded that the upper side of the estuary was having regions with highest potential. The relevance of identifying parameterized magnitudes those were not dependent on a specific instrument was demonstrated in this work, and also this study illustrated that in order to optimize hydrokinetic energy exploitation, estuarine hydrodynamic pattern should be properly characterized (Iglesias et al., 2021).

Hydrokinetic turbines with different designs operating on the principle the same as that of wind turbines have emerged in order to extract energy (hydrokinetic) from flowing streams. Savonius turbine is preferred for extracting hydrokinetic energy especially in the case of low velocity of the flow since it has good starting characteristics. Run-of-river scheme locations with low water velocities and low head or ultralow head (less than 1 meter) can also prefer Savonius turbine. Poor conversion efficiency was found to be one of the major drawbacks regarding Savonius turbine. In order to enhance its performance, turbine with improved design of blades is needed. In order to generate power, in addition to drag force if lift force can also be utilized, the performance of Savonius turbine can be improved. Recently a study was performed focusing on analyzing the performance of a customized Savonius turbine utilizing both the drag force and the lift force in an open flow water channel. The maximum power coefficient (0.268) was found to be higher than other modified Savonius turbines. Thus the study revealed significance of modifying Savonius turbine in order to have efficient and cleaner production of hydropower (Basumatary et al., 2021).

Recently a study was performed in which a simple barrier was utilized to deviate the flow of fluid from reversing bucket of the Savonius turbine in order to improve the power generation. Numerical modeling through computational fluid dynamics was performed to analyze the optimum barrier length. The continuity equation, Reynolds-averaged Navier-Stokes equation, and SST transition turbulence model were solved numerically. It was found that the result indicated were in accordance with the experimental data. The result showed that maximum power generation increased by 18% by utilizing the barrier with an optimum length (Alizadeh et al., 2020). A recent study was performed on single-stage Savonius hydrokinetic turbine; in this study the effect of blade arc angle and blade shape factor on the performance of the turbine were studied. A commercial unsteady Reynolds-averaged Navier-Stokes (URANS) solver in conjunction with realizable k-ε turbulence model was used for numerical analysis. The results showed that at a blade arc angle of 150° and a blade shape factor value of 0.6, maximum coefficient of power value 0.426 was obtained at tip speed ratio (TSR) value 0.9 (Kumar & Saini, 2017b). Another work was carried out in order to study and analyze the performance character sticks of a Savonius turbine rotor for the hydropower generation on a small scale. Irrigation channels of the rural areas which are having enough blade slopes were observed in order to generate hydrokinetic energy. A scale-down model (fabricated in house) of Savonius rotor was also tested. The aim was to analyze the performance under the well-developed controlled conditions. It was revealed that at a channel inclination of 0.5 and a tip speed ratio of 0.92, the power coefficient improved to 40% and coefficient of torque improved to 10%. At a bed slope of 2.0° the power and torque developed by the turbine were found to be maximum (Honnasiddaiah et al., 2021). In a recent study, performance analysis of Savonius hydrokinetic turbine was done which was having twisted blades. CFD study was carried out and flow distribution around the turbine was also discussed, and the result showed that at TSR value of 0.9 and velocity of water at 2 m/s, the turbine yields maximum coefficient of power as 0.39 at twist angle 12.5° (Kumar & Saini, 2017a). Another study was performed in which the effect of number of stages on the performance of modified Savonius turbine was studied. In this study, pressure and velocity distribution around the rotor were also analyzed. After numerical analysis, the result showed that for double-stage turbine at tip speed ratio (TSR) value of 0.9 and Reynolds number value at 37.53×10^4, the maximum power coefficient value of 0.9 was obtained. The numerical data obtained also helped in establishing a correlation for power coefficient as a function of number of stages, Reynolds number, and tip speed ratio (Kumar et al., 2020).

The performance of blade sections of the turbine is crucial because it directly affects the power coefficient of the rotor. Optimized sections of blades for wind turbines and aviation applications are only used since hydrokinetic is a relatively new technology. However it is better to consider hydrokinetics of water during the design process. Recently a study was aimed at optimizing blade sections for hydrokinetic turbines which are stall regulated. Cavitation, high value of hydro-dynamic force, ideal stalls behavior, and leading-edge contamination were the factors taken into consideration. For the optimization process, differential evolution algorithm was employed. Different performances of optimized sections such as blade, lift, drag, transition, and pressure coefficient were analyzed. It was concluded that for hydrokinetic turbines based on various objectives of design and constraints, the performance of optimized hydrofoils was found to be successful (Muratoglu et al., 2021).

A research was carried out aiming at enhancing and evaluating the operational performance of co-axial horizontal axis hydrokinetic turbines (mounted on a single shaft). The blade material was carbon fiber polymer composites. On increasing the number of rotors of the turbine from 1 to 2, it was noted that the efficiency increased by 75%, but the increase was not much when rotor increased to 3. The wake behavior and its effect were also examined. From the structural aspect, the composite materials were found to have the properties suitable for the water turbine blades (Abutunis et al., 2020). In a recent study, the performance of shrouded horizontal axis hydrokinetic turbines was evaluated corresponding to the yawed conditions. Results showed that for nearly all designs that were investigated, the power output in the case of off-axis flows was found to be decreasing. Till 10° value of yaw angle, it was found that the reduction was of negligible value, but as the angle was increased further, the reduction also saw the increasing trend. It was concluded that the convergent-divergent shroud was found to experience loss of performance with respect to other designs (Shahsavarifard & Bibeau, 2020). Another study was performed with an objective of analyzing the cavitation and hydrodynamic characteristics of the horizontal axis micro-hydrokinetic river turbine (HAMHRT). Analysis of the unsteady behaviors of the turbine was also the focus area taken into consideration. The results showed that the designed rotor was having good efficiency and also the stable output (Wang et al., 2019). In a recent study, a prototype of horizontal axis hydrokinetic turbine was manufactured and certain numerical simulations were performed. The results showed that cut in velocity of the turbine was 0.25 m/s with a maximum coefficient of power at 0.33. Thus this study revealed the feasibility of reducing the resistive torque problem in transmission parts with the application of magnetic couplings (Tian et al., 2018). In a recent study, geometric data of the broken blade of horizontal axis hydrokinetic turbine was obtained using reverse engineering approach. After reengineering, the rapid prototyping technique was adopted for manufacturing a new runner and data validation testing was also performed. Theoretical results with good precision were confirmed by the experimental data. Hydraulic power was found to be increased by 5% (Ciocănea et al., 2017).

7 Other Disruptive Technologies

Apart from the technologies discussed above, other disruptive technologies such as gravitational vortex hydroturbine and water wheels (as shown in Fig. 4) have also emerged in the field of hydropower and have great potential to further develop hydropower as the most accessible source of renewable energy. Few of them are discussed further.

7.1 *Gravitational Vortex Hydroturbine*

Gravitational water vortex turbine (GWVT) uses gravitational vortices generated upon water draining from the bottom of the tank, and power can be generated at low head and low flow rate. The typical layout is as shown in Fig. 4(a). Natural characteristic of water to form a free surface vortex is utilized in this hydropower system. A swirl flow is created by the vortex which has an ability to rotate a turbine. Water vortex that is formed under gravity is artificial, and through this, forces are generated which forms the basis of this type of hydropower system. Its major

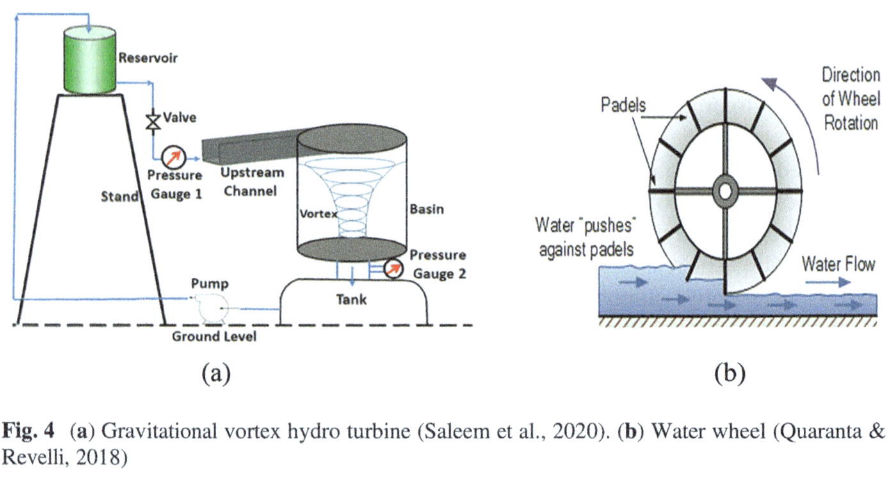

Fig. 4 (**a**) Gravitational vortex hydro turbine (Saleem et al., 2020). (**b**) Water wheel (Quaranta & Revelli, 2018)

components include an upstream channel, a cylindrical basin in which there is an orifice at the bottom, a rotor (with blades), and also a shaft (vertical). Water passes under gravity through a straight upstream channel, and in order to form a vortex, it takes the tangential entry into the cylindrical basin. Vortex is initiated at the free surface due to the effect of Coriolis force. The vortex gets intensified gradually towards the bottom surface of the orifice, and due to this, the water rotation starts getting speed up. The flow energy of the vortex generated can be converted to mechanical power with the help of a rotor. To produce electric power, the turbine rotor shaft is further coupled with the generator. Water passing through the bottom orifice gets return back to main river stream and canal and in this order the cycle is completed.

Recently a study of gravitational water vortex turbine (single staged) aiming at the parametric aspects was performed. This study presented different performance parameters like speed of rotation, torque, and the brake power. Various flow conditions and design aspects were taken into consideration in order to study the mechanical efficiency of the turbine. Systematic error analysis was also performed. The result showed that the parameters which affects the performance of GWVT are height of the vortex and a good shape of the vortex having a developed air core (Saleem et al., 2020).

It is considered as the emerging technology in the field of hydropower because it has less requirement of expertise, head required is low, and space for setup installation is also less. Issue of supplying electricity to the remote and rural areas can be solved with the proper utilization of this technology. Also it is minimally invasive to the ecosystem. GWVT can be operated under small rate of lows and low head, and this makes it, as compared to other conventional technologies, a more worthy hydropower project scheme (Rahman et al., 2017).

7.2 Water Wheels

Water wheels were scientifically investigated in the eighteenth century, but they were not much in use, also not much considered, as a potential technology in the twentieth century. Interest in the field of water wheels in the scientific community has evolved in the recent two decades.

In the field of hydropower, different machines are used to convert energy of water into mechanical energy (Williamson et al., 2014). Hydropower converting machines can be broadly classified as (1) action turbines, for example, stream water wheels, vertical axis water wheels, pelton turbine, turgo turbine, and cross flow turbines; (2) reaction-type hydropower turbines, for example, Francis turbine, Kaplan turbine, etc.; and (3) hydrostatic pressure converters like water wheels and Archimedes screw. Action turbines basically exploit the kinetic energy associated with the flowing water. Reaction turbines also exploit the water pressure. Hydrostatic force of water drives the hydrostatic pressure converters. The typical diagram of water wheel is shown in Fig. 4(b).

In the context of micro hydropower, the most significant and suitable options includes water wheels (gravity and streamed water wheels) and Archimedes screw. Stream water wheels (having maximum power coefficient to be 40%) generally finds its application in flowing water where the head differences are almost negligible. Stream water wheels are broadly categorized into three subcategories: stream wheels in shallow supercritical flow, shallow subcritical flow, and deep flow. Recently a study was published in which an analysis of experimental, numerical, and theoretical data regarding stream water wheels was performed. Guidelines for designing were also discussed focusing on wheel dimensions of wheel, design aspects of supporting structures, and also the blades and speed. Results illustrated that, in shallow water, generation of head difference takes place by the wheels which in turn increases the output power. Hydrostatic water force in addition to kinetic energy of the water flow can also be exploited force in the case of deep flow if hydrodynamic floating is accurate. Thus it can lead to power output improvement of more than 10 KW from 0.5 KW (per meter width), which makes the stream water wheel a potential energy supply technology in zero head sites (Quaranta, 2018).

Gravitational water wheels can be of three types: overshot in which the entry of water takes place from the top, breastshot, and undershot, in which entry of water takes place from the upstream side. A study was performed in order to optimize the performance characteristics of water wheels (gravity type). The results obtained illustrated that the undershot water wheels and overshot water wheels were having maximum efficiency of 85%, while in the case of breastshot water wheel, the range of maximum efficiency was around 75–80%, which was depending on the inflow configuration. Thus water wheels with better designed consideration can be a potential micro hydropower converters which can be both cost effective and efficient (Quaranta & Revelli, 2018).

8 Conclusion

In this book chapter, the focus was mainly on discussing the various aspects of hydropower energy. The worldwide potential of hydropower with Indian context has been studied. This study elaborated about the problems with large hydropower sector, and thus small hydropower can be the potential solution for the future. Different methods to extract hydropower have been discussed taking into consideration both the conventional and unconventional methods. Recent developments in the field of small hydropower sector were also studied. Other disruptive techniques to extract hydropower were also the point of our study.

References

Abutunis, A., et al. (2020). Experimental evaluation of coaxial horizontal axis hydrokinetic composite turbine system. *Renewable Energy, 157*, 232–245. https://doi.org/10.1016/j.renene.2020.05.010

Achebe, C. H., Okafor, O. C., & Obika, E. N. (2020). Design and implementation of a crossflow turbine for Pico hydropower electricity generation. *Heliyon, 6*(7), e04523. https://doi.org/10.1016/j.heliyon.2020.e04523

Adamkowski, A., Lewandowski, M., & Lewandowski, S. (2021). Fatigue life analysis of hydropower pipelines using the analytical model of stress concentration in welded joints with angular distortions and considering the influence of water hammer damping. *Thin-Walled Structures, 159*(October 2020). https://doi.org/10.1016/j.tws.2020.107350

Alizadeh, H., Jahangir, M. H., & Ghasempour, R. (2020). CFD-based improvement of Savonius type hydrokinetic turbine using optimized barrier at the low-speed flows. *Ocean Engineering, 202*(August 2019), 107178. https://doi.org/10.1016/j.oceaneng.2020.107178

Almeida, A. T., et al. (2005). Multi-impact evaluation of new medium and large hydropower plants in Portugal centre region. *Renewable and Sustainable Energy Reviews, 9*(2), 149–167. https://doi.org/10.1016/j.rser.2004.01.015

Alp, A., Akyüz, A., & Kucukali, S. (2020). Ecological impact scorecard of small hydropower plants in operation: An integrated approach. *Renewable Energy, 162*, 1605–1617. https://doi.org/10.1016/j.renene.2020.09.127

Basso, S., et al. (2020). Water-energy-ecosystem nexus in small run-of-river hydropower: Optimal design and policy. *Applied Energy, 280*(September), 115936. https://doi.org/10.1016/j.apenergy.2020.115936

Basumatary, M., Biswas, A., & Misra, R. D. (2021). Experimental verification of improved performance of Savonius turbine with a combined lift and drag based blade profile for ultra-low head river application. *Sustainable Energy Technologies and Assessments, 44*, 100999. https://doi.org/10.1016/j.seta.2021.100999

Bernad, S. *et al.* (2008) 'Flow investigations in Achard turbine', *Proceedings of the Romanian Academy Series A—Mathematics Physics Technical Sciences Information Science*, 9(2).

Brief History of Hydropower (2018). Available at: https://www.hydropower.org/discover/history-of-hydropower (Accessed: 13 January 2021).

Capik, M., Osman Ylmaz, A. and Cavusoglu, I. (2012). 'Hydropower for sustainable energy development in Turkey: The small hydropower case of the Eastern Black Sea Region', *Renewable and Sustainable Energy Reviews*. Pergamon, pp. 6160–6172. https://doi.org/10.1016/j.rser.2012.06.005.

Central Electricity Authority (2021). Available at: https://cea.nic.in/?lang=en (Accessed: 13 January 2021).

Chamorro, H. R., et al. (2019). A network control system for hydro plants to counteract the non-synchronous generation integration. *International Journal of Electrical Power and Energy Systems, 105*(September 2017), 404–419. https://doi.org/10.1016/j.ijepes.2018.08.020

Ciocănea, A., Nicolaie, S., & Băbuţanu, C. (2017). Reverse engineering for the rotor blades of a horizontal axis micro-hydrokinetic turbine. *Energy Procedia, 112*(October 2016), 35–42. https://doi.org/10.1016/j.egypro.2017.03.1056

Crespo Chacón, M., et al. (2020). Estimating regional potential for micro-hydropower energy recovery in irrigation networks on a large geographical scale. *Renewable Energy, 155*, 396–406. https://doi.org/10.1016/j.renene.2020.03.143

Dave, S. K., Parmar, A. A. and Parmar, D. K. (2015) 'International Journal of Advance Engineering and Research Small , Mini , Micro and Pico Hydro Power Plant : Scope , Challenges & Deployment in Indian Context', pp. 277–286.

Hennig, T., et al. (2013). Review of Yunnan's hydropower development. Comparing small and large hydropower projects regarding their environmental implications and socio-economic consequences. *Renewable and Sustainable Energy Reviews, 27*, 585–595. https://doi.org/10.1016/j.rser.2013.07.023

Hoerner, S., et al. (2019). Characteristics of the fluid–structure interaction within Darrieus water turbines with highly flexible blades. *Journal of Fluids and Structures, 88*, 13–30. https://doi.org/10.1016/j.jfluidstructs.2019.04.011

Honnasiddaiah, R., et al. (2021). Studies on application of vertical axis hydro turbine for sustainable power generation in irrigation channels with different bed slopes. *Renewable Energy, 163*, 845–857. https://doi.org/10.1016/j.renene.2020.09.015

Horynová, M., et al. (2017). Failure analysis of casing of draft tube of turbine used in hydropower plant. *Engineering Failure Analysis, 82*(September 2016), 848–854. https://doi.org/10.1016/j.engfailanal.2017.08.002

Hunt, J. D., et al. (2020). Existing and new arrangements of pumped-hydro storage plants. *Renewable and Sustainable Energy Reviews, 129*(April). https://doi.org/10.1016/j.rser.2020.109914

Iglesias, I., et al. (2021). Estuarine hydrodynamic patterns and hydrokinetic energy production: The Douro estuary case study. *Energy, 222*, 119972. https://doi.org/10.1016/j.energy.2021.119972

International Hydropower Association (Status Report) (2020). Available at: https://www.hydropower.org/resources/status-report (Accessed: 13 January 2021).

International Renewable Energy Agency (2012). Available at: https://cleanleap.com/2-hydropower-technologies-and-resources/24-large-and-small-hydropower-schemes (Accessed: 13 February 2021).

Janjua, A. B., et al. (2020). Static and dynamic computational analysis of Kaplan turbine runner by varying blade profile. *Energy for Sustainable Development, 58*, 90–99. https://doi.org/10.1016/j.esd.2020.07.008

Kaunda, C. S., Kimambo, C. Z., & Nielsen, T. K. (2012). Hydropower in the Context of Sustainable Energy Supply: A Review of Technologies and Challenges. *ISRN Renewable Energy, 2012*-(November 2015), 1–15. https://doi.org/10.5402/2012/730631

Kong, J., Skjelbred, H. I., & Fosso, O. B. (2020). An overview on formulations and optimization methods for the unit-based short-term hydro scheduling problem. *Electric Power Systems Research, 178*(April 2019), 106027. https://doi.org/10.1016/j.epsr.2019.106027

Kougias, I., et al. (2019). Analysis of emerging technologies in the hydropower sector. *Renewable and Sustainable Energy Reviews*. Elsevier Ltd, 109257. https://doi.org/10.1016/j.rser.2019.109257

Kumar, A., et al. (2020). Effect of number of stages on the performance characteristics of modified Savonius hydrokinetic turbine. *Ocean Engineering, 217*(September), 108090. https://doi.org/10.1016/j.oceaneng.2020.108090

Kumar, A., & Saini, R. P. (2016). Performance parameters of Savonius type hydrokinetic turbine—A Review. *Renewable and Sustainable Energy Reviews*. Elsevier Ltd, 289–310. https://doi.org/10.1016/j.rser.2016.06.005

Kumar, A., & Saini, R. P. (2017a). Performance analysis of a Savonius hydrokinetic turbine having twisted blades. *Renewable Energy, 108*, 502–522. https://doi.org/10.1016/j.renene.2017.03.006

Kumar, A., & Saini, R. P. (2017b). Performance analysis of a single stage modified Savonius hydrokinetic turbine having twisted blades. *Renewable Energy, 113*, 461–478. https://doi.org/10.1016/j.renene.2017.06.020

Mejbel Ali, A., Saadoon Algburi, S., & Abdelmajed Aljaradin, R. M. (2018). Design optimization of a hybrid hydro-wind micropower system for rural communities. *Journal of Engineering and Sustainable Development, 22*(02), 1–10. https://doi.org/10.31272/jeasd.2018.2.62

Menéndez, J., et al. (2020). Efficiency analysis of underground pumped storage hydropower plants. *Journal of Energy Storage, 28*(January), 101234. https://doi.org/10.1016/j.est.2020.101234

Ministry of Power (Government of India) (2019).

Muratoglu, A., Tekin, R., & Ertuğrul, Ö. F. (2021). Hydrodynamic optimization of high-performance blade sections for stall regulated hydrokinetic turbines using Differential Evolution Algorithm. *Ocean Engineering, 220*(November). https://doi.org/10.1016/j.oceaneng.2020.108389

Patel, V., Eldho, T. I., & Prabhu, S. V. (2017). Experimental investigations on Darrieus straight blade turbine for tidal current application and parametric optimization for hydro farm arrangement. *International Journal of Marine Energy, 17*, 110–135. https://doi.org/10.1016/j.ijome.2017.01.007

Patel, V., Eldho, T. I., & Prabhu, S. V. (2019). Performance enhancement of a Darrieus hydrokinetic turbine with the blocking of a specific flow region for optimum use of hydropower. *Renewable Energy, 135*, 1144–1156. https://doi.org/10.1016/j.renene.2018.12.074

Quaranta, E. (2018). Stream water wheels as renewable energy supply in flowing water: Theoretical considerations, performance assessment and design recommendations. *Energy for Sustainable Development, 45*, 96–109. https://doi.org/10.1016/j.esd.2018.05.002

Quaranta, E., & Revelli, R. (2018). Gravity water wheels as a micro hydropower energy source: A review based on historic data, design methods, efficiencies and modern optimizations. *Renewable and Sustainable Energy Reviews, 97*(November 2017), 414–427. https://doi.org/10.1016/j.rser.2018.08.033

Rahman, M. M., et al. (2017). A review on the development of gravitational water vortex power plant as alternative renewable energy resources. *IOP Conference Series: Materials Science and Engineering, 217*, 1. https://doi.org/10.1088/1757-899X/217/1/012007

Richards, S. K., et al. (2017). Governing the transition to renewable energy: A review of impacts and policy issues in the small hydropower boom. *Energy Policy, 101*(May 2016), 251–264. https://doi.org/10.1016/j.enpol.2016.11.035

Roy, N. C., & Roy, N. G. (2020). Risk management in small hydropower (SHP) projects of Uttarakhand: An innovative approach: Risk management in small hydropower projects. *IIMB Management Review, 32*(3), 291–304. https://doi.org/10.1016/j.iimb.2019.10.012

Saini, G., Kumar, A., & Saini, R. P. (2020). Assessment of hydrokinetic energy—A case study of eastern Yamuna canal. *Materials Today: Proceedings*, 2–6. https://doi.org/10.1016/j.matpr.2020.08.595

Saleem, A. S., et al. (2020). Parametric study of single-stage gravitational water vortex turbine with cylindrical basin. *Energy, 200*, 117464. https://doi.org/10.1016/j.energy.2020.117464

Sari, M. A., et al. (2018). Recent innovations and trends in in-conduit hydropower technologies and their applications in water distribution systems. *Journal of Environmental Management, 228*(May), 416–428. https://doi.org/10.1016/j.jenvman.2018.08.078

Shahsavarifard, M., & Bibeau, E. L. (2020). Performance characteristics of shrouded horizontal axis hydrokinetic turbines in yawed conditions. *Ocean Engineering, 197*(January), 106916. https://doi.org/10.1016/j.oceaneng.2020.106916

Sharma, S., Gandhi, B. K., & Pandey, L. (2021). Measurement and analysis of sediment erosion of a high head Francis turbine: A field study of Bhilangana-III hydropower plant, India. *Engineering Failure Analysis, 122*, 105249. https://doi.org/10.1016/j.engfailanal.2021.105249

Singal, S. K., & Saini, R. P. (2008). Cost analysis of low-head dam-toe small hydropower plants based on number of generating units. *Energy for Sustainable Development, 12*(3), 55–60. https://doi.org/10.1016/S0973-0826(08)60439-1

Singal, S. K., Saini, R. P., & Raghuvanshi, C. S. (2008). Cost optimisation based on electro-mechanical equipment of canal based low head small hydropower scheme. *The Open Renewable Energy Journal, 1*, 26–35.

The 10 biggest hydroelectric power plants in the world (2020). Available at: https://www.power-technology.com/features/feature-the-10-biggest-hydroelectric-power-plants-in-the-world/ (Accessed: 13 January 2021).

Tian, W., Mao, Z., & Ding, H. (2018). Design, test and numerical simulation of a low-speed horizontal axis hydrokinetic turbine. *International Journal of Naval Architecture and Ocean Engineering, 10*(6), 782–793. https://doi.org/10.1016/j.ijnaoe.2017.10.006

Tunio, I. A., et al. (2020). Investigation of duct augmented system effect on the overall performance of straight blade Darrieus hydrokinetic turbine. *Renewable Energy, 153*, 143–154. https://doi.org/10.1016/j.renene.2020.02.012

Varun, Prakash, R., & Bhat, I. K. (2010). Life cycle energy and GHG analysis of hydroelectric power development in India. *International Journal of Green Energy, 7*(4), 361–375. https://doi.org/10.1080/15435075.2010.493803

Venus, T. E., et al. (2020). The public's perception of run-of-the-river hydropower across Europe. *Energy Policy, 140*(July 2019). https://doi.org/10.1016/j.enpol.2020.111422

Wang, W. Q., Yin, R., & Yan, Y. (2019). Design and prediction hydrodynamic performance of horizontal axis micro-hydrokinetic river turbine. *Renewable Energy, 133*, 91–102. https://doi.org/10.1016/j.renene.2018.09.106

Williamson, S. J., Stark, B. H., & Booker, J. D. (2014). Low head pico hydro turbine selection using a multi-criteria analysis. *Renewable Energy, 61*, 43–50. https://doi.org/10.1016/j.renene.2012.06.020

World Small Hydropower Development Report (UNIDO) (2019). Available at: https://www.unido.org/our-focus-safeguarding-environment-clean-energy-access-productive-use-renewable-energy-focus-areas-small-hydro-power/world-small-hydropower-development-report (Accessed: 5 February 2021).

Zhang, C., et al. (2020). Small hydropower sustainability evaluation for the countries along the Belt and Road. *Environmental Development, 34*, 100528. https://doi.org/10.1016/j.envdev.2020.100528

Clean Energy Production Using Solar Energy Resources

Brajesh Kumar, Deepak Garg, Kurella Swamy, and Pradeep Kumar

Abstract Recently, a huge gap has been drawn between demand and supply of energy due to exorbitant consumption of conventional fuels which are not only in finite quantity on this globe but also adversely affecting the environment as a consequence of water and air pollution at large scale. Due to this reason, many researchers and scientists are focused in searching some renewable sources of energy as alternatives such as solar energy, wind energy, geothermal energy, ocean energy, etc. Among all of them, solar energy is recognized as the topmost renewable source as it is available free of cost and can be converted into useful variants of energy via many conversion techniques such as thermal conversion and photovoltaic (PV) conversion processes. This chapter deals with the necessary information regarding solar thermal conversion systems and solar photovoltaic conversion systems or the combination of both for commercial and domestic purposes. The more prevalent solar conversion techniques consist of solar water heating, solar air heating, solar thermal power plants, solar air conditioning and refrigeration, etc., while solar photovoltaic system (SPS) converts incident solar radiation into electrical energy by using solar cell consisting of semiconductor materials. In brevity to understand the depth of solar energy conversion process as incredible utility, some past and ongoing research details have also been mentioned. Moreover, many recent developments and future aspects have been discussed about solar energy in terms of applications, challenges, and requirement to focus on the utilization of sun's energy

B. Kumar · K. Swamy
Department of Chemical Engineering, National Institute of Technology Srinagar, Hazratbal, Srinagar, Jammu and Kashmir, India

D. Garg (✉)
University School of Chemical Technology, Guru Gobind Singh Indraprastha University, Dwarka, New Delhi, India
e-mail: deepak@ipu.ac.in

P. Kumar
Department of Chemical Engineering, Institute of Engineering & Technology, Lucknow, Uttar Pradesh, India

D. B. Pal, J. M. Jha (eds.), *Sustainable and Clean Energy Production Technologies*, Clean Energy Production Technologies,
https://doi.org/10.1007/978-981-16-9135-5_12

as promising, sustainable, economical, and environmental-friendly renewable source.

Keywords Solar energy · Solar cell · Renewable energy · Thermal conversion · Photovoltaic conversion

1 Introduction

Energy plays a vital role in any country's sustainable growth (industrial and agricultural growth) in terms of economics. Conventional fuels are reducing rapidly due to their excessive consumption because of their limited availability in the nature. Therefore, a great concern is rising towards a gap between demand and supply of energy for fulfilling our future needs. To overcome this huge serious threat, the search for renewable sources of energy is getting more attention nowadays worldwide by multitudinous academicians and researchers (Kumar et al., 2017).

Solar energy is more effective and prominent among various renewable sources of energy (wind, water, geothermal, etc.) quantity- and quality-wise because it is clean, free of cost, and available in great abundance in the atmosphere. It is the incident radiation from sun to earth which is enough capable to generate heat, responsible for the occurrence of chemical reactions and more importantly electricity generation. Solar energy is basically produced through thermonuclear fusion reactions occurring inside the sun, and its spectrum on earth's surface is spread in the form of visible and ultraviolet radiation.

There are several advantages and disadvantages of solar energy over nonrenewable sources of energy which makes it effective and efficient. Some are mentioned below:

1.1 Advantages of Solar Energy

(a) Solar energy conversion is considered as environmental friendly because it is clean, renewable, green, and sustainable source of energy.
(b) Unlike conventional fuels, solar energy has no participation in air pollution, water pollution, global warming, and acid rain. Because of these attributes, the utilization of solar energy minimizes the emissions of harmful greenhouse gases from atmosphere.
(c) Solar energy does not generate directly any waste like municipal waste, industrial waste, radioactive waste, etc.
(d) Solar energy conversion is considered as money saver because the energy obtained from sunlight is free of cost after recovery of initial investment cost. Therefore, it is out of concern of demand and supply of fuel and possesses negligible chances of fuel price hike in future.

(e) There is no transportation cost of solar energy as fuel. It lasts decades, if managed in proper ways, and requires low maintenance for solar energy conversion equipments.
(f) It has huge positive impact on the health of living beings in the long run.
(g) It reduces water usage for electricity generation and other solar energy conversion techniques.

1.2 Disadvantages of Solar Energy

(a) The initial investment cost is high.
(b) It is dependent on climatic conditions and availability of enough sunlight to produce heat.
(c) The solar energy storage is expensive.
(d) It requires space to install solar energy conversion equipments.

1.3 Utilization of Solar Energy

Solar energy is utilized by many developed and developing stage countries in advanced manner nowadays and trying to maximize the production and minimize the dependency on fossil fuel usage. Among these countries, the utilization of solar energy by various states in India is an appreciable step as developing stage country. The top five states of India which contribute significantly in solar energy sector to produce electricity as per the report of "Ministry of new and renewable energy" (March 2021) is given in Fig. 1 (National Informatics Centre, Govt. of India, 2021), in which Karnataka is the leading state and more aware about the importance of solar energy:

1.4 Solar Energy and Climate Change

It is considered by the environment experts that the reason of climate change is greenhouse gases emitted in the environment. Greenhouse gas emissions occur due to various sectors which contribute at large scale by human population such as conventional thermal power plants operated by coal, transportation sector, process industries, and domestic energy consumption. There is no other renewable source of energy except solar energy which can provide zero greenhouse gaseous emission and create positive impact on our climate. Sunlight produces radiation with less emission of greenhouse gases as compared to exploitation of conventional sources of energy. Although manufacturing and recycling of solar system may produce some emissions but generation of energy leads to almost negligible harmful emissions of greenhouse gases with zero impact.

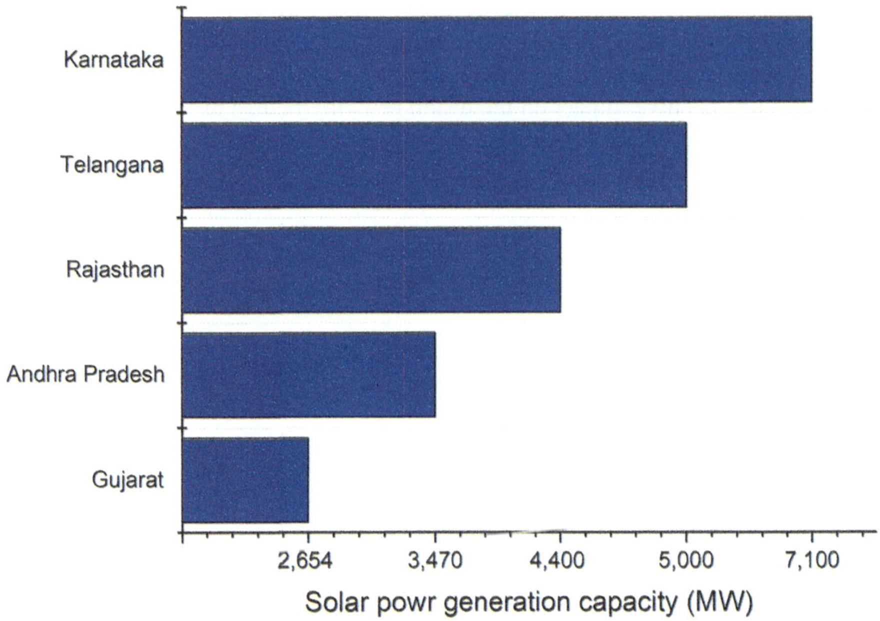

Fig. 1 Solar power generation capacity of top five states in India, 2021 (MNRE, GOI)

Solar energy is capable enough to mitigate harmful impact on the climate and to achieve some important climatic targets such as carbon-free climate, zero greenhouse gases, and minimization of occurrence of natural disasters. Except power generation, solar energy also reduces the wastage of definite natural resources such as coal, natural gas, etc. to heat space and water. Heating of space and water is quite a popular and successful technique by using solar radiation worldwide. Moreover, the purpose of clean and pollution-free (Chloroflourocarbon-free) cooling can also be accomplished by solar air conditioning instead of traditional air conditioning.

2 Solar Energy Conversion

Solar energy means energy from the sun. Solar photons through incident solar radiation on earth's surface convert itself into three forms of energy such as heat, chemical fuel, and power generation, which combines together and behaves like energy chain (Crabtree & Lewis, 2007). The discussion of all three forms is given below:

2.1 Heat

Normally, heat is produced by combustion of fossil fuels in conventional manner which is used to heat water, air, and space. Moreover, this heat is also converted into electricity. Solar energy replaces the requirement of this conventional heat with solar incident radiation which is also free from environmental pollution and also reduces the exploitation of finite sources of nonrenewable energy such as fossil fuels, coal, etc.

Unconcentrated solar incident radiation can produce a temperature around 200 °C which is capable enough to heat water and space for several domestic and industrial applications. Some countries have already adopted the concept of solar heating as per their demand and supply of energy. In opposite of this, concentrated solar incident radiation reaches up to 1500 °C or sometimes more than that. The high temperature obtained from solar energy is used in thermochemical water splitting and solar reforming of conventional fuels for the production of hydrogen. In opposite of this, concentrated solar incident radiation is used to produce heat for power heat engines and gas-fired electricity generators. The purpose for achieving the temperature more than 1500 °C requires large-scale plants for the generation of electricity by using power towers.

2.2 Chemical Fuel

There are many ways for converting solar photons into valuable chemical fuels such as photosynthesis and the production of starch and sugar from CO_2 and H_2O by catalytic endothermic reaction. Even after providing enough greenery by nature on earth to capture solar photons for converting solar energy into chemical fuel, the conversion efficiency is quite low to fulfill the demand of human being in terms of energy. It is because of emitting CO_2 on large scale by plants and trees during photosynthesis process. This problem can be overcome by three conversion pathways to enhance the production of photosynthesis-based solar fuel. The first is related to genetically engineered plants for the fast production of biomass materials. This increases plant yields and prominent pathways in the industries. Second is the search of new pathways to improve performance of solar energy-based chemical fuels. Third and last is the production of fuel from CO_2 and H_2O by using artificial bio-based nanoscale materials.

In the second and third pathways, the involvement of direct manipulation of photosynthesis process is found. Some reactions in green plants via solar energy divide H_2O into electron, proton, and oxygen. Artificial photosynthesis converts solar light into certain chemical fuels. In photoelectrochemical conversion, the charge separates holes and electrons separate H_2O to reduce CO_2 by using an electrolytic solution. The ultimate target of solar energy is to convert itself into chemical fuels to reduce the CO_2 amount and enhance the H_2 quantity with improved efficiency via more complex reaction system.

2.3 Power Generation

Power generation via solar radiation obtained from the sun is pollution-free and economical. Though electricity generation from solar incident radiation contributes less of global power production, the countries like the USA, Germany, China, etc. are more focused to raise solar energy-based power production in the world and presenting themselves as examples to increase awareness about solar energy attributes.

To get high intensity of sunlight for the generation of electricity, a variety of techniques are used such as photovoltaic (PV) cells and solar thermal power plants which harness solar energy in adequate and qualitative manner. Power generation through solar thermal power plant provides a promising approach to replace fuel-free way as compared to conventional power generating systems in which coal is burnt as fuel mainly. The list of top five countries that produce electricity with the capacity of using sunlight as per NS energy global report of 2019 is given below (NS Energy Staff Writer, 2021):

1. China – 205 GW.
2. USA – 76 GW.
3. Japan – 63.2 GW.
4. Germany – 49.2 GW.
5. India – 38 GW.

A detailed list of solar energy conversion techniques along with solar equipments is given below (Kothari et al., 2011):

I.	**Solar thermal energy conversion system (STECS)**				
	(A)	Solar water heating			
		(i)	Natural circulation solar water heater		
		(ii)	Forced-circulation solar water heater		
	(B)	Solar distillation			
		(i)	Passive distillation		
			(a)	Conventional solar still distillation with reflector	
			(b)	Multieffect solar still distillation with condenser	
			(c)	Inclined solar still distillation	
			(d)	Hybrid solar still distillation	
		(ii)	Active distillation		
			(a)	High-temperature distillation	
				(1)	Auxiliary heating distillation
				(2)	Distillation with collector panel
			(b)	Nocturnal distillation	

(continued)

	(C)	Liquid bath solar wax melter		
		(i)	Solar wax melter	
	(D)	Solar thermal power plant		
		(i)	Low-temperature solar power plant	
		(ii)	Medium-temperature solar power plant	
		(iii)	High-temperature solar thermal power generator	
	(E)	Solar pumping system		
		(i)	Submersible solar pumps	
		(ii)	Surface solar pumps	
		(iii)	DC solar pumps	
		(iv)	AC solar pumps	
	(F)	Solar ponds		
		(i)	Convective solar ponds	
		(ii)	Non-convective solar ponds	
			(a)	Membrane solar ponds
			(b)	Gel solar ponds
			(c)	Salt gradient solar ponds
	(G)	Solar crop drying		
	(H)	Solar kilns		
		(i)	Integrated solar dryers	
		(ii)	Distribution solar dryers	
	(I)	Solar cookers		
		(i)	Box solar cooker	
		(ii)	Dish solar cooker	
	(J)	Solar passive techniques		
		(i)	Direct heat gain solar passive system	
		(ii)	Indirect Trombe wall solar passive system	
	(K)	Solar air conditioning and refrigeration		
		(i)	Evaporative cooling	
		(ii)	Absorption cooling system	
		(iii)	Passive desiccant cooling	
	(L)	Solar greenhouses		
	(M)	Heating of swimming pool by solar energy		
	(N)	Solar air heaters		
		(i)	V-shaped absorber plate solar air heater	
		(ii)	Two-pass solar air heater	
		(iii)	Matrix air heater	
	(O)	Solar furnace		
		(i)	Direct solar furnace	
		(ii)	Heliostat solar furnace	
II.	**Solar photovoltaic system (SPS)**			
	(A)	Grid connected solar PV system		
	(B)	Off-grid solar PV system		

3 Solar Thermal Energy Conversion

Solar water heating is very popular among all solar thermal conversion technologies due to its ease of adaptive with minimal effort. It is mainly dependent on natural circulation of solar light, but for increasing intensity, sometime forced circulation is required. It is also known as solar domestic hot water system. The installation of solar water heater depends on several important factors such as adequate solar incident radiation, climate, and some nominal safety issues.

Safe drinking water is an essential requirement for arid, semiarid, and coastal regions to support life of living beings. For such places, saline water is available underground/oceans. To purify or desalinate this water, a device, namely, solar water still, based on distillation process is used to produce potable water by utilizing solar heat radiation. Solar distillation adopts two ways as passive and active distillation. In passive distillation, solar still are equipped with reflector and condenser which are well known for multiple effects to produce potable water from sea, river, lake, etc. at low cost. For getting more effect, inclined and hybrid systems are also prevalent. On the other hand, active distillation consists of high temperature and nocturnal distillation which are also equipped with some solar radiation collector panel for heating. Solar distillation is a technique used to purify or separate freshwater from sea or oceans, and this process is also known as desalination. It requires low maintenance and no energy cost.

Liquid bath solar wax melter is used to melt wax and reduces dependence on the conventional heating. Wax melted by solar energy is also a natural process and compensates with atmosphere as compared to conventional wax melting, but for getting higher efficiency, solar melter is used. For designing a liquid bath solar wax melter, the desired properties are as follows: melting point, 62.8 °C; specific heat for solid, 2.09 kJ/kg °C at 298 K, and for liquid, 2.3 kJ/kg °C at 338 K; and density for solid, 816 kg/m^3 at 298 K, and for liquid, 781 kg/m^3 at 338 K. Moreover, for melting around 5 kg of wax, 1 m^2 area of flat collector is required.

Solar thermal power plant has three types according to temperature variation as per requirement. It varies as low temperature (~45 °C), moderate temperature (~400 °C), and high temperature greater than 500 °C. Solar thermal power plant is a very good substitute of coal-fired thermal power plant due to many attributes such as carbon-free, fuel-free, harmful greenhouse gas emission–free, etc. These plants can sustain up to 25 years, and this period can also enhance with proper maintenance and upgradation. Although the initial cost of solar thermal power plant may be high, after return of investment, it produces power free of cost and is the proper utilization of epitome of renewable energy sources without wasting finite sources of energy. Solar thermal power plant works on the basis of utilization of concentrated solar incident radiation to heat water which is required to produce steam to feed into turbine for the electricity generation. Solar power generation is an essential requirement for those countries who are suffering from harmful greenhouse gaseous emission effects from their energy sectors. A representation of solar thermal power plant is shown in Fig. 2.

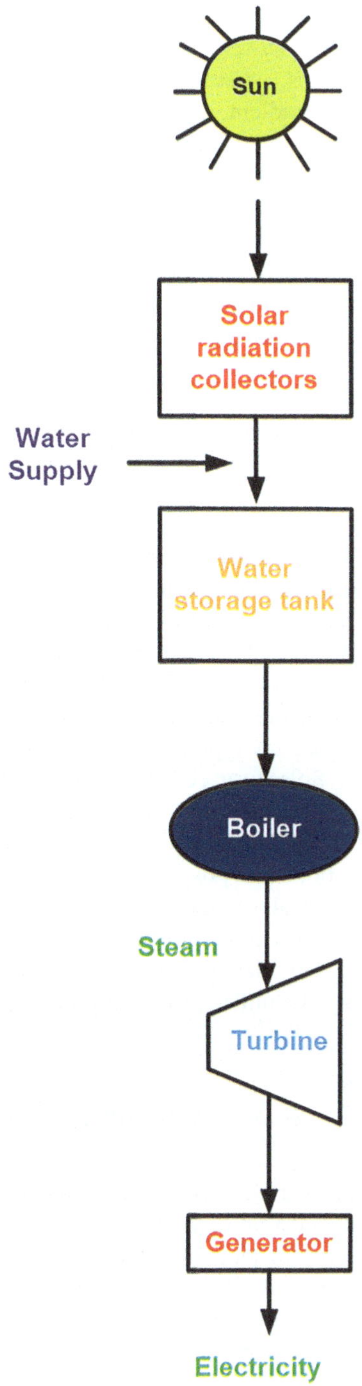

Fig. 2 Schematic diagram of solar thermal power plant

Pumping of water using solar energy instead of old method is found effective as direct current (DC) type, alternative current type, and surface and submersible pumps, whereas solar pumps require some special arrangements like membrane, salt gradient, and gel-type nonconvective solar ponds. The demand of solar pumps is high in rural and developing stage areas for irrigation in agricultural field which may become sustainable, safe, and viable option in renewable energy area. This technology is providing benefits in many continents such as Africa, South America, South Asia, and Southeast Asia.

In agriculture fields, drying of crops is a necessary task before and after storage to minimize the loss of crops during harvesting time. Solar crop drying is also found as an alternative to minimize the electricity requirement. This technique can be better option for farmers and reduces their expenditure and also enhances quality of crops in terms of hygiene without wasting essential nutrients. For fulfilling drying requirement expect solar crop drying, integrated and distribution type solar dryers are used in process industries. It reduces drying time and also provides protection from rain and wind.

In remote areas where electricity problem occurs, solar cookers are considered as better option for cooking rice, cereals, and other essential foods. The extra benefits of solar cooker are the following: no requirement of conventional fuel, no recurring charges, no soot formation, maintains quality of certain foods, and no health problem because of the absence of smoke like conventional methods. Solar greenhouse can be used as effective solar energy conversion technique to reduce greenhouse gas concentration and also minimizes the effect of global warming because it does not create any type of air pollution.

Solar furnace is used to attain high temperature through optical equipment which concentrates over solar light on a small area. In solar furnace, a temperature up to 3500 °C can be generated which is quite useful for the study of phase and vaporization, analysis of melting behaviors, and purification of ceramic and refractory materials. Solar furnace can be used for research using proper high-temperature measuring instruments.

For other applications such as conditioning and refrigeration, air heating, etc. can also be utilized by solar energy as per domestic and industrial requirement.

4 Solar Photovoltaic System

In solar photovoltaic system (SPS), the conversion of usable solar incident radiation directly into electricity is achieved by photovoltaic cells (PV cells) connected in series or parallel without creating any air and water pollution. The solar panel and solar cell are recognized as prime building components of solar photovoltaic system which are connected together with tiny wires. Solar cell is also called a developed and properly manufactured semiconductor device which is responsible to generate enough voltage to act an adequate manner when sunlight falls on it. Semiconductors are of two types such as n-type and p-type semiconductors.

n-type semiconductor consists of silicon crystals with four valance electrons doped with atoms (phosphorus, arsenic, antimony) having five valance electrons; then doped crystals carries excess-free electrons, whereas p-type semiconductor is made of silicon crystals which is doped with atom (boron, gallium, indium) of three valance electron and a hole with positive charge is created in which an electron can move freely within the crystals. Both type of semiconductor possesses high conductivity as compared to pure material (intrinsic material). Moreover, a platform where a junction is created between the boundaries of p-type and n-type semiconductors is known as p-n junction in which positive charged p-type semiconductor has an excess of holes and negative charged n-type semiconductor has an excess of electrons. This arrangement develops p-n junction diode which is used as photodiode and used as rectifiers in electrical circuits.

The electricity is generated by the PV cells which enables the photons to strike out the electrons from the molecular lattice, releasing a freed electron and hole pair which diffuses in an electric field where the charge carriers are separated, delivering a direct current (DC) to an outside circuit. Solar photovoltaic cell consists of semiconductor materials with layers of positive and negative charges. The semiconductor atoms absorb usable photons from the received light upon the solar cell in which electrons move from negative charge layer by an external source to the positive charge layer. This flow of electrons produces electric current. In a single module, individual photovoltaic cells in adequate number are interconnected to each other with water proof package in a sealed manner to enhance the utility.

The modules are connected through wires in series to increase the voltage at constant current and, on the other hand, when they are connected in parallel to increase the current at constant voltage (Denholm et al., 2010). The desired values of voltage and current are achieved by wiring this module as in series and parallel accordingly, and such setup is known as a PV array. The challenge for engineers in all the PV devices is to enhance the absorptivity of the cell material and the decrease in the rate of electron/hole recombination with cost considerations. Typical block diagram of an SPS is shown in Fig. 3.

4.1 Photovoltaic Effect

In a solar cell, p-n junction diode generates electron-hole pairs which developes electric current, and this electric current is equal to the difference between current generated by solar incident radiation and diode dark current. This current is denoted by symbol I and can be represented by the following standard relation:

$$I = I_L - I_j$$

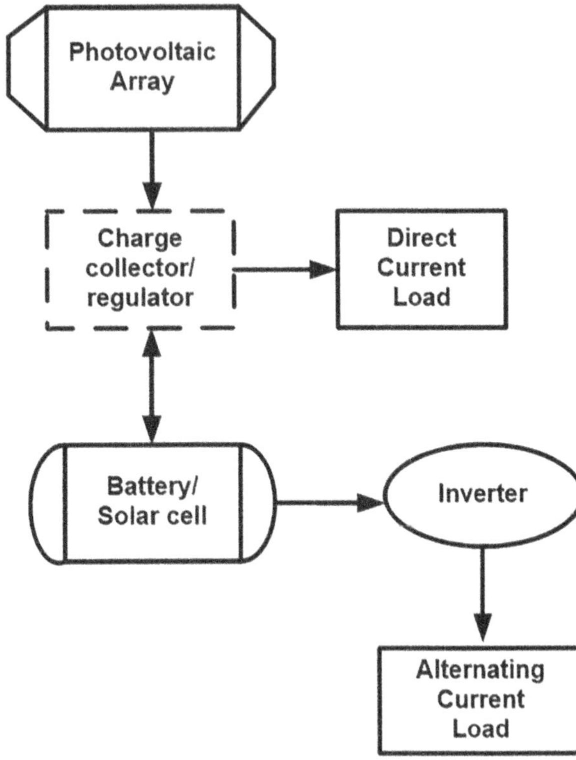

Fig. 3 Solar photovoltaic system block diagram

$$I = I_L - I_0\left[\exp\left(\frac{eV}{RT}\right) - 1\right]$$

where

I_L is the current generated by solar light, amp.
I_j is the diode dark current, amp.
V is the generated voltage, V.
e is the electronic charge,
I_0 is the saturation current generated due to dark current under reverse bias.
K is Boltzmann's constant.
T is the absolute temperature, K.

4.2 Semiconductors

The material through which the electrical current passes easily and has low resistivity is called as conductor. An insulator shows the opposite character of a conductor, i.e., high resistivity and opposes the electrical current flow. The materials that show the electrical properties in between of insulators and conductors are semiconductors. The conductors have plenty electrons that are not very tightly held within the molecules which upon getting very less energy become free, whereas, in an insulator, the electrons are primarily locked within the molecules at specified place and require sufficiently higher energy to become free. Semiconductors also show the properties like an insulator with a difference of less energy requirement to free some electrons.

They have very less free electrons as the atoms are grouped together closely. However, with the addition of certain dopants or impurities to the structure of semiconductor, charge carriers can be supplied that are very loosely held within the material and thus are easily movable. Removed electrons or additional electrons can be charge carriers which act like a positive charge in a nonstationary manner (Bhatia, 2014). Generally, dopants are distinguished as electron acceptors and electron donors. The electron donor dopants contribute additional electrons, and they are called as negative-type (n-type) semiconductor, and electron acceptor dopants withdraw the electrons from the material by making a "hole," and they are termed as positive-type (p-type) semiconductor. The dopant is added in very traceable concentration of the order of 1 ppm (Schwarz & Van Langenhove, 2013) . The semiconductors are generally of two types in which one is inorganic and the other one is an organic semiconductor. The silicon, carbon, metal oxides are inorganic semiconductors with high electrical performance and excellent stability, and the major drawback is that they require the temperature more than the thermal decomposition and/or glass transition temperatures of polymeric materials (Baca et al., 2008; Rogers et al., 2011) . The semiconductors of organic can be processed by printing techniques which is a low temperature deposition technique. The research efforts put in the direction of improvement for the electrical performance of organic semiconductors are promising. However, the obtained results are still lesser than that can be achieved by the inorganic semiconductors (Sun & Rogers, 2007).

4.3 Semiconductor Materials for Solar Cells

Solar cells are manufactured by materials in three physical forms such as single multicrystal, many small crystals (polycrystalline), and amorphous (nanocrystalline). The list of significant and prominent semiconductor materials which are used in solar cells is given below (Kothari et al., 2011):

(a) Single crystal silicon.
(b) Polycrystalline silicon cells.
(c) Bifacial crystalline cell over multicrystalline substrate.
(d) Amorphous silicon cells.

4.4 Advantages of Solar Photovoltaic Systems

(a) **Environmental friendly:** No harmful greenhouse gaseous emissions occur during power generation through solar photovoltaic systems. Therefore, SPV systems are known for clean and green systems as environmental friendly.
(b) **Abundance:** Solar energy as fuel for SPV system is provided by nature in unlimited quantity.
(c) **Easy installation:** Installation of solar panel on the ground surface pr rooftop is easy without any interference to residential lifestyle.
(d) **Noise-free system:** SPV system does not create noise pollution as they work silently and perfect solution for human habitation.
(e) **Stationary setup:** No moving part is required for the working of solar photovoltaic system. Therefore, less maintenance is required as it is a stationary setup.
(f) **Cost effective:** Operating and maintenance of SPV system are considered low as compared to the cost of other renewable sources of energy setup.
(g) **Direct electricity production:** Photoelectric effect generates electricity in a direct way by using solar panel.
(h) **No use of fuel and water:** There is a requirement of definite source as fuel and water during operation of SPV system to generate electricity.

4.5 Disadvantages of Solar Photovoltaic Systems

(a) **Intermittence issue:** In the absence of less intensity of sunlight or less shiny day, solar panels do not work in efficient manner as well as in cloudy and rainy season.
(b) **Large space requirement:** Large space is required to deploy or install solar panels on rooftop or ground's surface.
(c) **Additional requirement:** Requirement of additional equipment is inverter and other things to convert direct electricity into alternate electricity for proper functioning of power system.

4.6 Applications of PV Systems

The main applications of PV systems are given below in brevity (Kothari et al., 2011):

(a) Standalone solar photovoltaic systems.

 (i). Solar street light.
 (ii). Home lighting system.

(b) Solar photovoltaic water pumping systems.
(c) Solar photovoltaic cell for communication equipment in snow-bound areas.
(d) Sea water desalination.
(e) SPV cell for communication as equipment in snow-bound areas.

5 Combination of Solar Thermal and Solar Photovoltaic System

A solar photovoltaic system is not capable to supply of electricity by using storage cell or battery to supply of electricity alone in continuous way. This system is dependent on season especially high-intensive sunny days. Low intensity of solar incident radiation due to cloudy days and nights is not sufficient to convert into electricity. Therefore, a hybrid system containing solar thermal and solar photovoltaic has been developed to handle the adequate load requirements without any hindrance. Installing solar photovoltaic system with a diesel generator along with storage batteries is the most efficient and economical approach to enhance the capacity of hybrid system (Kothari et al., 2011). Different combinations other than diesel generator are also under research for trail to develop new hybrid system.

A combination of solar thermal and solar photovoltaic systems makes a hybrid system which leads to both thermal and electrical output. In 1997, the International Energy Association's (IEA) Photovoltaic Power Systems Program (PVSP) started IEA Task 7 to evaluate the technical status of combined solar-thermal conversion systems and solar photovoltaic system to establish a strategy for sustainable future needs (Collins, 2005). Due to this hybrid system, the demand of solar collectors is also increasing. Therefore, search for different feasible materials to manufacture solar collectors is getting more attention by researchers who are working in the fields of solar energy with a wide range of applications such as agricultural field, process industries, and building (Chow et al., 2012).

In recent times, many theoretical and experimental studies on the performance of solar collector panels have been done. In earlier stage, the approach was based on collecting maximum solar radiation, but later on, focus was shifted to design modification, testing procedure, innovative system, fluid flow pattern, and energy analysis of the whole system. The evaluation of system design and applications is based on customer feedback, availability of space and materials, and payback period. The modifications to attain the stability in power supply with conditioner and integration of power quality through solar energy as renewable source have also become new research areas. Nowadays, various powerful simulation tools including chemical fluid dynamics are also used for comprehensive study on solar energy. This study is extended on weather conditions and geographical conditions to enhance the performance of solar energy conversion (Chow et al., 2012).

Hybrid PV-T systems are used for the important future aspects such as increase sustainability, minimization of pollution, and enhancement of energy security. These systems produce 70% and sometimes higher efficiency including electrical and thermal efficiency in which thermal output is around 50–55% and electrical output is approximately 20%. The PV-T systems are well known as heating and cooling systems for domestic purpose on small scale. Some countries such as China, the USA, and Europe (Ramos et al., 2017) have already adopted the combination of solar photovoltaic and thermal conversion system. The progress on this combination is getting more attention by academics and as well as industries in terms of cost reduction, sustainability, availability, environmental friendly, reliability, etc.

Table 1 compiles important information of some significant solar conversion techniques with their attributes which are very prominent nowadays (Kothari et al., 2011).

6 Conclusion

To overcome the concern of depleting fossil fuels, an alternative in the form of solar energy conversion technologies is discussed and compiled in terms of its operating temperature, materials of construction for such equipments for converting solar energy into useful output, payback period, and other important aspects. Solar thermal conversion techniques along with solar photovoltaic systems are also summarized for many useful applications. Solar thermal energy conversion has a variety of techniques such as solar pumping, solar air drying, solar thermal power plants, etc., whereas solar photovoltaic consists of power generation and storage. Although initial investment is high for solar energy conversion technologies but payback period is good as a replacement of limited available conventional sources of energy fuels to avoid pollution and scarcity of valuable fuels. The present chapter is an attempt to gather some useful informative data at a place related to utilization of incident solar radiation over the surface of earth which is free of cost in nature. Moreover, solar photovoltaic system description is presented with semiconducting materials information, advantages, disadvantages, and applications in brief. Hybrid system such as the combination of SPV and solar thermal conversion system advancement has also been mentioned in this chapter with essential information. Therefore, solar energy conversion via various aforementioned modes can be considered as an alternative of nonrenewable source of energy with great benefits such as environmental and economical benefits.

Table 1 Some significant solar conversion techniques with their attributes and important aspects

S. No.	Solar conversion technology	Manufacturing materials	Operating temperature	Output	Payback period	Important aspects
1.	Solar water heating	Copper, glass, aluminum, GI thermostat, etc.	~60 °C	10,000 liters water/300 days	3–4 years	✓ Hot water obtained from solar water heating system can save 30,000 liters of diesel/year ✓ A solar water heater is very economical and lasts for a longer period around 15–20 years ✓ It lowers carbon footprints
2.	Solar distillation	Glass, plastic film, aluminum, sawdust, etc.	~65 °C	3 liters water per m^2/day	Minimum 45 days	✓ This system can purify ocean water with salinity around 30,000 mg/l ✓ The cost of water is around 50 paise/liter ✓ It removes electricity consumption powered by coal-fired conventional power plants
3.	Liquid bath solar wax melting	Stainless steel, plastic, glass, etc.	60 to 100 °C	5 kg wax/ m^2	Around 4 years	✓ Candle and match industries use this technology for melting wax and reduce dependency on oil and firewood ✓ The melting point of solar wax melter is around 62.8 °C ✓ It possesses structural reliability, lack of impurities, and high-performance solar energy conversion system
4.	Solar pumping	Chromium, nickel, and alloys of niobium, titanium, zirconium, etc.	54 to 82 °C	Capacity from 2 to 4.6 HP	4 years	✓ Energy efficiency is around 14% of this system ✓ No fuel cost and electricity cost ✓ Highly reliable and durable with long operating life

(continued)

Table 1 (continued)

S. No.	Solar conversion technology	Manufacturing materials	Operating temperature	Output	Payback period	Important aspects
5.	Solar crop drying	Plywood, glass, thermocouple wire, aluminum, PVC, thermocole, etc.	50 °C to 75 °C	Drying time for crop products = 2–4 days	------	✓ This technique is used for drying of corn, fruits, and fish; seasoning of timber; tea processing; and tobacco curing, etc. ✓ Crop dried by solar dryer is hygienic in nature and its nutritional value is also preserved ✓ Solar crop drying minimizes contamination produced by dust, insects, etc.
6.	Solar furnace	Concave mirror, steel, aluminum, etc.	Up to 3500 °C	Up to 1 MW power generation	-------	✓ It is applicable for purification of ceramic and refractory materials ✓ This is used for highly endothermic chemical reactions, e.g., paint or plastic aging experiments ✓ It can generate high power up to 1 MWs
7.	Solar thermal power plant	Glass, steel, plastics, copper, etc.	Low-temp. Plant ~45 °C, Medium-temp. Plant ~400 °C, High-temp. Plant >500 °C	~150 kW	~25 years	✓ There is no carbon dioxide emission during power plant operation ✓ Limitless quantity of energy, free of cost ✓ 60% less energy is required to heat water and up to 35% less energy for space heating

8.	Solar air heating	Solar dryer box, plywood, thermocouple, thermocole, aluminum, PVC, transparent glass, and thermometer	Up to 100 °C	Not applicable	~ 15 years	✓ No corrosion in solar air heater ✓ Solar air heater depends on several factors such as high efficiency and low fabrication, installation, and operational cost ✓ Improves indoor air quality (for systems heating ventilation air)
9.	Solar cookers	Wood, aluminum, plastic, etc.	65 to 400 °C	Not applicable	>15 years	✓ It reduces air pollution and keeps environment clean and healthy as compared to conventional cooking ✓ Preserve natural fuels because of no use of conventional fuel like wood for burning ✓ No noise in solar cooking
10.	Solar photovoltaic systems	Crystalline and amorphous silicon, gallium arsenide, organometallics, etc.	15 to 65 °C	~30GW	10–25 years	✓ No harmful greenhouse gaseous emissions ✓ Photovoltaic panels, through photoelectric phenomenon, generate power ✓ Noise-free application of SPVS

References

Baca, A. J., et al. (2008). Semiconductor wires and ribbons for high-performance flexible electronics. *Angewandte Chemie – International Edition, 47*(30), 5524–5542.

Bhatia, S. C. (2014). Solar photovoltaic systems. *Advanced Renewable Energy Systems*, 144–157.

Chow, T. T., Tiwari, G. N., & Menezo, C. (2012). Hybrid solar: A review on photovoltaic and thermal power integration. *International Journal of Photoenergy, 2012*.

Collins, M. R. (2005). Combined photovoltaic and solar-thermal systems : Overcoming. *IGEC-1 Proceedings of the International Green Energy Conference 12-16 June 2005, Waterloo, Ontario, Canada*, (June).

Crabtree, G. W., & Lewis, N. S. (2007). Solar energy conversion. *Physics Today, 60*(3), 37–42.

Denholm, P., et al. (2010). *Solar energy: The largest energy resource* (1st ed.). Elsevier Inc. https://doi.org/10.1016/B978-1-85617-655-2.00010-9

Kothari, D. P., Singal, K. C. & Ranjan, R. (2011). *Renewable energy sources and emerging technologies 2/E eBook*. Retrieved August 24, 2021, from https://www.amazon.in/Renewable-Energy-Sources-Emerging-Technologies-ebook/dp/B00K7YGORW.

Kumar, B., Kumar, S., & Kumar, S. (2017). Butanol reforming: An overview on recent developments and future aspects. *Reviews in Chemical Engineering, 34*(1), 1–19.

National Informatics Centre (NIC). (2021). *Current status*. Ministry of New and Renewable Energy, Government of India. Retrieved August 24, 2021, from https://mnre.gov.in/solar/current-status/

NS Energy Staff Writer. (2021). *Top five countries with the largest installed solar power capacity*. Retrieved August 24, 2021, from https://www.nsenergybusiness.com/features/solar-power-countries-installed-capacity/#

Ramos, A., et al. (2017). Hybrid photovoltaic-thermal solar systems for combined heating, cooling and power provision in the urban environment. *Energy Conversion and Management, 150*, 838–850. https://doi.org/10.1016/j.enconman.2017.03.024

Rogers, J. A., Lagally, M. G., & Nuzzo, R. G. (2011). Synthesis, assembly and applications of semiconductor nanomembranes. *Nature, 477*(7362), 45–53. https://doi.org/10.1038/nature10381

Schwarz, A., & Van Langenhove, L. (2013). *Types and processing of electro-conductive and semiconducting materials for smart textiles*. Woodhead Publishing Limited. https://doi.org/10.1533/9780857093530.1.29

Sun, Y., & Rogers, J. A. (2007). Inorganic semiconductors for flexible electronics. *Advanced Materials, 19*(15), 1897–1916.

The manufacturer's authorised representative in the EU is Springer Nature Customer Service Centre GmbH, Europaplatz 3, 69115 Heidelberg, Germany. If you have any concerns regarding our products, please contact ProductSafety@springernature.com

Printed and bound by CPI Group (UK) Ltd, Croydon, CR0 4YY
30/09/2026
02225250-0007